T0348485

PULSE FOODS

Food Science and Technology
International Series

Series Editor

Steve L. Taylor
University of Nebraska – Lincoln, USA

Advisory Board

Ken Buckle
The University of New South Wales, Australia

Mary Ellen Camire
University of Maine, USA

Roger Clemens
University of Southern California, USA

Hildegarde Heymann
University of California – Davis, USA

Robert Hutkins
University of Nebraska – Lincoln, USA

Ron S. Jackson
Quebec, Canada

Huub Lelieveld
Bilthoven, The Netherlands

Daryl B. Lund
University of Wisconsin, USA

Connie Weaver
Purdue University, USA

Ron Wrolstad
Oregon State University, USA

A complete list of books in this series appears at the end of this volume.

Pulse Foods

Processing, Quality and Nutraceutical Applications

Edited by

Brijesh K. Tiwari
Department of Food and Tourism, Manchester Metropolitan University, Manchester, UK

Aoife Gowen
UCD School of Agriculture, Food Science and Veterinary Medicine, University College Dublin, Dublin, Ireland

Brian McKenna
UCD School of Agriculture, Food Science and Veterinary Medicine, University College Dublin, Dublin, Ireland

ELSEVIER

AMSTERDAM • BOSTON • HEIDELBERG • LONDON
NEW YORK • OXFORD • PARIS • SAN DIEGO • SAN FRANCISCO
SINGAPORE • SYDNEY • TOKYO
Academic Press is an imprint of Elsevier

Academic Press is an imprint of Elsevier
32 Jamestown Road, London NW1 7BY, UK
30 Corporate Drive, Suite 400, Burlington, MA 01803, USA
525 B Street, Suite 1800, San Diego, CA 92101-4495, USA

First edition 2011

Notice

No responsibility is assumed by the publisher for any injury and/or damage to persons
or property as a matter of products liability, negligence or otherwise, or from any use or
operation of any methods, products, instructions or ideas contained in the material herein.
Because of rapid advances in the medical sciences, in particular, independent verification of
diagnoses and drug dosages should be made

British Library Cataloguing-in-Publication Data
A catalogue record for this book is available from the British Library

Library of Congress Cataloging-in-Publication Data
A catalog record for this book is available from the Library of Congress

ISBN : 978-0-12-382018-1

For information on all Academic Press publications
visit our website at www.elsevierdirect.com

Typeset by Thomson

11 12 13 14   10 9 8 7 6 5 4 3 2 1

Working together to grow
libraries in developing countries

www.elsevier.com | www.bookaid.org | www.sabre.org

ELSEVIER   BOOK AID International   Sabre Foundation

Contents

List of contributors vii

1. **Introduction** 1
    Brijesh K. Tiwari, Aoife Gowen and Brian McKenna

2. **Chemistry of pulses** 9
    B. Dave Oomah, Ankit Patras, Ashish Rawson, Narpinder Singh and Rocio Campos-Vega

3. **Functional and physicochemical properties
    of pulse proteins** 57
    Vassilis Kiosseoglou and Adamantini Paraskevopoulou

4. **Functional and physicochemical properties of pulse starch** 91
    Narpinder Singh

5. **Functional and physicochemical properties of legume fibers** 121
    Uma Tiwari and Enda Cummins

6. **Functional and physicochemical properties of non-starch
    polysaccharides** 157
    Charles Brennan and Uma Tiwari

7. **Post-harvest technology of pulses** 171
    Rangarajan Jagan Mohan, Arumugam Sangeetha, Hampapur V. Narasimha and Brijesh K. Tiwari

8. **Pulse milling technologies** 193
    Jennifer A. Wood and Linda J. Malcolmson

9. **Emerging technologies for pulse processing** 223
    Jasim Ahmed

10. **Pulse-based food products** 249
    Nissreen Abu-Ghannam and Aoife Gowen

11. **Novel food and industrial applications of pulse flours
    and fractions** 283
    Zubair Farooq and Joyce I. Boye

12. **By-product utilization** 325
    Ankit Patras, B. Dave Oomah and Eimear Gallagher

13. **The nutritional value of whole pulses and
    pulse fractions** 363
    Emma Derbyshire

14. **Role of pulses in nutraceuticals** 385
    Marina Carbonaro

15. **Quality standards and evaluation of pulses** 419
    Mahesh Gupta, Brijesh K. Tiwari and Amarinder Singh Bawa

16. **Global pulse industry: state of production,
    consumption and trade; marketing challenges
    and opportunities** 437
    Peter Watts

Index 465

Series List 473

List of contributors

Nissreen Abu-Ghannam
School of Food Science and Environmental Health, Dublin Institute
of Technology, Dublin, Ireland

Jasim Ahmed
Polymer Source Inc., Dorval (Montreal), QC, Canada

Joyce I. Boye
Food Research and Development Centre, Agriculture and Agri-Food
Canada, Saint-Hyacinthe, QC, Canada

Charles Brennan
Department of Food and Tourism, Manchester Metropolitan University,
Manchester, UK

Rocio Campos-Vega
Kellogg Company, Querétaro, Qro., Mexico

Marina Carbonaro
Istituto Nazionale di Ricerca per gli Alimenti e la Nutrizione (INRAN), Rome,
Italy

Enda Cummins
UCD School of Agriculture, Food Science and Veterinary Medicine,
University College Dublin, Dublin, Ireland.

Emma Derbyshire
Manchester Food Research Centre, Manchester Metropolitan University,
Manchester, UK

Zubair Farooq
McGill IR Group, Department of Food Science and Agricultural Chemistry,
Macdonald Campus, McGill University, Sainte-Anne-de-Bellevue, QC,
Canada

Eimear Gallagher
Teagasc, Ashtown Food Research Centre, Ashtown, Dublin, Ireland

Aoife Gowen
UCD School of Agriculture, Food Science and Veterinary Medicine, University
College Dublin, Dublin, Ireland

Mahesh Gupta
School of Food Science and Environmental Health, Dublin Institute of
Technology, Dublin, Ireland

Vassilis Kiosseoglou
Laboratory of Food Chemistry and Technology, School of Chemistry, Aristotle
University of Thessaloniki, Thessaloniki, Greece

Linda J. Malcolmson
Canadian International Grains Institute, Winnipeg, MB, Canada

Brian McKenna
UCD School of Agriculture, Food Science and Veterinary Medicine, University College Dublin, Dublin, Ireland

Rangarajan Jagan Mohan
Department of Food Product Development, Indian Institute of Crop Processing Technology, Thanjavur, India

Hampapur V. Narasimha
Department of Grain Processing, Central Food Technological Research Institute, Mysore, India

B. Dave Oomah
National Bioproducts and Bioprocesses Program, Pacific Agri-Food Research Centre, Agriculture and Agri-Food Canada, Summerland, BC, Canada

Adamantini Paraskevopoulou
Laboratory of Food Chemistry and Technology, School of Chemistry, Aristotle University of Thessaloniki, Thessaloniki, Greece

Ankit Patras
University College Dublin, Dublin, Ireland

Ashish Rawson
Teagasc, Ashtown Food Research Centre, Ashtown, Dublin, Ireland

Arumugam Sangeetha
Department of Food Product Development, Indian Institute of Crop Processing Technology, Thanjavur, India

Narpinder Singh
Department of Food Science and Technology, Guru Nanak Dev University, Amritsar, India

Amarinder Singh Bawa
Defence Food Research Laboratory, Siddartha Nagar, Mysore, India

Brijesh K. Tiwari
Department of Food and Tourism, Manchester Metropolitan University, Manchester, UK

Uma Tiwari
UCD School of Agriculture, Food Science and Veterinary Medicine, University College Dublin, Dublin, Ireland

Peter Watts
Pulse Canada, Winnipeg, MB, Canada

Jennifer A. Wood
Tamworth Agricultural Institute, Industry & Investment NSW, Calala, NSW, Australia

Introduction

1

Brijesh K. Tiwari[1], Aoife Gowen[2], Brian McKenna[2]
[1]Department of Food and Tourism, Manchester Metropolitan University, Manchester, UK
[2]UCD School of Agriculture, Food Science and Veterinary Medicine, University College Dublin, Dublin, Ireland

1.1 Pulses: what are they?

First cultivated by humans over 3000 years ago, the family Leguminosae consists of 650 genera and more than 18000 species. Members of the family, often referred to as legumes or pulses, are the second most important food source in the world after cereal grains. Food legumes are those species of the plant family Leguminosae that are consumed by human beings or domestic animals commonly as dry matter seeds, i.e. the grain legumes. The terms "legumes" and "pulses" are used interchangeably because all pulses are considered legumes *but* not all legumes are considered pulses. The Codex Alimentarius Commission defines pulses as "dry seeds of leguminous plants which are distinguished from leguminous oil seeds by their low fat content". The term "pulse", as used by the Food and Agriculture Organization (FAO), is exclusively for crops harvested solely for the dry seed of leguminous plants. This also excludes green beans and green peas which are consumed and considered as vegetables. A few oil-bearing seeds like groundnut (*Arachis hypogaea*) and soybean (*Glycine max*) are also excluded from pulses which are grown primarily for edible oil extraction. More than 80 different pulse species are consumed by humans, including beans, lentils, lupins, peas and peanuts. However, the FAO recognizes 11 primary pulses as listed in Table 1.1.

Pulse Foods: Processing, Quality and Nutraceutical Applications. DOI: 10.1016/B978-0-1238-2018-1.00007-0

Table 1.1 Commonly consumed pulses

Pulse class	Common/local names	Botanical name
1. Dry beans		*Phaseolus* spp., *Vigna* spp.
	Kidney bean, haricot bean, pinto bean, navy bean	*Phaseolus vulgaris*
	Lima bean, butter bean	*Phaseolus lunatus*
	Azuki bean, adzuki bean	*Vigna angularis*
	Mung bean, golden gram, green gram	*Vigna radiata*
	Black gram, urad	*Vigna mungo*
	Scarlet runner bean	*Phaseolus coccineus*
	Ricebean	*Vigna umbellata*
	Moth bean	*Vigna acontifolia*
	Tepary bean	*Phaseolus acutifolius*
2. Dry broad beans		*Vicia faba*
	Horse bean	*Vicia faba equina*
	Broad bean	*Vicia faba*
	Field bean	*Vicia faba*
3. Dry peas		*Pisum* spp.
	Garden pea	*Pisum sativum* var. *sativum*
	Protein pea	*Pisum sativum* var. *arvense*
4. Chickpea	Garbanzo, Bengal gram	*Cicer arietinum*
5. Dry cowpea	Black-eyed pea, black-eye bean	*Vigna unguiculata*
6. Pigeon pea	Arhar/Toor, cajan pea, Congo bean	*Cajanus cajan*
7. Lentil		*Lens culinaris*
8. Bambara groundnut	Earth pea	*Vigna subterranea*
9. Vetch	Common vetch	*Vicia sativa*
10. Lupins		*Lupinus* spp.
11. Minor pulses		
	Lablab, hyacinth bean	*Lablab purpureus*
	Jack bean	*Canavalia ensiformis*
	Sword bean	*Canavalia gladiata*
	Winged bean	*Psophocarpus teragonolobus*
	Velvet bean, cowitch	*Mucuna pruriens* var. *utilis*
	Yam bean	*Pachyrrizus erosus*

Grain legumes or pulses are important foodstuffs in tropical and subtropical countries, where they are second in importance only to cereals as a source of protein. In addition to their value as a foodstuff, the food legumes are important in cropping systems because of their ability to fix atmospheric nitrogen and increase the overall fertility of soil, reducing the need for expensive nitrogenous fertilizers. The dominant food legumes of any region may vary from country to country or even from region to region, but most of them can be grown under a reasonably wide range of ecological conditions and many legumes can be grown reasonably well on poor soils even without the application of fertilizers.

Pulses are regarded as a beneficial source of nutrients and are recommended as a staple food by health organizations and dieticians. They are rich sources of vitamins, minerals and carbohydrates in the human diet. They represent an important source of protein for vegetarians and are a low glycemic index food (Rizkalla et al., 2002). Pulses are also recognized as a food choice with significant potential health benefits. Pulses contain complex carbohydrates (dietary fibers, resistant starch and oligosaccharides), protein with a good amino acid profile (high lysine), important vitamins and minerals (B vitamins, folates and iron) as well as antioxidants and polyphenols.

1.2 Pulse processing and utilization

Grain processing, including cereal and pulse processing, is one of the oldest and most important of all food technologies and forms a large and important part of the food production chain. Grain pulses are grown widely throughout the world and their dietary and economic importance is globally appreciated and recognized. Today, the grain processing industry is as diverse as its range of products. Practically every meal produced contains grains in some form, while the range of non-food applications is increasing daily, all of which presents enormous processing challenges to food manufacturers.

Pulses undergo several primary and secondary processes such as dehulling, puffing, grinding and splitting prior to their consumption. The primary processing methods vary from country to country. Unlike cereals, processing of pulses also varies with cultivar. The oldest and most common home-scale technique for hulling grain legumes is to pound them in a mortar with a pestle, either after spreading the grains in the sun for a few hours, or after mixing them with a little water. The hull is then winnowed off to produce clean cotyledons. The traditional stone *chakki* design was used as a template for the attrition-type mills as commercial-scale dehulling and splitting of pulses emerged. Methods followed in the home, village industry or in commercial mills are usually similar in principle, but differ in the use of techniques for better yield, operational efficiency and large-scale application.

Novel food processing techniques have been introduced to improve microbial safety and nutritional quality, to improve or modify physicochemical properties, and to increase production and process efficiency. Among various emerging technologies, radiofrequency, microwave, irradiation and high-pressure processing have found potential application for storage and processing of pulses. The introduction of novel technologies has improved the processing and utilization of pulses in certain countries. Research studies show some promising results but industrial application of these novel processing techniques is likely to take some time due to several reasons, including the cost of the equipment, which remains the main challenge to overcome before food processors adopt the technology. Other processing challenges that are associated with the processing and utilization of pulses include:

1. Low profitability of pulse production
2. Post-harvest losses primarily during storage
3. Inadequate supply of high-quality and reasonably priced raw materials
4. Lack of sustainable and efficient processing and packaging technologies
5. Lack of internationally recognized quality standards and common nomenclature.

Interest in the utilization of whole pulses, and their milled constituents in food formulations, is growing in many developed countries. The processing of pulses into ingredients such as flours and fractions (e.g. protein, starch and fiber) and utilizing them in food products is virtually non-existent in Western-style food products, apart from a few specialty or niche markets, and only exists in a limited way in a few other countries. However, more recently, pulse flour and fractions have been used successfully as ingredients in the formulation of several meat products to improve functionality. Extrusion cooking has attracted the attention of researchers and food manufacturers to produce a variety of specialty foods from pulse flour including pasta products, ready-to-eat breakfast cereals, baby foods, snack foods, texturized vegetable protein, pet foods, dried soups and dry beverage mixes. Extrusion cooking not only improves digestibility but also improves bioavailability of nutrients compared to conventional cooking.

1.3 Challenges in pulse processing

Pulses are nutritionally diverse crops that could be successfully utilized as a food ingredient or a base for new product development. The incorporation of pulses with cereals through the application of different technologies is discussed in this book with numerous pulse-based food products given as examples. These have far reaching nutritional benefits to a wide range of the world's population and have the potential to raise significantly the profile of pulses as a highly nutritious and globally available food product base. Market forces have led to greater opportunities for product differentiation and added value to raw commodities because of:

1. Increased consumer demands regarding health, nutrition and convenience
2. Efforts by food processors to improve their productivity
3. Technological advances that enable producers to produce what consumers and processors desire (Siebert et al., 1997).

Following scientific evidence to recommend increased consumption of pulses from different species to improve health, future research needs to be focused on bioavailability and bioefficacy of bioactive compounds, upon technological processing and in novel food formulations, in order to increase their potential benefits. The pulse industry must advance its knowledge of the processing of pulses into ingredients and the impact of that processing on the functionality of the ingredients in food product formulations. The optimization of processing in terms of quality and functionality, in addition to other factors, such as yield and energy use, will be needed to introduce successfully more value-added pulse processing and the incorporation of these ingredients into foods. Ultimately, this will open the door to creating new ingredient markets for pulses leading to new food products and reformulated food products that address consumer needs.

By-products generated from the processing of pulses are promising sources of nutrients, including bioactive compounds (e.g. phytochemicals) which may be used for their favorable technological or beneficial nutraceutical properties. In recent years, many food companies have devoted effort to find value-added applications for these food by-products. As a result of much research, some biologically active compounds have been identified, which can be incorporated into food material to increase its functionality. The exploitation of by-products of pulse processing as a source of functional compounds and their application in food is a promising field which requires interdisciplinary research by food technologists, food chemists, nutritionists and toxicologists. In the near future, we are challenged to respond to the following research needs: first, food processing technology should be optimized in order to minimize the amounts of waste at the outset; secondly, methods for the complete utilization of by-products resulting from pulse processing on a large scale and at affordable cost need to be developed.

Lack of uniform quality standards for international trade is another major challenge to be addressed in the pulse industry. Current efforts for developing uniform standards are sporadic and limited in comparison to cereals and oil seeds. Such initiatives will provide new and valuable tools which will assist all sectors of the pulse processing industry. International cooperation in developing quality standards would channel new and exciting developments for the world pulse industry into the future. Pulse-growing countries will need to strengthen the necessary systems for quality certification in order to improve the competitiveness of the pulse industry. Such systems will become more important with the increasing globalization trend and growing concern for food safety and sustainability issues.

1.4   Relevance of this book

Research into the use of pulses and their components in food formulations is growing, and several factors are contributing to this drive. These include the reported nutritional and health benefits, changes in consumer lifestyles and demographics, increasing demand for variety/balance, rise in the incidence of food allergies, and novel production and processing technologies. This book brings together essential information on the processing and utilization of pulses. It also addresses processing challenges relevant to pulse grain processors, providing a dedicated and in-depth reference for pulse processing within the food industry. Pulse processing and utilization are expected to expand in the future as further economic development takes hold in Asia and the sub-Saharan African region, and as changing lifestyles compel more and more people to consume healthier foods. To sustain this trend, up-to-date information is provided in this book. This book also delivers an insight into the current state of art and emerging pulse-processing technologies of whole pulses, techniques for fractionating pulses into ingredients, their functional and nutritional properties, as well as their potential nutraceutical applications, so that the food industry can use this knowledge to incorporate pulses into new food products.

This book provides a comprehensive assessment of the current state of chemistry, nutrition and health aspects of pulses. It highlights the increasing range of pulse-based products and current best manufacturing processes while also discussing new and emerging technologies in pulse processing. This book also provides in-depth coverage of developments in nutraceutical applications of pulse protein- and carbohydrate-based foods. It is hoped that the book will guide researchers to the most appropriate pulse solution based on their application, providing insights for improving final food product results and assisting in identifying new product development opportunities based on nutritional properties.

References

Rizkalla, S.W., Bellisle, F., Slama, G., 2002. Health benefits of low glycaemic index foods, such as pulses, in diabetic patients and healthy individuals. Br. J. Nutr. 88, 255–262.

Siebert, J.W., Jones, R., Sporleder, T.L., 1997. The VEST model: an alternative approach to value added. Agribusiness 13, 561–567.

Chemistry of pulses

2

B. Dave Oomah[1], Ankit Patras[2], Ashish Rawson[3], Narpinder Singh[4], Rocio Compos-Vega[5]
[1]National Bioproducts and Bioprocesses Program, Pacific Agri-Food Research Centre, Agriculture and Agri-Food Canada, Summerland, BC, Canada
[2]University College Dublin, Ireland
[3]Teagasc, Ashtown Food Research Centre, Ashtown, Dublin, Ireland
[4]Department of Food Science and Technology, Guru Nanak Dev University, Amritsar, India
[5]Kellogg Company, Querétaro, Mexico

2.1 Introduction

Pulse grains are an excellent source of protein, carbohydrates, dietary fiber, vitamins, minerals and phytochemicals (phenolic acid, anthocyanins) (Tharanathan and Mahadevamma, 2003) and their consumption and production increasing worldwide. Researchers have demonstrated that pulses could prevent or manage chronic health issues such as diabetes, cardiovascular disease and obesity and contribute to overall health and wellness (Bassett et al., 2010). Pulses contain a number of bioactive substances including enzyme inhibitors, lectins, phytates, oligosaccharides and phenolic compounds that play metabolic roles in humans or animals that frequently consume these foods (Campos-Vega et al., 2010). These effects may be regarded as positive, negative or both (Champ, 2002). Some of these substances have been considered as antinutritional factors due to their effect on

Pulse Foods: Processing, Quality and Nutraceutical Applications. DOI: 10.1016/B978-0-1238-2018-1.00007-0

diet quality. Frequent legume consumption (four or more times compared with less than once a week) has been associated with 22% and 11% lower risk of coronary heart disease (CHD) and cardiovascular disease (CVD), respectively (Flight and Clifton, 2006). In an earlier study of 9632 participants free of CVD at their baseline examination in the First National Health and Nutrition Examination Survey (NHANES 1) Epidemiological Follow-up Study (NHEFS), Bazzano et al. (2001) found that legume consumption was significantly and inversely associated with risk of CHD and CVD.

Considerable genetic variation has been reported in the chemical composition of pulses both between and within species. In addition, chemical composition is modified by environmental factors during plant development, and many of the phytochemicals are secondary metabolites produced during seed development and seed maturation (Rochfort and Panozzo, 2007).

In many regions of the world, legume seeds are the unique supply of protein in the diet and regarded as versatile functional ingredients or as biologically active components more than as essential nutrients. Legumes are claimed to improve overall nutritional status (Guillon and Champ, 1996) and the needs of the food industry, respectively. This chapter focuses on the current knowledge around certain classes of proteins, carbohydrates, lipids and pulse phytochemicals including phytosterols, phenolic compounds, saponins and oxalate and phytic acid. The potential for these metabolites to influence human health is also briefly discussed.

2.2 Overview

Table 2.1 illustrates different pulses commonly consumed around the world because of their high protein content compared to other grains (Singh et al., 2004). This elevates pulses as a significant food source for developing countries, low-income people (Bressani and Elias, 1979) and even as animal feed. In fact, pulses are an important part of the human diet in many parts of the world, particularly in Latin America, Africa and the Asian subcontinent because they are both a rich and inexpensive source of protein and also a good source of B-complex vitamins, minerals and carbohydrates (Jood et al., 1988). Due to high cost and limited availability of animal proteins in the developing countries, attention has increased on the

Table 2.1 Protein content (g 100 g^{-1} db) and amino acid (mg g^{-1} N) content of different pulses

Common name	Protein	Isoleucine	Leucine	Lysine	Phenyl-alanine	Tyrosine	s-cont (Total)	Methionine	Cystine	Threonine	Tryptophan	Valine
Pigeon pea (*Cajanus cajan*)	20.9	380	490	450	540	210	160	70	90	240	30	330
Chickpea (*Cicer arietinum*)	20.1	280	570	370	390	240	180	110	80	310	–	330
Lentil (*Lens esculenta*)	24.2	340	480	400	310	200	200	80	110	250	90	330
Adzuki bean (*Phaseolus angularis*)	25.3	280	490	440	340	210	180	110	70	240	–	340
Mung bean (*Phaseolus aures*)	23.9	350	560	430	300	100	110	70	40	200	50	370
Black bean (*Phaseolus mungo*)	22.7	270	490	460	410	210	140	90	60	230	–	370
Kidney bean (*Phaseolus vulgaris*)	22.1	360	540	460	350	240	120	60	60	270	60	380
Pea (*Pisum sativum*)	22.5	350	520	460	320	250	160	80	80	240	70	350
Cow pea (*Vigna unguiculata*)	23.4	260	450	410	340	210	230	120	110	220	–	340

Adapted from www.fao.org

utilization of seed and legumes as potential sources of low-cost dietary proteins for food use (Wang et al., 1997). Thus, legumes are recognized as an important source of food protein, calories and other nutrients, minerals and vitamins (Salunkhe et al., 1985).

It is quite evident that protein–energy malnutrition is among the most serious problems developing countries are facing today. Some pulses play an important role in providing needed protein in these countries. For example, cowpeas have now become an important grain legume in East and West African countries as in other developing countries (Dovlo et al., 1976; McWatters, 1983; Philips and McWatters, 1991; Prinyawiwatkul et al., 1996). In Nigeria, cowpeas are grown extensively, with Niger and Nigeria producing 49.3% of the annual world crop (Rachie, 1985). This legume crop forms an important part of the diets of Nigerians (Oyenuga, 1968; Uriyo, 2001) as it is prepared and eaten as porridge, cake ("akara") and delicacies such as "moinmoin" (Faboya and Aku, 1996). Despite the benefits associated with cowpeas, such as good sources of low-cost vegetable protein, calcium, magnesium, zinc and B vitamins (Oke, 1967), they have been underutilized (Prinyawiwatkul et al., 1997).

2.3   Major constituents

2.3.1 Proteins

Legumes are widely recognized as important sources of food proteins. Plant proteins can now be regarded as versatile functional ingredients or as biologically active components more than as essential nutrients. This evolution towards health and functionality is mainly driven by consumer demands and health professionals (the partial replacement of animal foods with legumes is claimed to improve overall nutritional status (Guillon and Champ, 1996) and the needs of the food industry, respectively. Legume seeds accumulate large amounts of proteins during their development. Most are devoid of catalytic activity and play no structural role in the cotyledon tissue. They are stored in membrane-bound organelles (protein bodies) in the cotyledonary parenchyma cells, survive desiccation on seed maturation and undergo hydrolysis at germination, thus providing ammonia and carbon skeletons to the developing seedlings.

Food proteins not only are a source of constructive and energetic compounds such as the amino acids, but also may play bioactive roles by themselves and/or can be the precursors of biologically active peptides with various physiological functions. Out of the 20 amino acids, eight are essential and must be present in the diet. Unlike animal proteins, plant proteins may not contain all the essential amino acids in the required proportions. The nine essential amino acids required by humans are: histidine, leucine, isoleucine, valine, threonine, methionine, phenylalanine, tryptophan and lysine. Legume seeds also contain many anti-nutritional compounds (ANCs), which can be of proteinous (hydrolase inhibitors and lectins) and non-proteinous nature.

2.3.1.1 Structure and classification of proteins

Proteins can be classified according to source, solubility, physiological role and structure. Dry bean proteins can be broadly classified as metabolic and storage proteins. The metabolic proteins include enzymatic as well as non-enzymatic proteins. Many of the dry bean storage proteins are often referred to as globular proteins because of their globular shapes as well as the requirement of a certain amount of ionic strength (μ) for their solubilization in aqueous media. However, certain globular proteins in beans are also partly soluble in water (Deshpande and Nielsen, 1987), perhaps due to ionic salt(s) associated with the protein(s). Osborne (1924) classified proteins into five classes based on solubility in a range of solvents. Classification of proteins is based on sequential extraction using distilled water, dilute salt solution, dilute alkali and 70% ethanol. The following are five distinct classes of protein:

1. *Albumin*: soluble in water
2. *Globulins*: soluble in dilute salt solution
3. *Prolamin*: soluble in 70% ethanol solution
4. *Glutelin*: soluble in dilute alkali
5. *Residue*: left-over protein.

The most abundant class of storage proteins in grain legumes are the globulins. They are generally classified as 7S and 11S globulins according to their sedimentation coefficients (S). The 7S and 11S globulins of pea are named vicilin and legumin, respectively, so that the corresponding proteins of other seeds are often indicated as vicilin- and legumin-like globulins. The 7S proteins are oligomeric proteins (usually trimers). The 11S proteins are also oligomers, but usually they form hexamers (Duranti and Gius, 1997). Larger aggregates of 15–18S have also been reported for soybean legumin-like proteins (Koshiyama, 1983). Under dissociating conditions, both the 7S and 11S globulins liberate their constituent subunits. These polypeptide chains are naturally heterogeneous, in both size and charge levels (Brown et al., 1981) arising from a combination of various factors, including the multigene origin of each storage globulin and the post-translational modifications of relatively few expression products (Wright, 1986). The mutual contribution of these factors varies significantly. Characteristics of 11S and 7S globulins are discussed below briefly.

Characteristics of 11S globulins

- Major storage proteins of most legumes.
- Typically hexamers consisting of six subunit pairs that interact non-covalently.
- Each of these subunit pairs consists of an acidic subunit of M 40 000 and a basic subunit of M 20 000, linked by a single disulfide bond.
- Each subunit is cleaved after disulfide bond formation.

Characteristics of 7 S globulins

- Typically trimeric proteins of M 150 000 to 190 000.
- Lack cysteine residues and hence cannot form disulfide bonds.

2.3.1.2 Protein composition

2.3.1.2.1. Pigeon pea The protein content of the pigeon pea (Table 2.2) varies from 15.5 to 28.8% (Vilela and El-Dash, 1985; Salunkhe et al., 1986; Oshodi and Ekperigin, 1989) and depends on genetic and environmental factors (Salunkhe et al., 1986). Similar to other legumes, pigeon pea protein is deficient in sulfur-containing amino acids (methionine and cystine) and contains a surplus of lysine (limiting amino acid in cereals). Pigeon pea germ (embryo) is nutritionally better in terms of amino acid composition than that of the cotyledons. The amino acid composition of the cotyledon affects the overall nutritional quality since it constitutes approximately 85% of whole grain and is the finished edible portion of raw grain. Wide variation in protein fractions is also observed in pigeon pea protein, for instance albumin (15–27%), globulins (50–72%), prolamin (0.2–3.0%) and glutelin (5–23%). Non-protein nitrogen and glutelin are located mainly in the seed coat where prolamin is negligible. According to several reports, the prolamin fraction does not exist in pigeon pea and the reported values are merely due to

Table 2.2 Fractions of major storage protein (globulins) of pigeon pea

Globulins fraction	Molecular weight (Dalton)	Solubility	Sedimentation value
α-Globulins	186 000	Insoluble at	9.35
	294 000	pH 4.7	12.6
β-Globulins	227 000	Soluble at pH 4.7	10.6
γ-Globulins	50 000		3.9

contamination from other sources. The globulin fraction of pulses is often referred to as 7S globulins (vicilin) and 11S globulins (legumins). Fractionation of pigeon pea globulins yields three fractions: α-, β- and γ-globulins. α-Globulins consist of two subfractions with molecular weights of 186 000 and 294 000, with sedimentation values of 9.35 and 12.6 (Singh and Jambunathan, 1982) (Table 2.2). This fraction corresponds to legumins (11S) of other legumes. α-, β- and γ-Globulins have been characterized as glycoproteins (Singh and Jambunathan, 1982). Singh et al. (1981) fractioned pigeon pea proteins using water-solubility properties (albumins), salts (globulins), alcohol (prolamins) and acid/alkali (glutelins) as well as residual proteins and non-protein nitrogen. Pigeon pea globulin stored the highest amount of proteins (60–70%), similar to other legumes (Salunkhe et al., 1986). Within the protein fraction, globulins are deficient in sulfur-containing amino acids compared to albumins and glutelin. The albumin fraction of protein contains the highest amount of methionine, cystine, lysine, aspartic acid, glycine and alanine.

2.3.1.2.2. Chickpeas Chickpeas (*Cicer arietinum* L.) are one of the most widely consumed pulses in the world and protein content varies from 21.7 to 23.4% (El-Adawy, 2002). Chickpea protein is rich in lysine and arginine but most deficient in the sulfur-containing amino acids, methionine and cystine (Manan et al., 1984). Chickpea contains twice the amount of protein than that of cereals; hence, it can balance the amino acid and may improve the nutritive value of a cereal-based diet (Khalil et al., 1983; Singh et al., 1988). Similar to pigeon peas, the embryo amino acid profile of chickpeas deficient in sulfur-containing amino acids is nutritionally better than the cotyledons. Chickpea, in common with other pulses, contains relatively high concentrations of globulins, representing almost 60% of the total protein (Chavan et al., 1988; Clemente et al., 1998). The legumin-like globulin (11S) is the major globulin fraction in chickpea, whereas the vicilin (7S) constitutes about 30% of total globulins (Williams and Singh, 1987; Chavan et al., 1988).

2.3.1.2.3. Green gram Green gram is a protein-rich staple food. It contains about 25–30% protein. Like other pulses, green gram also contains globulins as a major seed protein. These globulins are broadly classified into legumin and vicilin as discussed earlier.

It should be noted that pulses also contain substantial amounts of non-protein nitrogen, in the 8.3–14.5% range of the total bean nitrogen (Deshpande and Nielsen, 1987).

2.3.2 Carbohydrates

Pulses generally contain about 60–65% carbohydrates, slightly lower compared to cereals (70–80%). Pulse carbohydrates mainly contain monosaccharides, disaccharides, oligosaccharides and polysaccharides. The primary storage carbohydrate is starch, which constitutes a major fraction of the total carbohydrates of almost all the pulses. Many health benefits are attributed to the carbohydrate components of pulse seeds. Pulse starch contributes to slow glucose release, inducing a low glycemic index (Rizkalla et al., 2002; Winham et al., 2007), whereas dietary fiber is involved in gastrointestinal health (Marlett et al., 2002). The soluble sugar fraction of pulses also includes monosaccharides (ribose, glucose, galactose and fructose) and disaccharides (sucrose and maltose). The major oligosaccharides of pulses belong to the α-galactoside group where galactose is present in an α-D-1,6-linkage. Galactosides derived from sucrose, such as raffinose, stachyose and verbascose, represent the most studied sugars in pulses. Other groups of α-galactosides in pulses include the glucose galactosides (melibiose and manninotriose) and inositol galactosides (galactinol, galactopinitol and ciceritol). Ciceritol is a trisaccharide (D-galactopyranosyl-6-α-D-galactopyranosyl-2-(1D)-4-*O*-methyl-chiro-inositol) most abundant in chickpea (Quemener and Brillouet, 1983; Bernabe et al., 1993; Sanchez-Mata et al., 1998).

2.3.2.1 Starch yield and chemical composition

Starch consists mainly of amylose and amylopectin (Fig. 2.1). Amylose is an essentially linear polymer of α-(1→4)-linked D-glucopyranosyl units with few (<0.1% according to Ball et al., 1996) α-(1→6) linkages. It has a number average degree of polymerization

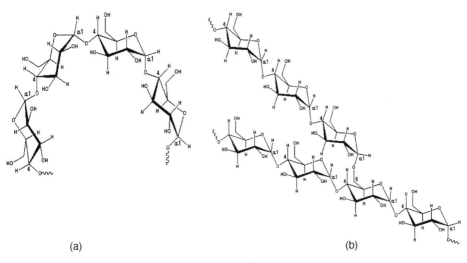

(a)   (b)

Figure 2.1 Structures of amylose (a) and amylopectin (b).

(DPn) of 800–4920, average chain lengths (CL) of 250–670 and β-amylolysis limits of 73–95% (Morrison and Karkallas, 1990). The starch yield, total lipid, total amylose and nitrogen contents of pulse starches (Table 2.3) generally range from 18.0 to 49.0%, 0.01 to 0.43%, 11.6 to 88.0% and 0.01 to 0.43%, respectively (Hoover et al., 2010). Figure 2.2 shows the starch granule structure at different levels of magnification. The wide range in amylose content within a starch source could be attributed to the following:

1. In many instances, the amylose content has been determined by colorimetric procedures without prior defatting and/or by not taking into account the iodine complexing ability of long external amylopectin chains (this leads either to an underestimation resulting from the failure to remove amylose complexed lipids or to an overestimation resulting from the failure to determine amylose content from a standard curve containing mixtures of amylose and amylopectin in various ratios).
2. Different methods that have been employed for determination of amylose content (Hoover and Ratnayake, 2002; Huang et al., 2007; Chung et al., 2008).
3. Cultivar differences.
4. The physiological state of the seed. Among pea mutants, the amylose content has been shown to range from 8.0 to 65.0% (Bogracheva et al., 1999).

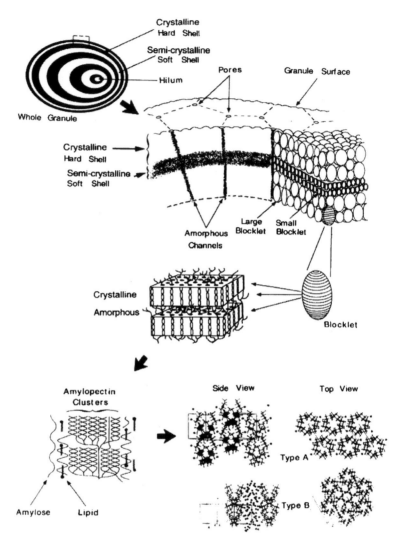

Figure 2.2 Overview of starch granule structure at different levels of magnification. (Adapted from Gallant et al., 1997.)

Variation in total lipid content within a starch source could be attributed to:

1. The method used for lipid determination (acid hydrolysis vs. solvent extraction)
2. Cultivar differences
3. The physiological state of the seed.

Table 2.3 Yield and composition of pulse starches

Starch source	Yield (%)	Total amylose (%)	Nitrogen (%)
Adzuki bean[a]	21.5	17.6–34.9	0.01–0.07
Beach pea[a]	12.3	29.0	0.08
Black bean[a]	16.4–22.2	27.2–39.3	0.04–0.07
Black gram[a]	18.0–45.0	30.7–34.6	0.10–0.15
Chickpea[a]	29.1–46.0	30.4–35.0	0.08–0.10
Chickpea (desi)[b]	–	28.3–52.8	–
Chickpea (kabuli)[b]	–	43.2–47.2	–
Cowpea[a]	37.0	25.8–33.0	0.06–0.09
Faba bean[a]	39.9	17.0–42.0	0.33–0.43
Field pea[c]	–	21.4–58.3	–
Grasspea[a]	21.1–25.5	35.2–38.3	0.04–0.09
Horsegram[a]	28.0	34.0	0.08
Jackbean[a]	18.4	37.5	0.09–0.16
Kidney bean[a]	25–45	34.0–41.5	0.02–0.05
Lentil[a]	27.4–47.1	23.5–32.3	0.03–0.09
Lima bean[a]	22.0	32.7–34.5	0.03–0.07
Mung bean[a]	31.1	33–45.3	0.02–0.05
Moth bean[a]	–	27.0	0.06
Navy bean[a]	24.0–25.0	28.6–41.4	0.02–0.05
Northern bean[a]	18–31	31.6–41.0	0.06–0.16
Pea (smooth)[a]	35–40	24.0–49.0	0.02–0.07
Pea (wrinkled)[a]	21.6	60.5–88.0	0.05–0.08
Pigeon pea[a]	29.7–49.3	27.0–46.4	0.02–0.05
Pinto bean[a]	25–30.1	31.3–37.4	0.05–0.07
Velvet bean[a]	–	39.2	0.11
Yam bean[a]	–	11.6	–

[a]Hoover et al. (2010); [b]Singh et al. (2010a); [c]Singh et al. (2010b).

Starch can be further classified according to digestibility as soluble, insoluble or resistant starch (RS). Until recently, starch was thought to undergo complete breakdown and absorption upon digestion. In 1992, RS was referred to as the proportion of starch that is not hydrolyzed or digested as it passes through the gastrointestinal tract (Englyst et al., 1992). RS that reaches the large intestine has a physiological function similar to that of dietary fiber. RS can be considered a probiotic and acts as a substrate for microbiological fermentation, producing short-chain fatty acids (SCFAs), methane and carbon dioxide, conferring benefits to human colonic health, and to a lesser extent can impact lipid and glucose metabolism. The production of these fermentation products from the consumption of RS is less than that from the consumption of non-digestible oligosaccharides (Christl et al., 1992). It is believed that the SCFAs produced mediate the benefits of RS rather than RS exerting a physical bulking effect (Topping et al., 2003). Resistant starch can be quantified by the following equation:

$$RS = TS - (RDS + SDS) \tag{2.1}$$

Table 2.4 shows starch content, rapidly digested starch (RDS), slowly digestible starch (SDS), resistant starch (RS) and free glucose (FG) of some common pulses (white bean, pinto bean, pea, chickpea, lentils). Worldwide, the dietary intake of RS varies considerably, with the intake between 30 and 40 g day^{-1} in developing countries (Baghurst and Baghurst, 2001), between 3 and 6 g day^{-1} in the EU (Dyssler and Hoffmann, 1994), and a similar intake in Australia (Baghurst and Baghurst, 2001). These values represent the total amount from all sources including fruit and vegetables.

2.3.2.2 Dietary fiber

Dietary fiber (DF) is generally defined as the macromolecules present in the diet that resist digestion by human endogenous enzymes and is essentially composed of plant cell wall remnants, such as cellulose, hemicelluloses, pectic polysaccharides and lignin. The American Association of Cereal Chemists (AACC) defines DF as "the edible parts of plants or analogous carbohydrates that are resistant to digestion and absorption in the human small intestine with complete or partial fermentation in the large intestine" and DF includes polysaccharides, oligosaccharides, lignin and associated plant substances (McCleary, 2008). During the last two decades, DF has played an important role in decreasing the risk of many diseases (Tharanathan and Mahadevamma, 2003). DF contents range from 23 to 32, 18 to 22, 18 to 20 and 14 to 26 (%) for dry bean, chickpeas, lentils and peas, respectively (Table 2.5). It does not constitute a defined chemical group; instead it represents a combination of chemically heterogeneous substances such as cellulose, hemicelluloses, pectins, gums, mucilages,

Table 2.4 Starch content, rapidly digested starch (RDS), slowly digestible starch (SDS), resistant starch (RS) and free glucose (FG) of some legumes

Legumes	Total starch	RDS	SDS	RS	FG
White bean	39.96	6.23	15.84	17.89	2.46
Pinto bean	40.99	4.11	12.86	24.02	2.36
Lentil	51.50	13.01	16.06	22.44	0.97
Chickpea	46.47	4.97	10.21	31.60	0.22
Pea	20.33	12.00	2.01	6.32	1.00

Table 2.5 Dietary fiber contents (total, insoluble and soluble) of raw pulse seeds

| Pulses | Dietary fiber content (g 100 g^{-1}) | | | |
	Total dietary fiber	Insoluble fiber	Soluble fiber	
Beans, dry	23–32	20–28	3–6	Granito et al. (2002), Kutos et al. (2003), Perez-Hidalgo et al. (1997)
Chickpeas	18–22	10–18	4–8	Dalgetty and Baik (2003), Perez-Hidalgo et al. (1997)
Lentils	18–20	11–17	2–7	Dalgetty and Baik (2003), Perez-Hidalgo et al. (1997)
Peas, dry	14–26	10–15	2–9	Borowska et al. (1998)

Adapted from Tosh and Yada (2010).

resistant starch, other polysaccharides and lignin, a non-carbohydrate polymer of phenyl propane residues. DF is classified mainly into two types, insoluble and soluble. The insoluble DF includes cellulose, lignin and some hemicelluloses, whereas soluble DF includes natural gel-forming fibers like pectins, gums, mucilages and some hemicellulosic fractions. Each one of these exerts specific physiological effects (Stephen, 1995). Pulses contain appreciable amounts of crude fiber (1.2–≈13.5). Rather large variations in crude fiber content were observed in black gram, Bengal gram, mung bean and red gram. Cellulose is the major component of crude fiber in smooth and wrinkled peas, red kidney beans, navy beans, pinto beans, pink beans and black- eye beans, while in other pulses (lupin seeds, lentil, broad beans, red gram, black gram), hemicellulose is the major component of fiber. Several researchers reported that glucose is the major sugar in hemicelluloses of *Vicia faba* (Pritchard et al., 1973), cowpeas (Longe, 1981), mung beans (Buchala and Franz, 1974), wrinkled peas (Cerning-Beroard and Filiatre, 1976) and winged beans (Sajjan and Wankhede, 1981). Hemicelluloses of horse beans contain essentially xylose, small amounts of arabinose and traces of galactose and rhamnose (Cerning et al., 1975). Sajjan and Wankhede (1981) hydrolyzed hemicellulose A and B fractions (extracted with alkaline solution and precipitated with acetic acid and ethanol to isolate A and B fractions) of winged beans in order to establish the proportion of hexose to pentose. They found that hemicellulose A consists of glucose, xylose and arabinose in ratios of 15.5:9:1 and hemicellulose B contained glucose and xylose in the proportion 15:1.

2.3.2.3 Non-starch polysaccharides and minor carbohydrates

The non-starch polysaccharides (NSPs) include various polysaccha-ride molecules excluding α-glucans (starch). The classification of NSP was based originally on the methodology used for extraction and isolation of polysaccharides. The residue remaining after a series of alkaline extractions of cell wall materials was called cellulose, and the fraction of this residue solubilized by alkali was called hemicellulose. The complexity in the structure and confusion in the nomenclature have made it almost impossible to draw a clear-cut classification of NSPs, however, NSPs fall into three main groups, namely cellulose, non-cellulosic polymers and pectic polysaccharides (Bailey, 1973). Grain pulses are used in monogastric diets mainly to supply protein. In addition to the protein, they contain substantial amounts of NSP. Cellulose and xylans, which are the major NSPs in cereal grains, are generally found in the hulls or husks of most pulses. The NSPs in the cotyledon of legumes are pectic polysaccharides. Total sug-ars (mono- and oligosaccharides) represent only a small percentage of total carbohydrates in dry legume seeds (Table 2.6). Among the sugars, oligosaccharides of the raffinose family (raffinose, stachyose, verbascose and ajugose) predominate in most pulses and account for a significant percentage (31–76%) of the total sugars (Becker et al., 1974; Nene et al., 1975; Cerning-Beroard and Filiatre, 1976; Akpapunam and Markakis, 1979; Kon, 1979; Rockland et al., 1979; Ekpenyong and Borchers, 1980; Reddy and Salunkhe, 1980; Flem-ing, 1981; Sathe and Salunkhe, 1981). Certain pulses such as smooth peas, wrinkled peas, black gram and red gram contain higher amounts of total oligosaccharides than others (Table 2.6). The predominance of a particular oligosaccharide depends on the type of pulses. For ex-ample, verbascose is the major oligosaccharide in black gram, Bengal gram, red gram, mung bean and broad beans (faba beans), whereas stachyose is the major oligosaccharide in smooth and wrinkled peas, great northern beans, California small white beans, red kidney beans, navy beans, pinto beans, pink beans, black eye beans, Bengal gram, lentils, cowpeas and lupin seeds. Ajugose is the other higher molecu-lar weight oligosaccharide of the raffinose family of sugar, present in small amounts in smooth and wrinkled peas and lupin seeds. Raffino-se is present in moderate to low amounts in most legumes. These oligosaccharides cause flatulence when incompletely digested by the stomach and are fermented by microflora in the lower intestine. Soak-ing, fermentation, processing and certain spices have been reported to reduce the flatulence problem.

Table 2.6 Carbohydrate content of dry beans

Pulses	Total soluble sugars (%)	Sucrose (%)	Raffinose (%)	Stachyose (%)	Verbascose (%)	Ajugose (%)	Crude fiber (%)	Lignin (%)	Cellulose (%)	Hemicellulose (%)
Red kidney beans	8	1.6	0.3–0.9	2.4–4.0	0.1–0.5		3.7	2.7–3.1	2.5–5.9	0.3
Navy beans	5.6–6.2	2.2–3.5	0.4–0.7	2.6–3.5	0.1–0.4		3.4–6.6	0.1	3.2	0.5–4.9
Pinto beans	6.7	2.8	0.4–0.6	2.9–3	0.1–0.2		4.3–7.2	1.8–3.0	9	4
Pink beans		1.4	0.2–0.4	0.2–0.4				0.2	6	
Black eye beans		2.6	0.4–1.0	0.4–0.9			3.1	0.1	4.9	
Black gram	3.0–7.1	0.7–1.5	0.0–1.3	0.9–3.0	3.4–3.5		1.2–7.1	3.8	5	10.7
Bengal gram	3.5–9.0	0.7–2.9	0.7–2.4	2.1–2.6	0.4–4.5		1.2–13.5	2.9–7.1	1.1–13.7	0.6–8.4
Mung bean	3.9–7.2	0.3–2.0	0.3–2.6	1.2–2.8	1.7–3.8		1.2–12.8	2.2–7.2	2.5–4.6	0.3–9.1
Red gram	3.5–10.2	2.7	1.0–1.1	2.7–3.0	4.0–4.1		1.2–8.1	2.9	7.3	10.1
Winged bean seeds	3.4	0.3–8.2	0.2–2	0.1–3.6	0.04–0.9		3.4–12.5	0.7–1		1.36
Smooth peas	5.3–8.7	2.3–2.4	0.3–0.9	2.2–2.9	1.7–2.3	0.06	4.6–7	0.5–0.9	0.9–4.9	1.0–5.1
Wrinkled pea	10.2–15.1	2.3–4.2	1.2–1.6	2.9–5.5	2.2–4.2	0.13	7.6	0.3–1	1.2–4.2	0.9–6.6
Great northern beans	9.9	2.0–3.8	0.3–0.7	2.3–3.8	ND		4.5–6.7			
California small white beans	7.7	3	0.3–0.7	2.9–3.7	0.1					
Broad beans	3.1–7.1	1.4–2.7	0.1–0.5	0.5–2.4	1.6–2.1		8	0.7–1.1	1	4.0–4.6
Lentil	4.2–6.1	1.8–2.5	0.4–1.0	1.9–2.7	1.0–3.1		3.8–4.6	2.6	4.1	6

(*Continued*)

Table 2.6 Carbohydrate content of dry beans – continued

Pulses	Total soluble sugars (%)	Sucrose (%)	Raffinose (%)	Stachyose (%)	Verbascose (%)	Ajugose (%)	Crude fiber (%)	Lignin (%)	Cellulose (%)	Hemicellulose (%)
Cowpea	6.0–13.0	1.8–3.1	0.4–1.2	2.0–3.6	0.6–3.1		1.7–4.0	0.6–1.8		
Lupin seeds	7.4–9.5	1.0–2.6	0.5–1.1	0.9–7.1	0.6–3.4	0.3–2.0	3	0.7–0.8		9.3–9.9

Data compiled from: Nigam and Giri (1961); Patwardhan (1962); Schoch and Maywald (1968); Kawamura (1969); Becker et al. (1974); Cristofaro et al. (1974); Nene et al. (1975); Olson et al. (1975); Satterlee et al. (1975); Cerning-Beroard and Filiatre (1976); Kumar and Venkataraman (1976); Morton (1976); Rao (1976); Iyengar and Kulkarni (1977); Watson (1977); Claydon (1978); Shahen et al. (1978); Akpapunam and Markakis (1979); Bhatty and Slinkard (1979); Biliaderis et al. (1979); Garcia (1979); Kon (1979); Lorenz (1979); Naivikul and D'Appolonia (1979); Rockland et al. (1979); Silva and Luh (1979); Sosulski and Youngs (1979); Colonna et al. (1980); Ekpenyong and Borchers (1980); Eskin et al. (1980); Goel and Verma (1980); Iyer et al. (1980); Kamath and Belavady (1980); Monte and Maga (1980); Morad et al. (1980); Reddy and Salunkhe (1980); Vose (1980); Chen and Anderson (1981); Fleming (1981); Geervani and Theophilus (1981); Halaby et al. (1981); Jaya and Venkataraman (1981); Labaneiah and Luh (1981); Longe (1981); Saijan and Wankhede (1981); Sathe and Salunkhe (1981); Sathe et al. (1982).
ND, not detected.

2.3.3 Fatty acids

Pulse seeds contain 2–21% fat with a beneficial composition of exogenic unsaturated fatty acids: linoleic (21–53%) and linolenic (4–22%) acids (Campos-Vega et al., 2010). They also contain significant levels of vitamin E: 12–187 i.u. kg^{-1} feed and 470–2560 i.u. kg^{-1} oil. Common beans are vital sources of free unsaturated fatty acids, accounting for 61.1% of total fatty acids (FA): palmitic acid (16:0), oleic acid (18:1) and linoleic acid (18:2) (Campos-Vega et al., 2010). Linolenic (18:3) acid accounts for 43.1% of total fatty acids in common beans (Grela and Gunter, 1995). Chickpeas, butter beans and kidney beans, respectively, have the highest content of monosaturated fatty acids (MUFAs) (34.2 g $100 g^{-1}$), saturated fatty acids (SFAs) (28.7 g $100 g^{-1}$) and polysaturated fatty acids (PUFAs) (71.1 g $100 g^{-1}$) (Table 2.7) (Ryan et al., 2007).

Azuki beans grown in Japan contained 2.2% fat, consisting mainly of phospholipids (63.5%), triglycerides (21.2%), steryl esters (7.5%), hydrocarbons (5.1%), diacylglycerols (1.3%), free fatty acids (0.9%), and other minor components. The fatty acid composition of azuki bean lipids was linoleic, palmitic and linolenic acids representing 45%, 25%, and 21%, respectively, of the total lipid content (Yoshida et al., 2008). The lipid contents of seeds of eight *Vicia* species collected in Turkey varied between 2.5% and 3.9%, with palmitic (7–23%) and stearic (15–35%) acids as the major fatty acids (Akpinar et al., 2001). Food pulses generally contain only 1–2% lipids with the unsaponifiable fraction of the oil, ranging from 0.5% to 4% (Harborne et al., 1971).

2.4 Minor components

2.4.1 Vitamins

Pulses are also good sources of many vitamins such as thiamine, riboflavin, niacin, pyridoxamine, pyridoxal and pyridoxine. The vitamin composition of different pulses is given in Tables 2. 8 and 2.9. The raw bean contained 0.99 mg of thiamine, 0.20 mg riboflavin, 1.99 mg niacin, 0.49 mg vitamin B_{12} and 0.30 mg folic acid $100 g^{-1}$ d.m. Pulses are excellent sources of folates which are not readily available to humans due to complex binding with biomolecules (Kadam and Salunkhe, 1989). Folate contents between 400 and 600 μg representing 95% of daily requirements have been reported. Higher folate intake

Table 2.7 Total oil (g 100g^{-1}) and fatty acid composition (% of total) of pulses

Pulse	Total oil	Fatty acid												
		16:0	16:1	17:0	18:0	18:1	18:2	18:3	20:0	20:1	22:0	SFA	MUFA	PUFA
Butter bean	0.9	23.68	0.20	0.37	3.62	10.35	42.43	18.64	ND	ND	0.30	28.7	10.5	69.8
Chickpea	5.0	10.87	0.23	0.06	1.85	33.51	49.74	2.41	0.60	0.39	0.21	13.7	34.2	52.1
Kidney bean	1.2	14.20	0.16	0.22	1.30	11.97	26.04	45.69	0.24	ND	0.51	16.5	12.1	71.7
Lentil	1.4	14.57	0.09	0.13	1.24	22.95	47.17	11.67	0.44	0.70	0.28	16.7	23.7	58.8
Pea	1.5	10.65	0.07	0.19	3.11	28.15	47.59	9.29	0.22	0.21	ND	14.7	28.4	56.9

SFA, saturated fatty acids; MUFA, monosaturated fatty acids; PUFA, polysaturated fatty acids; ND, not detected.
Adapted from Ryan et al. (2007).

Table 2.8 Vitamin B$_{12}$ content of selected pulses

Local name	Botanical name	Vitamin B$_{12}$ (µg 100g^{-1})
Kalai	Phaseolus mungo	0.42±0.02
Mung	Phaseolus radiatus	0.61±0.01
Musur	Lens esculenta	0.43±0.01
Mator	Pisum sativum	0.36±0.02
Khesari	Lathyrus sativus	0.37±0.0
Arhar	Cajunus indicus	0.53±0.01
Chhola	Cicer arietinum	0.35±0.01

Adapted from Rohatgi et al. (1955).

Table 2.9 Thiamine, riboflavin and niacin (mg $100\,g^{-1}$ d.m.), pyridoxamine, pyridoxal and pyridoxine (μg $100\,g^{-1}$ d.m.) contents in pulses

Legume	Thiamine	Riboflavin	Niacin	Pyridoxamine	Pyridoxal	Pyrodoxine
Lens culinaris var. *vulgaris*	0.433±0.005	0.061±0.004	2.0 ±0.05			
Lens culinaris var. *variabilis*	0.647± 0.006	0.062±0.001	0.93±0.05			
Vicia faba var. *major*	0.253±0.003	0.123±0.008	2.23±0.008			
Faba beans			2.79±0.01			
Peas			3.15±0.08			
Lupins			2.05±0.009			
Lentils				63.3±2.6	94.2±2.8	44.2±2.5
Chickpeas				26.9±0.8	74.8±2.3	23.5±1.3
Haricot beans				64.5±1.2	128.8±1.6	24.0±0.2
Beans raw[a]	0.81–1.32	0.112–0.411	0.85–3.21			

[a]Augustin et al. (1981).
Adapted from Prodanova and Vidal-Valverde (1997), Vidal-Valverde et al. (2001).

has been inversely associated with the risk of colon cancer (Giovannucci et al., 1998). Chickpeas have a higher folate content of 149.7 μg $100\,g^{-1}$ compared with peas 101.5 μg $100\,g^{-1}$ (Dang et al., 2000).

Lentils *Lens culinaris* var. *variabilis* have thiamine, riboflavin, and niacin contents of 0.647, 0.062 and 0.93 mg $100\,g^{-1}$ d.m., respectively, against 0.253, 0.123 and 2.233 mg $100\,g^{-1}$ d.m., respectively, for *Vicia faba* var. *major* (Prodanov et al., 1997; Vidal-Valverde et al., 2001).

Most plant-derived foods contain low to moderate levels of vitamin E activity. However, owing to the abundance of plant-derived foods in our diets, they provide a significant and consistent source of vitamin E (Eitenmiller and Lee, 2004). Tocopherol content is higher in seeds and legumes than cereals. Peas contain greater amounts of α- than $\beta + \gamma$-tocopherols (10.4 and 5.7 mg $100\,g^{-1}$, respectively) and chickpeas contain similar levels of α- and $\beta + \gamma$-tocopherols (6.9 and 5.5 mg $100\,g^{-1}$, respectively) (Table 2.10) (Ryan et al., 2007).

2.4.2 Minerals

Mineral contents of pulses (Table 2.11) indicate that beans and lentils have the highest iron (110 and 122 μg g^{-1}, respectively) and zinc contents (44 and 48 μg g^{-1}, respectively). The levels of minerals in

Table 2.10 Tocopherol, phytosterols and squalene (mg 100 g^{-1}) in legumes

Legume	α-Tocopherol	β+γ-Tocopherol	β-Sitosterol	Campesterol	Stigmasterol	Squalene
Butter bean	0.7±0.18	4.7±0.40	85.1±7.3	15.2±2.9	86.2±5.7	0.4±0.02
Chickpea	6.9±0.04	5.5±0.72	159.8±7.1	21.4±0.7	23.4±0.7	0.5±0.03
Kidney bean	1.2±0.16	2.6±0.13	86.5±2.6	6.5±0.8	41.4±1.6	0.7±0.05
Lentil	1.6±0.43	4.5±0.11	123.4±4.1	15.0±0.4	20.1±0.6	0.7±0.15
Pea	10.4±0.09	5.7±0.64	191.4±0.4	25.0±6.9	26.0±0.6	1.0±0.07

Adapted from Ryan et al. (2007).

Table 2.11 Copper, iron and zinc content (μg g^{-1}) in legumes and food composition

Sample	Origin	Copper	Iron	Zinc	Reference
Bean	Spain	–	62.0	35.0	Mataix and Manas (1998)
	Spain	–	–	35.4±2.8	Terres et al. (2001) Ministerio de Sanidad Consumo (1999)
	Germany	0.14[b]	0.83[b]	0.18[b]	Souci et al. (2000)
	India	9–22	108–150	50–109	Vadivel and Janardhanan (2000)
	UK	3.2±0.7	68±1.6	44±1.2	Elhardallon and Walker (1992)
	UK	10.9	42.0	50.5	Holland et al. (1992)
	Italy	–	57.9±2.0	32.9±4.0	Lombardi-Boccia et al. (1994)
Broad bean	Spain	–	55.0	31.0	Mataix and Manas (1998)
	UK	9.1±0.7	110±3.2	58±2.7	Mataix and Manas (1998)
Chickpea	Spain	–	–	33.5±3.6	Terres et al (2001)
	Spain	–	68.0	10.0	Ministerio de Sanidad Consumo (1999)
	Spain	3.51	72.0	8.0	Souci et al (1994), Jimenez et al. (1998)
Lentil	Spain	–	–	45.1±14.2	Terres et al (2001)
	Spain	–	82.0	37.0	Mataix and Manas (1998)
	Spain	2.5	70.0	55.0	Jimenez et al. (1998)
	France	–	80.0	–	Feinberg et al. (1995)
	UK	9.1±0.7	122±4.1	48±1.0	Elhardallon and Walker (1992)
	UK	10.2	111.0	39.0	Holland et al. (1992)
Green pea	Spain[a]	19.0[a]	7.0[a]	–	Mataix and Manas (1998)
	Spain	1.75[a]	19.0[a]	7.0[a]	Jimenez et al. (1998)

[a]Fresh weight.
[b]Results are expressed as mg MJ^{-1}.
Adapted from Campos-Vega et al. (2010).

Table 2.12 Mineral content ($\mu g\ g^{-1}$ of the edible portion) of legumes

Legume	Copper	Chromium	Iron	Zinc	Aluminum	Nickel	Lead	Cadmium
Lentil	2.5	0.31	71	56.5	30.2	0.24	0.51	0.009
Haricot bean	2.8	0.15	62.5	39.7	13.4	0.15	0.62	0.0005
Kidney bean	3.0	0.17	64.4	46.9	19.0	0.17	0.69	0.007
Broad bean	4.3	0.28	80.0	41.2	6.7	0.17	0.4	0.012
Chickpea	3.5	0.12	68.8	39.2	10.2	0.26	0.48	0.01
Green peas	1.7	0.08	20.2	38.9	6.5	0.05	0.37	ND

ND, not detected.
Adapted from Cabrera et al. (2003).

legumes generally range from 1.5 to 5.0 μg Cu g^{-1}, 0.05 to 0.60 μg Cr g^{-1}, 18.8 to 82.4 μg Fe g^{-1}, 32.6 to 70.2 μg Zn g^{-1}, 2.7 to 45.8 μg Al g^{-1}, 0.02 to 0.35 μg Ni g^{-1}, 0.32 to 0.70 μg Pb g^{-1} and not detectable to 0.018 μg Cd g^{-1} (Table 2.12) (Cabrera et al., 2003). The content of Fe and other minerals is generally high in pulses, with beans having the highest mineral content (Table 2.13). However, Iqbal et al. (2006) reported that cowpea had the highest concentrations of potassium, magnesium and phosphorus among four legumes: chickpea, cowpea, lentil and green pea (Table 2.14). Sodium was found in an appreciable concentration in green pea. Chickpea contained good amounts of calcium, zinc and copper. These results revealed that legumes may provide sufficient amounts of minerals to meet the human mineral requirement (recommended dietary allowance, RDA) (NRC, 1980).

2.4.3 Phytic acids

Phytic acid (myo-inositol hexakisphosphate or IP6), a major phosphorus storage form in plants, and its salts known as phytates, constitutes 1–3% of cereal grains, pulse seeds and nuts. In particular, wholegrain cereals and legumes have a high content of phytate (Sandberg, 2002). Phytates are located in the protein bodies in the endosperm portion of pulses. Phytate occurs as a mineral complex, which is insoluble at the physiological pH of the intestine (Fredlund et al., 2006). The phytate content of selected pulses is shown in Table 2.15 (Morris and Hill, 1996; Rochfort and Panozzo, 2007). InsP6 and InsP5 account for more than 95% of the total inositol

Table 2.13 Mineral content (mg kg^{-1}) of dry bean cultivars

Cultivar	MK	Ca	Mg	K	P	B	Cu	Zn	Fe	Mn
AC Cruiser	Navy	1381 de	2002 b	17286 f	6207 b	13.7 b	8.8 a	25.0 c	55.0 b	13.3 d
AC Earlired	Small red	13344 ef	1677 f	17305 f	5727 d	11.3 c	0.4 h	18.9 e	34.1 e	13.2 d
AC Mast	Navy	1649 b	1952 c	17944 d	6564 a	11.1 c	3.1 e	18.3 e	44.1 de	15.8 c
AC Ole	Pinto	1328 f	1758 e	18638 b	6194 b	10.2 e	3.5 d	21.5 d	43.6 d	13.3 d
CDC Jet	Black	1229 g	1843 d	16166 h	5044 f	10.5 de	2.4 f	26.6 b	46.1 cd	13.8 d
Envoy	Navy	2025 a	1841 d	16945 g	5695 d	9.4 f	6.1 c	27.1 ab	53.3 bc	17.2 b
Galley	Navy	2035 a	2078 a	19464 a	6447 a	10.9 cd	1.0 g	21.2 d	42.2 d	19.4 a
Onyx	Black	1408 d	1746 e	18428 c	5483 e	14.7 a	0.1 l	18.8 e	49.4 bcd	13.3 d
Resolute	Great northern	1516 c	1693 f	17585 e	6034 c	10.2 e	0.9 g	18.5 e	28.0 e	13.5 d
ROG 802	Dark red kidney	823 h	1525 g	17090 g	5660 d	11.1 c	7.1 b	28.3 a	66.6 a	10.8 e
Means	Black	1300 x	1804 x	17070 y	5220 y	12.2 x	1.5 y	23.5 x	47.4 x	13.6 x
Means	Navy	1749 x	1958 x	17768 x	6209 x	11.3 x	5.1 x	23.1 x	49.2 x	16.2 x
Overall mean		1456	1804	17595	5901	11.2	3.6	22.6	46.3	14.2

Means in a column with different letters are significantly different ($P<0.05$).
Adapted from Oomah et al. (2008).

Table 2.14 Mineral constituent of important grain legumes

Mineral (mg 100 g^{-1})	Chickpea	Cowpea	Lentil	Green pea
Sodium	101±3.51 [b]	102±5.29 [ab]	79±2.65 [c]	111±2.65 [a]
Potassium	1155±5.00 [b]	1280±8.62 [a]	874±6.43 [d]	1021±12.49 [c]
Phosphorus	251±6.11 [b]	303±7.94 [a]	294±3.61 [a]	283±3.00 [a]
Calcium	197±3.61 [a]	176±4.58 [b]	120±6.24 [c]	110±3.61 [c]
Iron	3.0±0.20 [a]	2.6±0.20 [ab]	3.1±0.26 [a]	2.3±0.05 [b]
Copper	11.6±0.20 [a]	9.7±0.20 [b]	9.9±0.10 [b]	10.0±0.40 [b]
Zinc	6.8±0.26 [a]	5.1±0.20 [a]	4.4±0.20 [a]	3.2±0.56 [a]
Manganese	1.9±0.10 [a]	1.7±0.04 [a]	1.6±0.03 [a]	2.2±0.02 [a]
Magnesium	4.6±0.04 [ab]	4.8±±0.10 [a]	4.5±0.04 [b]	4.2±0.04 [c]

Means in each row for each mineral followed by the same letter are not significantly different ($P \geqslant 0.05$). Each column contains mean and SD of mean.
Adapted from Iqbal et al. (2006).

phosphates in most raw grains and legumes. InsP3 is undetectable in most of these seeds and InsP4 is usually less than 5% of the total. Raw lentils contained 0.3 mmol kg^{-1} of InsP3. The highest InsP4 concentration in raw legumes was 0.26 mmol kg^{-1} in black-eye peas and accounted, on average, for only about 1% of the total inositol phosphates in raw, dry legumes. The mean InsP5 concentration in

Table 2.15 Phytate content (μmol g^{-1}) in pulses

Food legume	IP$_4$	IP$_5$	IP$_6$
Baby lima bean	0.23	2.13	9.96
Black bean	0.13	1.87	14.2
Black pea (cowpea)	0.26	2.52	12.6
Chickpea (Garbanzo bean)	0.04	1.76	6.00
Great northern bean	0.19	2.19	12.7
Green split pea	0.17	1.36	6.48
Lentil	0.21	1.39	8.37
Navy bean	0.14	1.80	12.4
Pigeon pea	0.04	2.41	7.96
Pinto bean	0.17	2.05	11.7
Red chili bean	0.02	2.18	11.9
Red kidney bean	0.16	1.84	13.5
Roman bean	0.02	1.95	10.6
Yellow pea	0.12	1.49	8.82
Field pea	0.01	0.08	0.43
Pink bean	ND	0.60	13.07
Mung bean	ND	1.18	5.87

ND, not detected.
Adapted from Morris and Hill (1996), Phillippy (2003), Chen (2004), Rochfort and Panozzo (2007).

raw, dry pulses was 1.9 mmol kg^{-1} and $1.36–2.52 \text{ mmol kg}^{-1}$ in green split peas and black-eye peas, respectively, accounting for 16% of total inositol phosphates. The most abundant inositol phosphate in raw, dry pulses was InsP6, accounting for 83% of total inositol phosphates, 77% in chickpeas and 88% in black beans. Oomah et al. (2008) reported that phytic acid (InsP6) represents 75% of the total phosphorus in several Canadian bean varieties. Varietal and agronomic factors, alone and in combination, often result in wide variation in phytate content of mature pulse seeds (Dintzis et al., 1992; Mason et al., 1993). Chen (2004) reported InsPn contents in beans with only InsP6 and InsP5 detected in all beans. There was a wide variation in the InsP6 or InsP5 content among different types of raw dry black beans or red kidney beans. InsP6 content (kg^{-1}, adjusted by moisture) in raw dry beans ranged from 5.87 to 14.86 mmol in mung beans and black beans, respectively. InsP6 was the predominant inositol phosphate of the total InsPn determined in raw dry beans, ranging from 63.9% in red kidney beans to 97.5% in pinto beans. Of the four possible InsP5 isomers (excluding enantiomers), DL-Ins(1,2,4,5,6)P5 was dominant in raw dry beans, followed by Ins(1,3,4,5,6)P5, DL-Ins(1,2,3,4,5)P5 and Ins(1,2,3,4,6)P5, if present, which indicated that there probably were some common profiles at least for raw dry beans.

2.4.4 Polyphenols

The potential health benefits of common beans are attributed to the presence of secondary metabolites such as phenolic compounds that possess antioxidant properties (Cardador-Martinez et al., 2002; Azevedo et al., 2003; Lazze et al., 2003). The major polyphenolic compounds of pulses consist mainly of tannins, phenolic acids and flavonoids (Fig. 2.3). Pulses with the highest polyphenolic content have dark and highly pigmented grains, such as red kidney beans (*Phaseolus vulgaris*) and black gram (*Vigna mungo*). Condensed tannins (proanthocyanidins) have been quantified in hulls of several varieties of field beans (*Vicia faba*) and are also present in pea seeds of colored-flower cultivars. Tannin-free and sweet seeds have been selected among broad beans, lentils and lupins (Smulikowska et al., 2001). Pulses vary based on their total phenolic contents and antioxidant activities (Table 2.16). Lentils have the highest phenolic, flavonoid and condensed tannin content (6.56 mg gallic acid equivalents

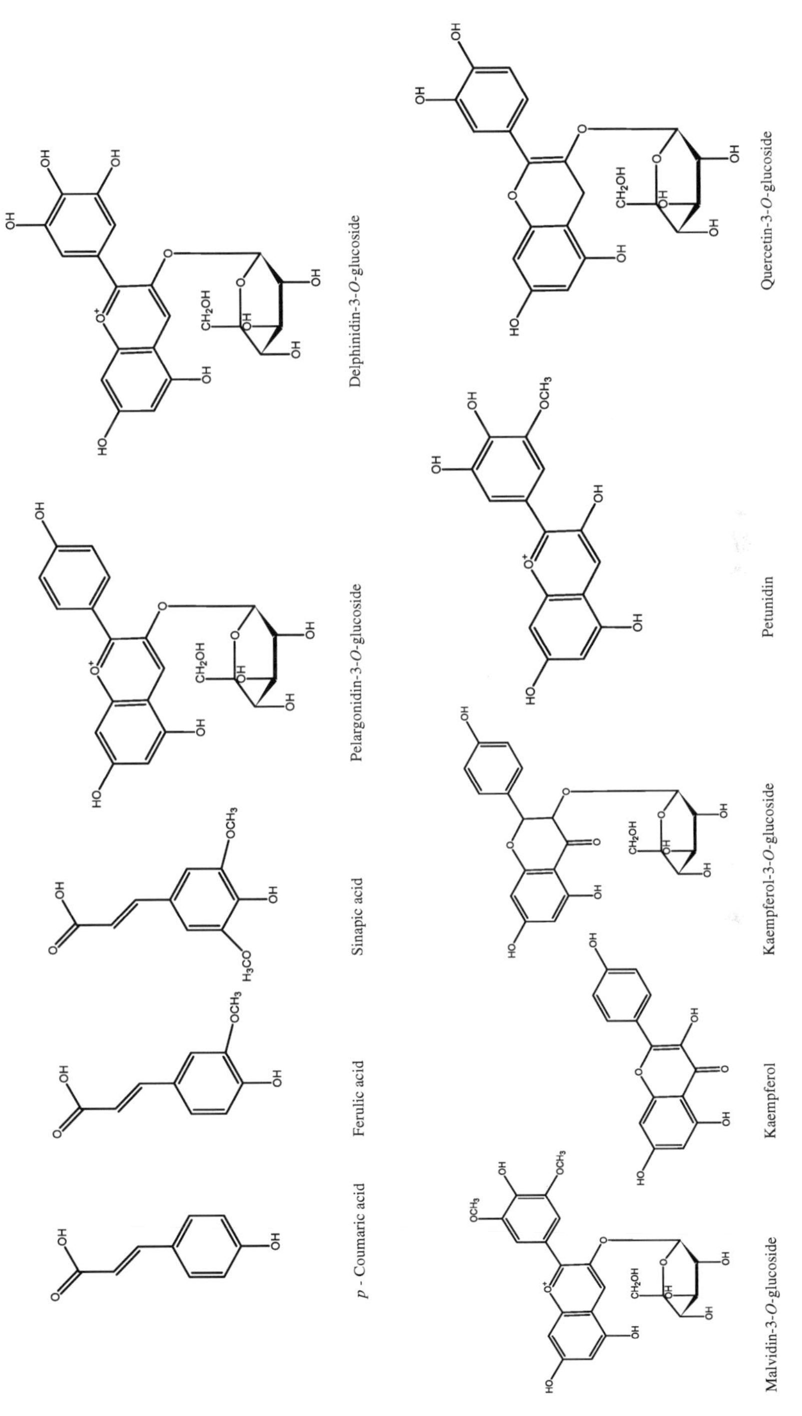

Delphinidin-3-glucoside

Pelargonidin-3-O-glucoside

Quercetin-3-O-glucoside

Petunidin

Sinapic acid

Ferulic acid

p - Coumaric acid

Kaempferol-3-O-glucoside

Kaempferol

Malvidin-3-O-glucoside

Figure 2.3 Some common phytochemicals present in pulses.

(Continued)

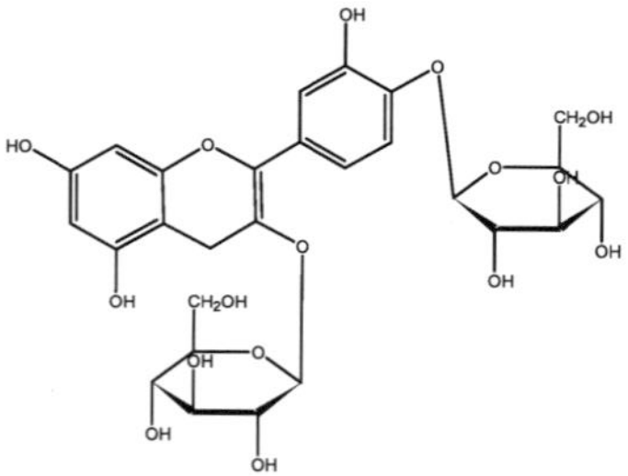

Biochanin A

Stigmasterol

Formononetin

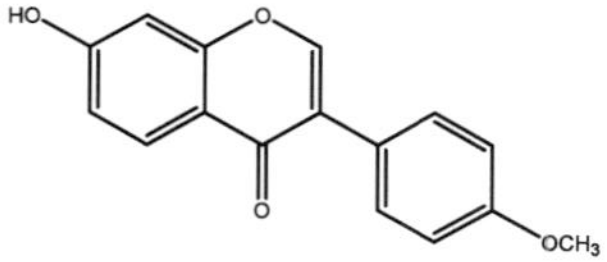

β-Stosterol

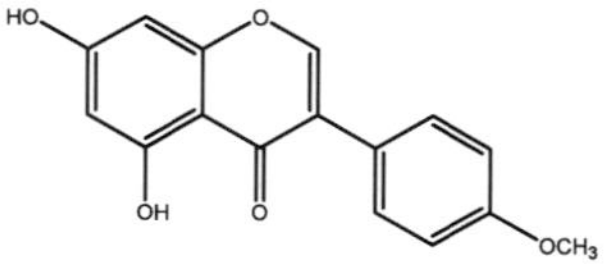

Daidzein

Genistein

Quercetin-3,4-di-glucoside

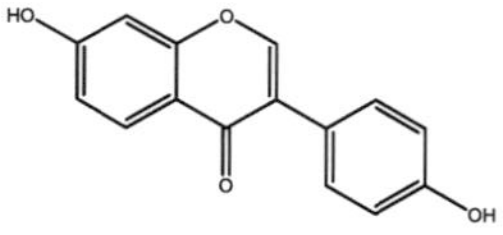

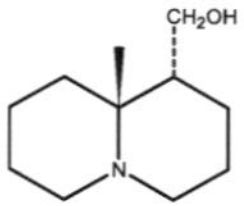

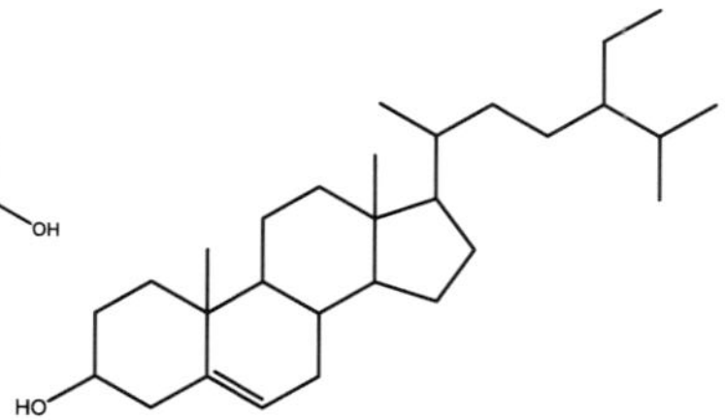

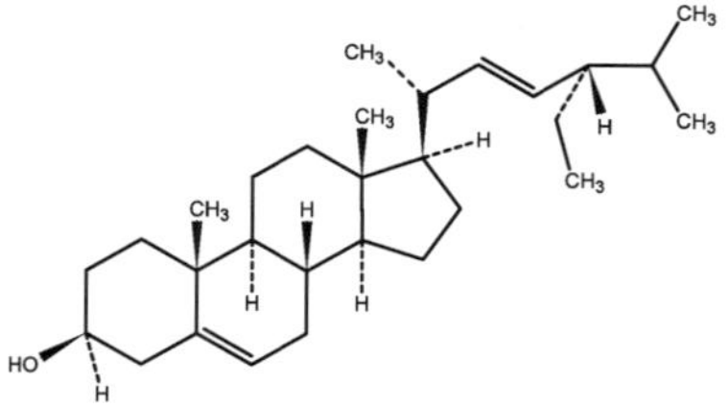

(−)-Lupinine

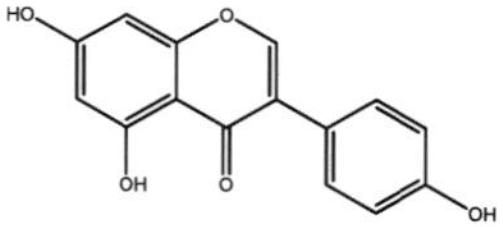

Figure 2.3 (Continued) Some common phytochemicals present in pulses.

(Continued)

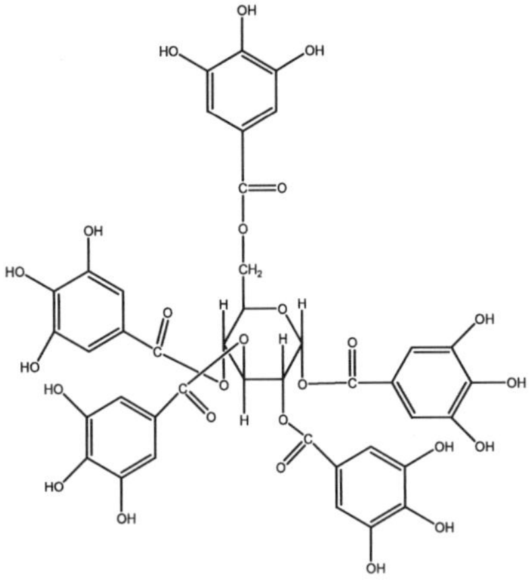

Oxalate ion

Tannins

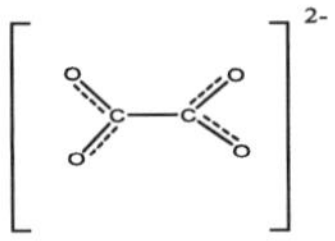

Figure 2.3 (Continued) Some common phytochemicals present in pulses.

g^{-1}, 1.30 and 5.97 mg catechin equivalents g^{-1}, respectively), followed by red kidney and black beans (Xu and Chang, 2007).

Ferulic acid is the most abundant phenolic acid in common beans and intermediate levels of p-coumaric and sinapic acids are also present (Table 2.17) (Luthria and Pastor-Corrales, 2006). Oomah et al. (2005) reported a fivefold variation (3.3–16.6 mg catechin equivalents g^{-1}) in total phenolic content of six Canadian bean varieties, while variations in flavonoids, anthocyanins, flavonols and tartaric esters were minimal. Twenty-four common bean samples analyzed recently by Long-Ze et al. (2008) contained the same hydroxycinnaminic acids, but the flavonoid components showed distinct differences. Black beans contained primarily the 3-O-glucosides of delphinidin, petunidin

Table 2.16 Phenolic contents of pulses

Legume	Total phenolic content (mg gallic acid equivalents g^{-1})	Total flavonoid content (mg catechin equivalents g^{-1})	Condensed tannin content (mg catechin equivalents g^{-1})
Green pea	1.53	0.08	0.26
Yellow pea	1.67	0.18	0.42
Chickpea	1.81	0.18	1.05
Lentil	6.56	1.30	5.97
Red kidney	4.98	2.02	3.85
Black bean	5.04	2.49	3.40

Adapted from Xu et al. (2007).

Table 2.17 Phenolic acid content of commonly consumed dry beans in the USA

Bean class	Cultivars	Mean phenolic acid concentration (mg 100 g^{-1})			Total phenolic acid content (mg g^{-1})
		p-Coumaric	Ferulic	Sinapic	
Pinto	Maverik	4.5	22.9	8.5	36.0
	Buster	4.5	16.0	9.0	29.5
	Othello	5.6	15.2	5.9	26.7
Great northern	Norstar	4.0	17.0	9.4	30.4
Matterhorn		6.3	17.2	9.0	32.5
Navy	Vista	12.4	26.6	9.2	48.3
Black	T–39	11.6	25.5	9.0	47.1
	Jaguar	7.0	11.7	5.7	24.4
	Eclipse	9.8	24.7	6.8	42.5
Dark red kidney	Red Hawk	1.8	15.3	3.8	20.9
Light red kidney	Cal Early	7.0	14.8	5.7	27.4
Red Mex	UI 239	5.8	17.4	5.4	28.6
Cranberry	Taylor Cranberry	1.7	14.0	3.5	19.1
Pink	UI 537	6.8	19.4	8.2	34.4
Alubia	Beluga	5.3	10.6	4.0	19.8

Adapted from Luthria and Pastor-Corrales (2006).

and malvidin, while kaempferol and its 3-O-glycosides were present in pinto beans (see Fig. 2.3). Light red kidney beans had traces of quercetin 3-O-glucoside and its malonates, but pink and dark red kidney beans contained the diglycosides of quercetin and kaempferol. Small red beans contained kaempferol 3-O-glucoside and pelargonidin 3-O-glucoside, while flavonoids were undetected in alubia, cranberry, great northern and navy beans. Total anthocyanin content in whole grain and seed coat ranged from 37.7 to 71.6 mg g^{-1} and 10.1 to 18.1 mg g^{-1}, respectively, among 15 black bean cultivars grown in Mexico. The anthocyanins in seed coats of beans were identified as delphinidin 3-glucoside 65.7%, petunidin 3-glucoside 24.3% and maldivin 3-glucoside 8.7% (Salinas-Moreno et al., 2005). Chickpeas also contain a wide range of polyphenolic compounds, including flavonols, flavone glycosides, flavonols, and oligomeric and polymeric proanthocyanidins (Sarma et al., 2002; Singh et al., 2003). Total phenolic content in chickpea ranges from 0.92 to 1.68 mg gallic acid equivalents g^{-1} (Xu and Chang, 2007; Zia-Ul-Haq et al., 2008). Lignans, diphenolic compounds with a 2,3-dibenzylbutane skeleton, have both estrogenic and antiestrogenic properties (Orcheson et al., 1998).

Table 2.18 Lignan content (μg 100 g^{-1} dry wt) of legumes as SEC and MAT or as ED and EL

Legume	Direct analysis[a]		In vitro fermentation		
	SEC	MAT	ED[b]	EL[b]	Total[c]
Lentil	Nr	0	1092	864	1956
Kidney bean	69.9	0	266	377	643
Navy bean	85.8	0	144	399	543
Pinto bean	79.1	Traces	53	173	226
Yellow pea	8.2	Traces	48	185	233

ED, entreodiol; EL, enterolactone.
Mazur et al. (1998); Nr, not reported; SEC, secoisolariciresinol;
MAT, matairesinol; [b]Thompson et al. (1991); [c]Total=sum of ED + EL.
Adapted from Meagher and Beecher (2000).

The plant lignans, secoisolariciresinol (SEC) and matairesinol (MAT), are converted to the metabolites enterodiol (ED) and enterolactone (EL), known as the mammalian lignans, in the gastrointestinal tract. Most studies have only looked at the isoflavonoid content of legumes, only one study (Mazur et al., 1998) has analyzed the SEC and MAT content. The concentrations of SEC lignans in legumes ranged from 0 to 240 μg 100 g^{-1} and 13 to 273 μg 100 g^{-1} in soybean (Table 2.18) with trace or no MAT detected (Mazur et al., 1998).

2.4.4.1 Isoflavones

Flavones and isoflavones have been isolated from various plants, though the isoflavones are largely reported from the Fabaceae/Leguminosae family. According to the USDA survey on isoflavone content, lentils do not contain significant amounts of these isoflavones (USDA, 2002). Chickpeas contain daidzein, genistein and formononetin (0.04, 0.06 and 0.14 mg 100 g^{-1}, respectively) and approximately 1.7 mg 100 g^{-1} biochanin A. Soybeans have significantly higher levels of daidzein and genistein (47 and 74 mg 100 g^{-1}, respectively) but contain less formononetin and biochanin A compared to chickpeas, 0.03 and 0.07 mg 100 g^{-1}, respectively (see Fig. 2.3).

Total isoflavones in *L. mutabilis* range from 9.8 to 87, 16.1 to 30.8 and 1.3 to 6.1 mg 100 g^{-1} fresh weight of sample (expressed as genistein) in seed coat, cotyledon and hypocotyl fractions, respectively (Ranilla et al., 2009). Barcelo and Munoz (1989) identified isoflavones such as genistein, 2'-hydroxigenistein, luteon and wighteone in sprouted hypocotyls of *L. albus* CV multolupa, suggesting that these compounds are related to cell wall lignification. This may explain luteone (a tetrahydroxyisoflavone) which was detected in immature seeds of *L. luteus* (Fukui et al., 1973). Dini et al. (1998) detected two genistein derivatives, mutabilin (glycosylated form) and mutabilein (aglycon form), in seeds of *L. mutabilis*. Furthermore, formononetin, genistein and the phytoestrogen secoisolariciresinol were found in seeds of *L. mutabilis* (23, 2420 and 3.1 μg $100\,g^{-1}$, respectively) (Mazur et al., 1998). The cotyledon of Andean lupins has the highest content of total isoflavones (16–31 mg $100\,g^{-1}$ cotyledon FW) compared to the hypocotyls (1.3–6.1 mg $100\,g^{-1}$ hypocotyls FW) and seed coats (9.8–10 mg $100\,g^{-1}$ seed coat FW). Interestingly, the genistein derivative (GD) was the major isoflavone found in seed coats and cotyledons from *L. mutabilis* cultivars. Furthermore, the H-6 cultivar was remarkable because of its high total isoflavone content in seed coats (87 mg $100\,g^{-1}$ FW), cotyledon (30.8 mg $100\,g^{-1}$ FW) and hypocotyls (6.1 mg $100\,g^{-1}$ FW).

2.4.5 Saponins

Saponins are present in many edible legumes, though the detailed structures have not been established. These have been reported in lupins (Woldemichael et al., 2003), lentils (Morcos et al., 1976; Ruiz et al., 1996), chickpeas (El-Adawy, 2002), various beans and peas (Shi et al., 2004). Saponin content may vary even among the same species of edible beans, because of variations in cultivars (Khokhar and Chauhan, 1986), locations (Fenwick and Oakenfull, 1983; Price et al., 1987), irrigation condition, type of soil, climatic and environmental conditions. The saponin content in various legumes is listed in Table 2.19. Chickpeas, black grams, moth beans, broad beans and peas are reported to contain 3.6, 2.3, 3.4, 3.7 and 2.5 g kg^{-1} dry matter of saponins, respectively (Khokhar and Chauhan, 1986). Saponin content in dehulled light and dark colored peas ranges from 1.2 to 2.3 g kg^{-1} dry matter (Daveby et al., 1998).

2.4.6 Oxalate

The oxalate content (Fig. 2.3) of different pulses is shown in Table 2.20. Dietary oxalate is a potential risk factor for kidney stone development and lowers the availability of dietary minerals such as

Table 2.19 Saponin content of pulse seeds

Common name	Saponin content (g kg^{-1} dry matter)
Beans	
Broad	0.1–3.7
Butter	1.0
Field	0.03–3.5
Green moth	3.3
Haricot	2.3
Kidney	2.16
Moth	3.4
Mung	3.4
Navy	2–16
Red	0.02
Runner	3.5
Peas	
Black eyed	0.03
Green	1.8–11
Pea	2.5
Snow	0.01
Yellow split	1.1–11
Black gram	2.3
Chickpea	2.3–60

Adapted from Khokhar and Chauhan (1986), Price et al. (1987), Oakenfull and Sidhu (1989).

calcium and magnesium. However, this needs to be confirmed by bioavailability investigations. Oxalates in food strongly chelate with dietary minerals such as calcium to form complexes and precipitate as insoluble salts accumulating in the renal glomeruli, and contribute to the development of renal disorders such as kidney stone formation (Noonan and Savage, 1999; Judprasong et al., 2006). About 75% of all kidney stones are composed primarily of calcium oxalate (Williams and Wandzilak, 1989; Chai and Liebman, 2005).

2.4.7 Phytosterols

Pulses contain small quantities of phytosterols, of which β-sitosterol, campesterol and stigmasterol are the most common (Benveniste, 1986) (Fig. 2.3). These compounds are also abundant as sterol glucosides and esterified sterol glucosides, with β-sitosterol representing 83% of the glycolipids in defatted chickpea flour (Sanchez-Vioque et al., 1998). Ryan et al. (2007) reported that the total phytosterol content in the pulses ranged from 134 mg 100 g^{-1} (kidney beans)

Table 2.20 Oxalate content of pulses	
Legumes	**Oxalate content (mg 100 g^{-1} wet weight)**
Beans	
Anasazi	80
Adzuki	25
Black	72
Garbanzo	9
Great northern	75
Large lima	8
Mung	8
Navy	57
October	28
Pink	75
Pinto	27
Red kidney	16
Small red	35
Small white	78
Peas	
Black eye	4
Green split	6
Yellow split	5

Adapted from Chai and Liebman (2005).

to 242 mg 100 g^{-1} (peas) while total β-sitosterol content ranged from 160 mg 100 g^{-1} (chickpeas) to 85 mg 100 g^{-1} (butter beans). Chickpeas and peas contained high levels of campesterol (21.4 and 25.0 mg 100 g^{-1}, respectively). Stigmasterol content is higher in butter beans (86 mg 100 g^{-1}) as is the squalene content in peas (1.0 mg 100 g^{-1}).

2.4.8 Alkaloids

Alkaloids constitute a group of very diverse compounds consisting of a heterocycle with a nitrogen atom within the cycle. This conformation confers a basic character on the molecule, which tends to acquire a proton in aqueous solution, except when the nitrogen atom is close to an electron acceptor in the molecule (e.g. ricinine). They are mainly present in lupins, but breeding of alkaloid-free varieties ("sweet varieties") has increased the lupin content of fodder for all classes of domestic livestock (Bond and Duc, 1993). Alkaloids are present in some other grain legumes, such as the jackbean, in which trace quantities of lupanine have been found (Oboh et al., 1998).

2.5 Conclusions and prospects

The key nutritional role of grain pulses is unquestionable, due to the massive presence of macro- and micronutrients. Pulses supply significant amounts of protein and calories for both rural and urban populations of developing and developed countries. These pulses contain up to 60% carbohydrates (mainly starch). Pulses are also a good source of major and minor (polyphenols, vitamins, minerals) compounds which may have important metabolic and/or physiological effects. More recent evidence provides potential information of their impact on health, so these secondary metabolites are currently marketed as functional foods and nutraceutical ingredients. In the frame of a reappraisal of the effects that grain legume components may have on human health and wellness, widely accepted claims on their beneficial activity in the prevention and treatment of various diseases have been made. Altogether, these claims strongly support the regular dietary intake of grain legumes as one of the ways to a healthy life. Nevertheless, many efforts and further studies are still needed in order to disclose the mechanism(s) underlying the legume proteins/peptides effects; to identify and characterize novel biological activities often "hidden" inside the polypeptide chains, and to establish clear dose–response relationships in order to calibrate the preparation and use of nutraceutically enhanced foods. Similar direction is needed for other bioactive compounds in pulses (phenolic acids, anthocyanins, vitamins, etc.). As parallel or further steps, the biotechnological approaches can be extremely useful as cognitive tools and in the design of novel, more effective biologically active molecules. Future research may determine whether they should be preserved or eliminated in each main nutritional situation.

References

Akpapunam, M.A., Markakis, P., 1979. Oligosaccharides of 13 American cultivars of cowpeas (*Vigna sinensis*). J. Food Sci. 44, 1317–1318.

Akpinar, N., Akpinar, M.A., Türkoglu, S., 2001. Total lipid content and fatty acid composition of the seeds of some *Vicia* L. species. Food Chem. 74, 449–453.

Augustin, J., Beck, C.B., Kalbfleish, G., Kagel, L.C., Matthews, R.H., 1981. Variation in the vitamin and mineral content of raw and cooked commercial *Phaseolus vulgaris* classes. J. Food Sci. 46, 1701–1706.

Azevedo, A., Gomes, J.C., Stringheta, P.C., 2003. Black bean (*Phaseolus vulgaris* L.) as a protective agent against DNA damage in mice. Food Chem. Toxicol. 41, 1671–1676.

Baghurst, K., Baghurst, P.A., 2001. Dietary fibre, non-starch polysaccharide and resistant starch intakes in Australia. In: Spiller, G.A. (Ed.), CRC handbook of dietary fibre in human health. CRC Press, Boca Raton, pp. 583–591.

Bailey, R.W., 1973. Chemistry and biochemistry of herbage, Vol. 1. Academic Press, London.

Ball, S., Guan, H.P., James, M., 1996. From glycogen to amylopectin: a model for the biogenesis of the plant starch granule. Cell 86, 349–352.

Barcelo, A.R., Munoz, R., 1989. Epigenetic control of a cell wall scopoletin peroxidase by lupisoflavone in *Lupinus*. Phytochemistry 28, 1331–1333.

Bassett, C., Boye, J., Tyler, R., Oomah, B.D., 2010. Molecular, functional and processing characteristics of whole pulses and pulse fractions and their emerging food and nutraceutical applications. Special Issue Food Res. Int. 43, 397–659.

Bazzano, L.H.J., Ogden, L.G., Loria, C., Vupputuri, S., Myers, L., Whelton, P.K., 2001. Legume consumption and risk of coronary heart disease in US men and women: NHANES I Epidemiologic Follow-up Study. Arch. Intern. Med. 161, 2573–2578.

Becker, R., Olson, A.C., Frederick, D.P., Kon, S., Gumbmann, M.R., Wagner, J.R., 1974. Conditions for the autolysis of alpha-galactosides and phytic acid in California small white beans. J. Food Sci. 39, 766–769.

Benveniste, P., 1986. Sterol biosynthesis. Annu. Rev. Plant Physiol. 37, 275–308.

Bernabe, M., Fenwick, R., Frias, J., Jimenez-Barbero, J., Price, K., Valverde, S., 1993. Determination, by NMR spectroscopy, of the structure of ciceritol, a pseudotrisaccharide isolated from lentils. J. Agric. Food Chem. 41, 870–872.

Bhatty, R.S., Slinkard, A.E., 1979. Composition, starch properties and protein quality of lentils. Can. Inst. Food Sci. Technol. J. 12, 88–92.

Biliaderis, C.G., Grant, D.R., Vose, J.R., 1979. Molecular weight distribution of legume starches by gel chromatography. Cereal Chem. 56, 475–480.

Bogracheva, T.Y., Cairns, P., Noel, T.R., 1999. The effect of mutant genes at the r, rb, rug3, rug4, rug5 and lam loci on the granular structure and physicochemical properties of pea seed starch. Carbohydr. Polym. 39, 303–314.

Bond, D.A., Duc, G., 1993. Plant breeding as a means of reducing antinutritional factors of grain legumes. In: Van der Poel, A., Huisman, J., Saini, H.S. (Eds.), Recent advances of research in antinutritional factors in legume seeds, Proceedings of the 2nd International Workshop on Antinutritional Factors (ANFs) in Legume Seeds. Wageningen Press, Wageningen, The Netherlands, pp. 379–396.

Borowska, J., Zadernowski, R., Borowski, J., Swiecicki, W.K., 1998. Intravarietal differentiation of pea (*Pisum sativum* L. subsp. *sativum*) seeds – their chemical composition and physical properties. Plant Breed. Seed Sci. 42, 75–85.

Bressani, R., Elias, L.G., 1979. The problems of legume protein digestibility. J. Food. Sci. 39, 61–67.

Brown, J.W.S., Bliss, F.A., Hall, T.C., 1981. Linkage relationships between genes controlling seed proteins in French bean. Theor. Appl. Genet. 60, 251.

Buchala, A.J., Franz, G., 1974. A hemicellulose beta-glucan from the hypocotyls of *Phaseolus aureus*. Phytochemistry 13, 1887–1899.

Cabrera, C., Lloris, F., Gimenez, R., Olalla, M., Lopez, M.C., 2003. Mineral content in legumes and nuts: contribution to the Spanish dietary intake. Sci. Total Environ. 308, 1–14.

Campos-Vega, R., Loarca-Pina, G., Oomah, B.D., 2010. Minor components of pulses and their potential impact on human health. Food Res. Int. 43, 461–482.

Cardador-Martinez, A., Loarca-Pina, G., Oomah, B.D., 2002. Antioxidant activity in common beans. J. Agric. Food Chem. 50, 6975–6980.

Cerning, J., Sapsonik, A., Guilbot, A., 1975. Carbohydrate composition of horsebeans (*Vicia faba* L.) of different origins. Cereal Chem. 52, 125–138.

Cerning-Beroard, J., Filiatre, A., 1976. A comparison of the carbohydrate composition of legume seeds: horsebeans, peas, and lupines. Cereal Chem. 53, 968–978.

Chai, W., Liebman, M., 2005. Oxalate content of legumes, nuts, and grain-based flours. J. Food Compos. Anal. 18, 723–729.

Champ, M.M., 2002. Non-nutrient bioactive substances of pulses. Br. J. Nutr. 88(Suppl. 3), S307–S319.

Chavan, J.K., Kadan, S.S., Salunkhe, D.K., 1988. Biochemistry and technology of chickpea (*Cicer arietinum* L.) seeds. Crit. Rev. Food Sci. Nutr. 25, 107–158.

Chen, Q.C., 2004. Determination of phytic acid and inositol pentakisphosphates in foods by high-performance ion chromatography. J. Agric. Food Chem. 52, 4604–4613.

Chen, W.J.L., Anderson, J.W., 1981. Soluble and insoluble plant fiber in selected cereals and vegetables. Am. J. Clin. Nutr. 34, 1077–1082.

Christl, S.U., Murgatroyd, P.R., Gibson, G.R., Cummings, J.H., 1992. Production, metabolism, and excretion of hydrogen in the large intestine. Gastroenterology 102, 1269–1277.

Chung, H.J., Liu, Q., Donner, E., Hoover, R., Warkentin, T.D., Vandenberg, B., 2008. Composition, molecular structure, properties and *in vitro* digestibility of starches from newly released Canadian pulse cultivars. Cereal Chem. 85, 471–479.

Claydon, A., 1978. Winged bean, a food with many uses. Plant Foods Man 2, 203–224.

Clemente, A., Sanchez-Vioque, R., Vioque, J., Bautista, J., Millan, F., 1998. Effect of cooking on protein quality of chickpea (*Cicer arietinum* L.) seeds. Food Chem. 62, 1–6.

Colonna, P., Gallant, D., Mercier, C., 1980. *Pisum sativum* and *Vicia. faba* carbohydrates: studies of fractions obtained after dry and wet protein extraction processes. J. Food Sci. 45, 1629–1636.

Cristofaro, E., Mottu, F., Wuhrmann, J.J., 1974. Involvement of raffinose family of oligosaccharides in flatulence. In: Sipple, H.L., McNutt, K.W. (Eds.), Sugars in nutrition. Academic Press, London, pp. 313–336.

Dalgetty, D.D., Baik, B.-K., 2003. Isolation and characterization of cotyledon fibers from peas, lentils, and chickpeas. Cereal Chem. 80, 310–315.

Dang, J., Arcot, J., Shrestha, A., 2000. Folate retention in selected processed legumes. Food Chem. 68, 295–298.

Daveby, Y.D.A.P., Betz, J.M., Musser, S.M., 1998. Effect of storage and extraction on ratio of soyasaponin I to 2,3-dihydro-2,5-dihydroxy-6-methyl-4-pyroneconjugated soyasaponin I in de-hulled peas (*Pisum sativum* L.). J. Sci. Food Agric. 32, 141–146.

Deshpande, S.S., Nielsen, S.S, 1987. *In vitro* enzymatic hydrolysis of phaseolin, the major storage protein of *Phaseolus vulgaris* L. J. Food Sci. 52, 1326.

Dini, I., Schettino, O., Dini, A., 1998. Studies on the constituents of *Lupinus mutabilis* (Fabaceae). Isolation and characterization of two new isoflavonoid derivatives. J. Agric. Food Chem. 46, 5089–5092.

Dintzis, F.R., Lehrfeld, J., Nelsen, T.C., Finney, P.L., 1992. Phytate content of soft wheat brans as related to kernel size, cultivar, location, and milling and flour quality parameters. Cereal Chem. 69, 577–581.

Dovlo, F.E., Williams, C.E., Zoaka, L., 1976. Cowpeas, home preparations and use in West Africa. Publication IDRC-055e. International Development Research Centre, Ottawa, Canada.

Duranti, M., Gius, C., 1997. Legume seeds: protein content and nutritional value. Field Crops Res. 53, 31.

Dyssler, P., Hoffmann, D., 1994. In: Asp, N.-G., Amelsvoort, J.M.M.v., Hautvat, J.G.A.J. (Eds.), Estimation of resistant starch intake in Europe. EURESTA, Wageningen, pp. 84-86.

Eitenmiller, R.R., Lee, J., 2004. Vitamin E: food chemistry, composition and analysis. Marcel Dekker, New York.

Ekpenyong, T.E., Borchers, R.L., 1980. Effect of cooking on the chemical composition of winged beans (*Psophocarpus tetragonolobus*). J. Food Sci. 45, 1559–1560.

El-Adawy, T.A., 2002. Nutritional composition and antinutritional factors of chickpeas (*Cicer arietinum* L.) undergoing different cooking methods and germination. Plant Foods Hum. Nutr. 57, 83–97.

Elhardallon, S.B., Walker, A.F., 1992. Binding of iron by three starchy legumes in the presence of iron alone, with calcium or with calcium, zinc, magnesium and copper. J. Food Sci. Nutr. 43, 61–68.

Englyst, H.N., Kingman, S.M., Cummings, J.H., 1992. Classification and measurement of nutritionally important starch fractions. Eur. J. Clin. Nutr. 46(Suppl. 2), S33–S50.

Eskin, N.A.W., Johnson, S., Vaisey-Genser, M., McDonald, B.E., 1980. A study of oligosaccharides in a select group of legumes. Can. Inst. Food Sci. Technol. J. 13, 40–42.

Faboya, O.O.P., Aku, U.U., 1996. Interaction of oxalic acid with divalent metals – calcium, magnesium and zinc – during the cooking of cowpea. Food Chem. 57, 365–369.

Feinberg, M., Favier, J.C., Ireland-Ripert, J., 1995. Repértoir e general des aliments: table de composition. Institut National de la Recherche Agronomique, Paris.

Fenwick, D.E., Oakenfull, D., 1983. Saponin content of food plants and some prepared foods. J. Sci. Food Agric. 34, 186–191.

Fleming, S.E., 1981. A study of relationships between flatus potential and carbohydrate distribution in legume seeds. J. Food Sci. 46, 794–798.

Flight, I., Clifton, P., 2006. Cereal grains and legumes in the prevention of coronary heart disease and stroke: a review of the literature. Eur. J. Clin. Nutr. 60, 1145–1159.

Fredlund, K., Isaksson, M., Rossander-Hulthen, L., Almgren, A., Sandberg, A.-S., 2006. Absorption of zinc and retention of calcium: dose-dependent inhibition by phytate. J. Trace Elem. Med. Biol. 20, 49–57.

Fukui, H., Egawa, H., Koshimizu, K., 1973. A new isoflavone with antifungal activity from immature fruits of *Lupinus luteus*. Agric. Biol. Chem. 37, 417–421.

Gallant, D.J., Bouchet, B., Baldwin, P.M., 1997. Microscopy of starch: evidence of a new level of granule organization. Carbohydr. Polym. 32, 177–191.

Garcia, V.V., 1979. Biochemical composition of mature winged beans (*Psophocarpus tetragonolobus* L.). DCPhD Dissertation, VPI and State University, Blacksburg, Virginia.

Geervani, P., Theophilus, F., 1981. Studies on digestibility of selected legume carbohydrates and its impact on the pH of the gastrointestinal tract in rats. J. Sci. Food Agric. 32, 71–78.

Giovannucci, E., Stampfer, M.J., Colditz, G.A., 1998. Multivitamin use, folate, and colon cancer in women in the Nurses' Health Study. Ann. Intern. Med. 129, 517–524.

Goel, R., Verma, J., 1980. Removal of flatulence factor of some pulses by microbial fermentation. Indian J. Nutr. Diet. 18, 215–217.

Granito, M., Frias, J., Doblado, R., Guerra, M., Champ, M., Vidal-Valverde, C., 2002. Nutritional improvement of beans (*Phaseolus vulgaris*) by natural fermentation. Eur. Food Res. Technol. 214, 226–231.

Grela, E.R., Gunter, K.D., 1995. Fatty acid composition and tocopherol content of some legume seeds. Anim. Feed Sci. Technol. 52, 325–331.

Guillon, F., Champ, M., 1996. Grain legumes and transit in humans. In: AEP. (Ed.), Grain legumes, 11, pp. 18.

Halaby, G.A., Lewis, R.W., Ray, C.R., 1981. Nutrient content of commercially prepared legumes. Food Technol. 35, 86–88.

Harborne, J., Boulter, D., Turner, B.E., 1971. Chemotaxonomy of the Leguminosae. Academic Press, London.

Holland, B., Unwin, I.D., Buss, D.H., 1992. First supplement to McCance and Widdowson's The composition of foods, 5th edn. Royal Society of Chemistry, Cambridge.

Honow, R., Hesse, A., 2002. Comparison of extraction methods for the determination of soluble and total oxalate in foods by HPLC-enzyme-reactor. Food Chem. 78, 511–521.

Hoover, R., Ratnayake, W.S., 2002. Starch characteristics of black bean, chickpea, lentil, navy bean and pinto bean cultivars grown in Canada. Food Chem. 78, 489–498.

Hoover, R., Hughes, T., Chung, H.J., Liu, Q., 2010. Composition, molecular structure, properties, and modification of pulse starches: a review. Food Res. Int. 43, 399–413.

Huang, J., Schols, H.A., Jin, Z., Sulmann, E., Voragen, A.G.J., 2007. Characterization of differently sized granule fractions of yellow pea, cowpea and chickpea starches after modification with acetic anhydride and vinyl acetate. Carbohydr. Polym. 67, 11–20.

Iqbal, A., Khalil, I.A., Ateeq, N., Khan, M.S., 2006. Nutritional quality of important food legumes. Food Chem. 97, 331–335.

Iyengar, A.K., Kulkarni, P.R., 1977. Oligosaccharide levels of processed legumes. J. Food Sci. Technol. 14, 222–223.

Iyer, V., Salunkhe, D.K., Sathe, S.K., Rockland, L.B., 1980. Quick cooking of beans (*Phaseolus vulgaris* L.): II. Phytates, oligosaccharides, antienzymes. Plant Foods Hum. Nutr. 30, 45–52.

Jaya, T.V., Venkataraman, L.V., 1981. Changes in the carbohydrate constituents of chickpea and green gram during germination. Food Chem. 7, 95–104.

Jimenez, A., Cervera, P., Bacardi, M., 1998. Tabla de composición de alimentos. Novartis Nutrition, Bacelona.

Jood, S., Chauhan, B.M., Kapoor, A.C., 1988. Contents and digestibility of carbohydrates of chick pea and black gram as affected by domestic processing and cooking. Food Chem. 30, 113–127.

Judprasong, K., Charoenkiatkul, S., Sungpuag, P., Vasanachitt, K., Nakjamanong, Y., 2006. Total and soluble oxalate contents in thai vegetables, cereal grains and legume seeds and their changes after cooking. J. Food Compos. Anal. 19, 340–347.

Kadam, S.S., Salunkhe, D.K., 1989. Minerals and vitamins. In: Salunkhe, D.K. (Ed.), Handbook of world food legumes. CRC Press, Boca Raton, pp. 117–121.

Kamath, M.V., Belavady, B., 1980. Unavailable carbohydrates of commonly consumed Indian foods. J. Sci. Food Agric. 31, 194–202.

Kawamura, S., 1969. Studies on the starches of edible legume seeds. J. Jpn. Soc. Starch Sci. 17, 19–40.

Khalil, J.K., Bashir, A., Khalil, I.A., Hussain, T., 1983. Nitrogen balance in human subjects as influenced by corn bread supplemented with peanut and chickpea flour. Pak. J. Sci. Ind. Res. 26, 83–86.

Khokhar, S., Chauhan, B.M., 1986. Antinutritional factors in moth bean: varietal differences and effects of methods of domestic processing and cooking. J. Food Sci. 51, 591–594.

Kon, S., 1979. Effect of soaking temperature on cooking and nutritional quality of beans. J. Food Sci. 44, 1329–1334, 1340.

Koshiyama, I., 1983. Storage proteins of soybean. In: Gottschalk, W., Muller, H.P. (Eds.), Seed proteins. Biochemistry, genetics, nutritive value. Martinus Nijhoff, The Hague, pp. 428.

Kumar, K.G., Venkataraman, L.V., 1976. Studies on the *in-vitro* digestibility of starch in some legumes before and after germination. Nutr. Rep. Int. 13, 115–124.

Kutos, T., Golob, T., Kac, M., Plestenjak, A., 2003. Dietary fibre content of dry and processed beans. Food Chem. 80, 231–235.

Labaneiah, M.E.O., Luh, B.S., 1981. Changes of starch, crude fiber, and oligosaccharides in germinating dry beans. Cereal Chem. 58, 135–138.

Lazze, M.C., Pizzala, R., Savio, L.A., Stivala, E.P., Bianchi, L., 2003. Anthocyanins protects against DNA damage induced by tert-butyl hydroperoxide in rat smooth muscle and hepatoma cells. Mutat. Res. 535, 103–115.

Lombardi-Boccia, G., Carbonaro, M., Di Lullo, G., Carnovale, E., 1994. Influence of protein components (G1, G2 and albumin) on Fe and Zn dialysability from bean (*Phaseolus vulgaris* L.). J. Food Sci. Nutr. 45, 183–190.

Longe, O.G., 1981. Carbohydrate composition of different varieties of cowpea (*Vigna sinensis* L.). Food Chem. 6, 153–161.

Long-Ze, L., Harnly, J.M., Pastor-Corrales, M.S., Luthria, D.L., 2008. The polyphenolic profiles of common bean (*Phaseolus vulgaris* L.). Food Chem. 107, 399–410.

Lorenz, K.J., 1979. The starch of the faba bean (*Vicia jaba* L.). Starke 31, 181–184.

Luthria, D.L., Pastor-Corrales, M.A., 2006. Phenolic acids content of fifteen dry edible bean (*Phaseolus vulgaris* L.). varieties. J. Food Compos. Anal. 19, 205–211.

Manan, F., Hussain, T., Iqbal, P., 1984. Proximate composition and minerals constituents of important cereals and pulses grown in NWFP. Pak. J. Sci. Res. 36, 45–49.

Marlett, A.J., McBurney, M.I., Slavin, J.L., 2002. Position of the American Dietetic Association: health implications of dietary fiber. J. Am. Diet. Assoc. 102, 993–1000.

Mason, A.C., Weaver, C.M., Kimmel, S., Brown, R.K., 1993. Effect of soybean phytate content on calcium bioavailability in mature and immature rats. J. Agric. Food Chem. 41, 246–249.

Mataix, F.J., Manas, M.T., 1998. Tabla de composición de alimentos españoles. Instituto de Nutricion y Tecnologia de los Alimentos–Universidad de Granada, Granada.

Mazur, W.M., Duke, J.A., Wahala, K., Rasku, S., Adlercreutz, H., 1998. Isoflavonoids and lignans in legumes: nutritional and health aspects in humans. J. Nutr. Biochem. 9, 193–200.

McCleary, B., 2008. Dietary fiber definition task force submits comments on nutrition labelling. Cereal Foods World 53, 91.

McWatters, K.H., 1983. Compositional, physical and sensory characteristics of "akara" processed from cowpea paste and Nigerian cowpea flour. Cereal Chem. 60, 332–336.

Meagher, L.P., Beecher, G.R., 2000. Assessment of data on the lignan content of foods. J. Food Compos. Anal. 13, 935–947.

Ministerio de Sanidad y Consumo, 1999. Tablas de composición de alimentos españoles. Ministerio de Sanidad y Consumo. Boletin Oficial del Estado, Madrid.

Monte, W.C., Maga, J.A., 1980. Extraction and isolation of soluble and insoluble fiber fractions from the pinto bean (*Phaseolus vulgaris* L.). J. Agric. Food Chem. 28, 1169–1174.

Morad, M.M., Leung, H.K., Hsu, D.L., Finney, P.L., 1980. Effect of germination on physiochemical and bread-baking properties of yellow pea, lentil, and faba bean flours and starches. Cereal Chem. 57, 390–396.

Morcos, S.R., Gabriel, G.N., El-Hafez, M.A., 1976. Nutritive studies on some raw and prepared leguminous seeds commonly used in the Arab Republic of Syria. Z. Ernaehrungswiss 15, 378–386.

Morris, E.R., Hill, A.D., 1996. Inositol phosphate content of selected dry beans, peas, and lentils, raw and cooked. J. Food Compos. Anal. 9, 2–12.

Morrison, W.R., Karkalas, J., 1990. Starch in carbohydrates. In: Dey, P.M. (Ed.), Methods in plant biochemistry, Vol. 2. Academic Press, London, pp. 323–352.

Morton, J.F., 1976. The pigeon pea (*Cajanus cqjan*): a high protein tropical bush legume. HortScience 11, 11–19.

Naivikul, O., D'Appolonia, B.L., 1979. Carbohydrates of legume flours compared with wheat flour: II. Starch. Cereal Chem. 56, 24–28.

National Research Council, 1980. Recommended dietary allowances, 9th edn. National Academy of Science, Washington, DC.

Nene, S.P., Vakil, U.K., Sreenivasan, A., 1975. Effect of gamma radiation on physico-chemical characteristics of red gram (*Cajanus cajan*) starch. J. Food Sci. 40, 943–947.

Nigam, V.N., Giri, K.V., 1961. Sugars in pulses. Can. J. Biochem. Physiol. 39, 1847–1853.

Noonan, S.C., Savage, G.P., 1999. Oxalic acid content of foods and its effect on humans. Asia Pacif. J. Clin. Nutr. 8, 64–74.

Oakenfull, D., Sidhu, G.S., 1989. Saponins. Toxicants of plant origin-CRC Press, Boca Raton.

Oboh, H.A., Muzquiz, M., Burbano, C., Pedrosa, M.M., Ayet, G., Osagie, A.U., 1998. Anti-nutritional constituents of six underutilized legumes grown in Nigeria. J. Chromatogr. A 823, 307–312.

Oke, O.L., 1967. Chemical studies of some Nigerian pulses. W. Afr. J. Biol. Appl. Chem. 9, 52–55.

Olson, A.C., Becker, R., Miers, J.C., Gumbmnn, M.R., Wagner, J.R., 1975. Problems in the digestibility of dry beans. In: Friedman, M. (Ed.), Protein nutritional quality of foods and feeds, Part II. Marcel Dekker, New York, pp. 551–563.

Oomah, B.D., Blanchard, C., Balasubramanian, P., 2008. Phytic acid, phytase, minerals, and antioxidant activity in Canadian dry bean (*Phaseolus vulgaris* L.) cultivars. J. Agric. Food Chem. 56, 11312–11319.

Oomah, B.D., Cardador-Martinez, A., Loarca-Pina, G., 2005. Phenolics and antioxidative activities in common beans (*Phaseolus vulgaris* L.). J. Sci. Food Agric. 85, 935–942.

Orcheson, L.J., Rickard, S.E., Seidl, M.M., Thompson, L.U., 1998. Flaxseed and its mammalian lignan precursor cause a lengthening or cessation of estrous cycling in rats. Cancer Lett. 125, 69–76.

Osborne, T.B., 1924. The vegetable proteins. Longmans, Green and Co., London.

Oshodi, A.A., Ekperigin, M.M., 1989. Functional properties of pigeon pea (*Cajanus cajan*) flour. Food Chem. 34, 187–191.

Oyenuga, V.A., 1968. Nigeria's food and feeding stuffs. Ibadan University Press, Ibadan, pp. 79–83.

Patwardhan, V.N., 1962. Pulses and beans in human nutrition. Am. J. Clin. Nutr. 11, 12–30.

Perez-Hidalgo, M.A., Guerra-Hernandez, E., Garcia-Villanova, B., 1997. Dietary fiber in three raw legumes and processing effect on chick peas by an enzymatic–gravimetric method. J. Food Compos. Anal. 10, 66–72.

Phillips, R.D., McWatters, K.H., 1991. Contribution of cowpeas to nutrition and health. Food Technol. 45, 127–130.

Phillippy, B.Q., 2003. Inositol phosphates in food. Adv. Food Nutr. Res. 45, 1–60.

Price, K.R., Johnson, I.T., Fenwick, G.R., 1987. The chemistry and biological significance of saponins in food and feeding stuffs. Crit. Rev. Food Sci. Nutr. 26, 27–135.

Prinyawiwatkul, W., Beuchat, L.R., Mcwatters, K.H., Philips, R.D., 1996. Changes in fatty acid, simple sugar, and oligosaccharide content of cowpea flour as a result of soaking boiling and fermentation with *Rhizopus microsporus* var. *oligosporus*. Food Chem. 57, 405–415.

Prinyawiwatkul, W., McWatters, K.H., Beuchat, L.R., Phillips, R.D., 1997. Functional characteristics of cowpea flour and starch as affected by soaking, boiling and fungal fermentation before milling. Food Chem. 58, 361–372.

Pritchard, P.J., Dryburgh, E.A., Wilson, B.J., 1973. Carbohydrates of spring and winter field beans (*Vicia faba* L.). J. Sci. Food Agric. 24, 663–669.

Prodanova, M., Sierra, I., Vidal-Valverde, C., 1997. Effect of germination on the thiamine, riboflavin and niacin contents in legumes. Z. Lebensm. -Unters. -Forsch. A 205, 48–52.

Quemener, B., Brillouet, J.M., 1983. Ciceritol, a pinitol digalactoside form seeds of chickpea, lentil and white lupin. Phytochemistry 22, 1745–1751.

Rachie, K.O., 1985. Introduction. In: Singh, S.R., Rachie, K.O. (Eds.), Cowpea research, production and utilization. John Wiley, New York, pp. XXI–XXI10.

Ranilla, L.G., Genovese, M.I., Lajolo, F.M., 2009. Isoflavones and antioxidant capacity of Peruvian and Brazilian lupin cultivars. J. Food Compos. Anal. 22, 397–404.

Rao, P.S., 1976. Nature of carbohydrates in pulses. J. Agric. Food Chem. 24, 958–961.

Reddy, N.R., Salunkhe, D.K., 1980. Changes in oligosaccharides during germination and cooking of black gram and fermentation of black gram/rice blend. Cereal Chem. 57, 356–360.

Rizkalla, S.W., Bellisle, F., Slama, G., 2002. A health benefits of low glycaemic index foods, such as pulses, in diabetic patients and healthy individuals. Br. J. Nutr. 88, 255–262.

Rochfort, S., Panozzo, J., 2007. Phytochemicals for health, the role of pulses. J. Agric. Food Chem. 55, 7981–7994.

Rockland, L.B., Zaragosa, E.M., Oracca-Tetteh, R., 1979. Quick-cooking of winged beans (*Psophocarpus tetragonolobus*). J. Food Sci. 44, 1004–1007.

Rohatgi, K., Banerjee, M., Banerjee, S., 1955. Effect of germination on vitamin B_{12} values of pulses (leguminous seeds). J. Nutr. 56, 403–408.

Ruiz, R.G., Price, K.R., Arthur, A.E., Rose, M.E., Rhodes, M.J.C., Fenwick, R.G., 1996. Effect of soaking and cooking on the saponin content and composition of chickpeas (*Cicer arietinum*) and lentils (*Lens culinaris*). J. Agric. Food Chem. 44, 1526–1530.

Ryan, E., Galvin, K., O'Connor, T.P., Maguire, A.R., O'Brien, N.M., 2007. Phytosterol, squalene, tocopherol content and fatty acid profile of selected seeds, grains, and legumes. Plant Foods Hum. Nutr. 62, 85–91.

Sajjan, S.U., Wankhede, D.B., 1981. Carbohydrate composition of winged bean (*Psophocarpus tetragonolobus*). J. Food Sci. 46, 601–602.

Salinas-Moreno, Y., Rojas-Herrera, L., Sosa-Montes, E., Perez-Herrera, P., 2005. Anthocyanin composition in black bean (*Phaseolus vulgaris* L.) varieties grown in Mexico. Agrociencia 39, 385–394.

Salunkhe, D.K., Kadam, S.S., Chawan, J.K., 1985. Post-harvest biotechnology of food legumes. CRC Press, Boca Raton, FL, pp. 29–52.

Salunkhe, D.K., Chavan, J.K., Kadam, S.S., 1986. Pigeonpea as an important food source. Crit. Rev. Food Sci. Nutr. 23, 103–145.

Sanchez-Mata, M.C., Penuela-Teruel, M.J., Camara-Hurtado, M., Diez-Marques, C., Torija-Isasa, M.E., 1998. Determination of mono-, di-, and oligosaccharides in legumes by high-performance liquid chromatography using an amino-bonded silica column. J. Agric. Food Chem. 46, 3648–3652.

Sanchez-Vioque, R., Clemente, A., Vioque, J., Bautista, J., Millan, F., 1998. Polar lipids of defatted chickpea (*Cicer arietinum* L.) flour and protein isolates. Food Chem. 63, 357–361.

Sandberg, A.S., 2002. Bioavailability of minerals in legumes. Br. J. Nutr. 88, 281–285.

Sarma, B.K., Singh, D.P., Mehta, S., Singh, H.B., Singh, U.P., 2002. Plant growth promoting rhizobacteria-elicited alterations in phenolic profile of chickpea (*Cicer arietinum*) infected by *Selerotium rolfsii*. J. Phytopathol. 150, 277–282.

Sathe, S.K., Salunkhe, D.K., 1981. Isolation, partial characterization and modification of the great northern bean (*Phaseolus vulgaris* L.) starch. J. Food Sci. 46, 617–621.

Sathe, S.K., Rangnekar, P.D., Deshpande, S.S., Salunkhe, D.K., 1982. Isolation and partial characterization of black gram (*Phaseolus mungo* L.) starch. J. Food Sci. 47, 1524–1527.

Satterlee, L.D., Bembers, M., Kendrick, J.G., 1975. Functional properties of the great northern bean (*Phaseolus vulgaris* L.) protein isolate. J. Food Sci. 40, 81–84.

Schoch, T.J., Maywald, E.C., 1968. Preparation and properties of various legume starches. Cereal Chem. 45, 564–569.

Shahen, N., Roushi, M., Hassan, R.A., 1978. Studies on lentil starch. Starch 30, 148–150.

Shi, J., Arunasalam, K., Yeung, D., Kakuda, Y., Mittal, G., Jiang, Y., 2004. Saponins from edible legumes: chemistry, processing, and health benefits. J. Med. Food 7, 67–78.

Silva, H.C., Luh, B.S., 1979. Changes in oligosaccharides and starch granules in germinating beans. Can. Inst. Food Sci. Technol. J. 12, 103–105.

Singh, D.K., Rao, A.S., Singh, R., 1988. Amino acid composition of storage protein of a promising chickpea (*Cicer arietinum* L.) cultivar. J. Sci. Food Agric. 55, 37–46.

Singh, N., Sandhu, K.S., Kaur, M., 2004. Characterization of starches separated from Indian chickpea (*Cicer arietinum* L.) cultivars. J. Food Eng. 63, 441–449.

Singh, N., Kaur, S., Isono, N., Noda, T., 2010a. Genotypic diversity in physico-chemical, pasting and gel textural properties of chickpea (*Cicer arietinum* L.). Food Chem. 122, 65–73.

Singh, N., Kaur, N., Rana, J.C., Sharma, S.K., 2010b. Diversity in seed and flour properties in field pea (*Pisum sativum*) germplasm. Food Chem. 122, 518–525.

Singh, U., Jambunathan, R., Gurtu, S., 1981. Seed protein fraction and amino acid composition of some wild species of pigeon pea. J. Food Sci. Technol. 18, 83–85.

Singh, U.P., Sarma, B.K., Singh, D.P., 2003. Effect of plant growth promoting rhizobacteria and culture filtrate of *Selerotium rolfsii* on phenolic and salicylic acid contents in chickpea (*Cicer arietinum*). Curr. Microbiol. 46, 131–140.

Smulikowska, S., Pastuszewska, B., Swiech, E., Ochtabinska, A., Mieczkowska, A., Nguyen, V.C., 2001. Tannin content affects negatively nutritive value of pea for monogastrics. J. Anim. Feed Sci. (Pol.) 10, 511–523.

Sosulski, F., Youngs, C.G., 1979. Yield and functional properties of air classified protein and starch fractions from eight legume flours. J. Am. Oil Chem. Soc. 56, 292–295.

Souci, S.W., Fachmann, W., Kraut, H., 1994. Food composition and nutrition tables. Medpharm Scientific Publications, CRC Press, Stuttgart.

Stephen, A.M., 1995. Resistant starch. In: Kritchevsky, D., Bonfield, C. (Eds.), Dietary fibre in health and disease. Eagan Press, St. Paul, MN, pp. 453–458.

Terres, C., Navarro, M., Martin-Lagos, F., Gimenez, R., Lopez, H., Lopez, M.C., 2001. Zinc levels in foods from southeastern Spain: relationship to daily dietary intake. Food Addit. Contam. 18, 687–695.

Tharanathan, R.N., Mahadevamma, S., 2003. Legumes – a boon to human nutrition. Trends Food Sci. Technol. 14, 507–518.

Thompson, L.U., Serraino, R.M., Cheung, F., 1991. Mammalian lignan production from various food. Nutr. Cancer 16, 43–52.

Topping, D.L., Fukushima, M., Bird, A.R., 2003. Resistant starch as a prebiotic and symbiotic: state of art. Proc. Nutr. Soc. 62, 171–176.

Tosh, S.M., Yada, S., 2010. Dietary fibres in pulse seeds and fractions. characterization, functional attributes, and applications. Food Res. Int. 43, 450–460.

Uriyo, M.G., 2001. Changes in enzyme activity during germination of cowpeas. Food Chem. 73, 7–10.

US Department of Agriculture, Agricultural Research Service, 2002. USDA–Iowa state University Database on the Isoflavone Content of Foods, Release 1.3-2002. Nutrient Data Laboratory website: http://www.nal.usda.gov/fnic/foodcamp/Data/isoflav/isoflav.html.

Vadivel, V., Janardhanan, K., 2000. Nutritional and anti-nutritional composition of velvet bean: an under-utilized food legume in South India. Int. J. Food Sci. Nutr. 51, 279–287.

Vidal-Valverde, C., Sierra, I., Diaz-Pollan, C., Blazquez, I., 2001. Determination by capillary electrophoresis of total and available niacin in different development stage of raw and processed legumes: comparison with high-performance liquid chromatography. Electrophoresis 22, 1479–1484.

Vilela, E.R., El-Dash, A.A., 1985. Producëäo de farinha de guandu (*Cajanus cajan*, Millsp.): moagem por via seca. Bol. Soc. Brasil. e Cie. Tecnol. Alim. 19, 101–108.

Vose, J.R., 1980. Production and functionality of starches and protein isolates from legume seeds. Cereal Chem. 57, 406–410.

Wang, N., Lewis, M.J., Brennam, J.G., Westby, A., 1997. Effect of processing methods on nutrients and anti nutritional factors in cowpea. Food Chem. 58, 59–68.

Watson, J.D., 1977. Chemical composition of some less commonly used legumes in Ghana. Food Chem. 2, 267–271.

Williams, H.E., Wandzilak, T.R., 1989. Oxalate synthesis, transport and the hyperoxaluric syndromes. J. Urol. 141, 742–747.

Williams, P.C., Singh, U., 1987. Nutritional quality and the evaluation of quality in breeding programmes. In: Saxena, M.C., Singh, K.B. (Eds.), The chickpea. CAB International, Wallingford, pp. 329–356.

Winham, D.M., Hutchins, A.M., Melde, C.L., 2007. Pinto bean, navy bean, and black-eyed pea consumption do not significantly lower the glycemic response to a high glycemic index treatment in normoglycemic adults. Nutr. Res. 27, 535–541.

Woldemichael, G.M., Montenegro, G., Timmermann, B.N., 2003. Triterpenoidal lupin saponins from the Chilean legume *Lupinus oreophilus*. Phil. Phytochem. 63, 853–857.

Wright, D.J., 1986. The seed globulins. In: Hudson, B.J.F. (Ed.), Development in food proteins, Vol. 5. Elsevier, Amsterdam.

Xu, B.J., Chang, S.K.C., 2007. A comparative study on phenolic profiles and antioxidant activities of legumes as affected by extraction solvents. J. Food Sci. 72, S159–S166.

Yoshida, H., Tomiyama, Y., Yoshida, N., Saiki, M., Mizushina, Y., 2008. Lipid classes, fatty acid compositions and triacylglycerol molecular species from adzuki beans (*Vigna angularis*). J. Food Lipids 15, 343–355.

Zia-Ul-Haq, M., Iqbal, S., Ahmad, S., Bhanger, M.I., Wiczkowsi, W., Amarowicz, R., 2008. Antioxidant potential of desi chickpea varieties commonly consumed in Pakistan. J. Food Lipids 15, 326–342.

Functional and physicochemical properties of pulse proteins

3

Vassilis Kiosseoglou, Adamantini Paraskevopoulou
Laboratory of Food Chemistry and Technology, School of Chemistry,
Aristotle University of Thessaloniki, Thessaloniki, Greece

3.1 Introduction

A common feature of pulses such as dry pea, chickpea, lentil or the various types of dry bean, is their high protein content, ranging between 17 and 30% (Boye et al., 2010a). In lupin, which may also be included in this group of legumes since the average crude fat content of lupin seed of various varieties is found in the range of 5.8–14.1% (Peterson, 1998) and is not, therefore, exploited as in the case of soybean for its oil (Roy et al., 2010), the protein content may exceed 40%, depending on lupin variety. Dry pea, chickpea, lentil and bean are also low in fat which, depending on pulse type and variety, may reach a level as low as 0.83 and as high as 6.6% (Boye et al., 2010a).

Pulse Foods: Processing, Quality and Nutraceutical Applications. DOI: 10.1016/B978-0-1238-2018-1.00007-0

The proteins of pulses are mainly storage proteins belonging to the groups of albumins, globulins and glutelins, with the salt-soluble globulins constituting the main proteins found in the seeds. There are also a number of proteins, other than the water-soluble storage proteins, mainly enzymes, enzyme inhibitors and lectins which constitute part of the defensive mechanism of the seed but are considered as antinutritional factors for the human diet (Roy et al., 2010). As in the case of all proteins found in nature, the differences in physicochemical properties between the various protein constituents of pulses reflect differences in their primary molecular structure. Pulse proteins belonging to the albumin fraction are characterized by a relatively low to medium molecular weight and a hydrophilic surface that renders the proteins water soluble. On the other hand, the globulins are multisubunit molecules of high molecular weight and relatively hydrophobic surface that limits their solubility in aqueous media. One important difference between the various types of pulses, with regard to their protein fraction, is an appreciable variation encountered in terms of the albumin/globulin ratio. Thus, the albumin/globulin ratio for lentil, black gram, French bean and chickpea was reported by Gupta and Dhillon (1993) to be 1/3, 1/6.3, 1/>3 and 1/4, respectively. This variation in albumin/globulin ratio, along with possible variation in vicillin/legumin ratio, could be responsible to some extent for the differences in functionality reported by investigators for rich-in-protein materials originating from different types of pulses.

Pulses are exploited for the preparation of various meals, either on their own or in admixture with other food materials. In the latter case, the main target is to prepare meals that are nutritionally superior to either of the mixture's ingredients. In addition, such mixtures may possess superior functional properties due to the enrichment of the mixture in biomacromolecules such as proteins and starch. An alternative way to exploit the nutritional and functional potential of biopolymeric constituents of pulses is their isolation from the seeds by applying a number of well-known methods and their use as ingredients in many foods. This approach is particularly useful considering that an appreciable percentage of the raw material is damaged during harvesting or processing and may, therefore, constitute a low-cost by-product that could be exploited as a source for the extraction of proteins and other useful seed constituents (Sanchez-Vioque et al., 1999).

The application of various processing techniques to extract proteins from pulse seeds usually leads to the preparation of rich-in-protein concentrates and isolates. These materials may differ in their protein constituent composition, depending mainly on the conditions prevailing during protein extraction in terms of temperature, salt content and pH. In addition, the protein extraction and recovery processes may influence the secondary, tertiary or quaternary structure of the extracted protein molecules and hence protein functionality. As a result, the functional properties of the protein molecules in these materials may differ depending on the method of preparation. The systematic investigation of the functional properties of enriched in pulse protein materials should, therefore, be connected with their compositional and physicochemical characteristics and the extent to which these characteristics are affected by the extraction process applied. Some recent studies on pulse protein functionality have addressed this point and the main findings of these studies, as well as their critical discussion in accordance with the fundamentals of modern food colloids science, constitute the basis of this chapter. In addition, a number of potential food applications of pulse protein isolates and concentrates are presented at the end of the chapter, focusing on possible interactions of protein molecules with other food ingredients and their effect on macroscopic properties such as physical stability, texture and mechanical characteristics.

3.2 Preparation of protein concentrates and isolates

The techniques employed for legume exploitation in preparing rich-in-protein materials, such as concentrates or isolates with a protein content of about 70 and 90%, respectively, are air classification and the so-called wet protein extraction methods. Air classification is based on the separation of legume flours into protein- and starch-rich fractions by taking advantage of the existence of particles differing in size and density, the light fine particles of proteins and the heavy coarse ones of starch. The process involves milling of whole or dehulled seeds and subsequent classification in a spiral air stream to separate the two fractions.

The application of air classification is not possible in the case of legume seeds such as soybean since the high fat content of the flour originating from soybean results in particle agglomeration and hence poor separation efficiency (Gueguen and Cerleti, 1994). Taking into account that pulses are relatively low in fat, this method is suitable for the preparation of protein materials from pulses (Kohnhorst et al., 1990). According to a number of reports (Tyler et al., 1981; Elkowicz and Sosulski, 1982; Tyler and Panchuk, 1982; Tyler, 1984), the purity of the separated protein particle fractions may depend on the type of pulse or the flour moisture content. As a rule, rich-in-protein flours or protein concentrates with contents as low as 40 and as high as 75.1%, but not isolates, have been obtained by applying air classification to a number of pulses. In addition, the type of pulse may influence the yield of protein extraction. Very satisfactory protein separation efficiencies (PSEs), as high as 88.9%, were reported for mung bean by Tyler et al. (1981). These investigators suggested that this technique was more suitable for legumes such as mung bean and lentil and less suitable for lima bean and cowpea.

In spite of its relative usefulness in obtaining moderately enriched-in-protein materials from pulses, air classification is not suitable for the preparation of protein isolates. The latter are more suitable than the flours or the concentrates in a number of food applications, especially when emulsification or gelation by the protein is required. In addition, the presence in flours, and to a lesser extent in protein concentrates, of constituents such as polysaccharides, at relatively high concentrations, does not allow an in-depth investigation of the functionality of the pulse proteins in food systems. Protein isolates with a protein content of 90%, or even higher, are prepared from pulses by applying one of the wet methods available. These methods involve an initial aqueous extraction step of the proteins from full-fat or defatted with hexane ground pulse meal, followed by the application of a protein recovery step and dehydration. The wet methods are differentiated in terms of the conditions used for extracting the protein from the meal or the technique employed for recovering the protein from the extract.

As most proteins in general and legume proteins in particular exhibit a relatively high solubility at alkaline or acidic conditions, extraction of protein may be easily effected by dispersing the ground legume flour in water at pH 8–10, in alkaline extraction, or below 4, in acidic extraction, followed by prolonged agitation. Extraction at these conditions results in a mixed solution of proteins belonging to the globulin and albumin fractions. Recovery of the extracted proteins is very often performed either by isoelectric precipitation, involving pH adjustment to a value around 4.5, where the globulins exhibit minimum solubility, or by concentrating the extract by ultrafiltration. Following pH adjustment of the recovered protein to pH 7, in order to improve solubility, materials in the form of powder are obtained by applying spray or freeze-drying (Boye et al., 2010a).

Depending on extraction conditions, but mainly on the application or not of a number of washing steps at the end of the process, the resulting protein material could be either a protein concentrate or an isolate. Thus, protein contents close to 90% have been reported in a number of papers for powdered materials obtained from pulses such as chickpea, pea, lupin or faba bean, by applying the so-called isolate process (Fan and Sosulski, 1974; Chakraborty et al., 1979; Paredes-Lopez et al., 1991; Kiosseoglou et al., 1999; Papalamprou et al., 2008; Zhang et al., 2009). In other studies, the protein content of the materials obtained from a number of pulses was lower than 90% and in some cases even lower than 70% (Sanchez-Vioque et al., 1999; Makri and Doxastakis, 2006a; Boye et al., 2010b).

The isolates obtained by isoelectric precipitation or ultrafiltration are expected to differ with respect to their protein composition. Thus, the proteins of isolates obtained by isoelectric precipitation should mainly belong to the globulin fraction, while in the case of the isolates prepared by ultrafiltration, both globulin as well as albumin constituents are expected to be found. Domination of proteins belonging to the globulin fraction over those of the albumin fraction in a chickpea protein isolate obtained by isoelectric precipitation was reported by Sanchez-Vioque et al. (1999). Papalamprou et al. (2008, 2010), on the other hand, reported that the chickpea protein isolate obtained by ultrafiltration was made up of a mixture of proteins belonging to both the albumin and the globulin fractions but the albumins were practically absent from the isolate obtained by isoelectric precipitation.

One way for the preparation of albumin- and globulin-rich isolates constitutes a method based on an initial extraction of the albumins in a slightly acidic environment, followed by extraction at an alkaline pH of the proteins belonging to the globulin fraction (Tomoskozi et al., 2001; Papalamprou et al., 2008). Alternatively, an initial extraction of both albumins and globulins may be effected by employing a salt solution at neutral pH, followed by dilution of the resulting protein extract with water to salt out and separate the globulins (micellization) from the more water-soluble albumins. Depending on protein extraction and recovery conditions, either protein concentrates or isolates have been obtained by applying the micellization method to various pulse flours (Satterlee et al., 1975; Sathe and Salunkhe, 1981; Paredes-Lopez et al., 1991).

3.3 Functional properties of pulse proteins

3.3.1 Solubility, water- and oil-absorption capacity

Protein solubility depends on the hydrophilicity/hydrophobicity balance of the protein molecule but mainly on the composition of molecular surface in terms of polar/non-polar amino acids that in turn affects the thermodynamics of protein–protein and protein–solvent interactions. Since a number of protein functional properties, such as emulsification, foaming or gelation, are closely associated with solubility, this property may determine to a large extent the protein suitability as an ingredient in various food applications. Legume globulins, which constitute the main proteins of concentrates or isolates prepared from pulses, are relatively hydrophobic in nature and tend to exhibit reduced solubility at pH environments close to the protein isoelectric point, where electrostatic repulsion and ionic hydration of molecules reach a minimum. In general, the solubility of a number of pulse protein materials (flours, concentrates or isolates), recovered from alkaline protein extracts by isoelectric precipitation or ultrafiltration, is very low at a pH range between 4 and 6 but exhibits a sharp rise when the pH is moved either to more acidic or to neutral and alkaline environments.

The method of protein recovery appears to have an effect on protein solubility, with the concentrates or isolates obtained by ultrafiltration usually exhibiting higher solubility compared to those recovered by isoelectric precipitation, although the type of pulse may also have an effect. For example, protein isolates of faba bean or pea prepared by ultrafiltration, exhibited, respectively, 22 and 15% higher solubility compared to those obtained by isoelectric precipitation (Vose, 1980). Papalamprou et al. (2008) reported that protein solubility of chickpea protein isolates, obtained by either isoelectric precipitation or ultrafiltration, was practically the same at pH 3 or above 6.5. The isolate prepared by ultrafiltration, on the other hand, exhibited higher solubility by about 20% at pH 4 or 5, compared to that recovered by isoelectric precipitation. Although such a difference could be partly attributed to more extensive molecular structure changes resulting from protein precipitation, the isolate prepared by ultrafiltration was also richer in water-soluble albumins which are much more soluble than the globulins at the isoelectric point. Boye et al. (2010b) reported that the solubility of a lentil protein concentrate was much higher at pH 4, or at a more acidic pH, compared to the respective isolate obtained by isoelectric precipitation and suggested that this property could be useful in the preparation of beverages with an acidic character.

Incorporation of ionic polysaccharides in the solution may improve the solubility of pulse proteins. As was reported by Braudo et al. (2001) the solubility of faba bean legumin at the isoelectric point increases in the presence of chitosan, possibly due to the formation of an ionic complex between the two biomolecules. In a recently published paper, Liu et al. (2010) studied the effect of gum Arabic addition on the functionality of pea protein isolate and observed that the protein solubility was considerably increased at the pH range of 4–2.5, an effect attributed by the authors to a soluble coacervate formation (Fig. 3.1).

Water- and oil-absorption capacity (WAC, OAC) are defined as the amount of water and oil, respectively, that can be bound per unit weight of the protein material and constitute useful indices of the ability of the protein to prevent fluid leakage from a product during food storage or processing. WAC values as low as 0.6 and as high as $4.9\,g\,g^{-1}$ have been reported in a number of studies (Fernandez-Quintela et al., 1997; Kaur and Singh, 2007; Boye et al., 2010b) for protein isolates or concentrates prepared from pulses such as chickpea, pea, faba bean or lentil, indicating that both the type of pulse and variety may have an effect on WAC. In addition to the influence of the type of pulse on WAC, it appears that the technique employed for protein recovery may also influence the water-holding capacity (WHC) value, with the protein materials obtained by isoelectric precipitation, for example, from pea or chickpea exhibiting higher water-binding ability than those prepared by ultrafiltration (Boye et al., 2010b). By analogy to WAC, OAC values reported for protein materials prepared from various pulses range from 1.0 to $3.96\,g\,g^{-1}$,

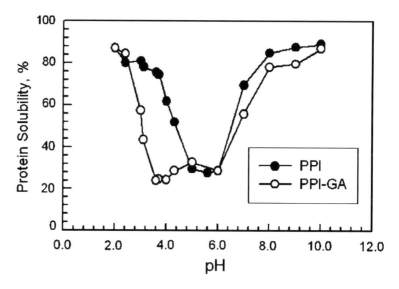

Figure 3.1 Percent protein solubility as a function of pH for a homogeneous PPI and mixed PPI–GA (2:1) system (data represent mean values (*n*=3) ± one standard deviation). (From Liu et al., 2010, with permission.)

depending on the type of pulse, pulse variety and method of preparation (Paredes-Lopez et al., 1991; Fernandez-Quintela et al., 1997; Kaur and Singh, 2007).

Although WHC and OAC data are useful for assessing the technological suitability of pulse protein materials in food applications, their fundamental importance is rather limited considering that they are not usually accompanied in the published papers by data on protein composition and molecular characteristics that would enable a search for relationships between structure and functionality.

3.3.2 Emulsifying and foaming properties

During the course of mixing/homogenization of oil with protein solutions, continuous creation of an oil–water interface and oil droplet formation takes place, aided by adsorption of protein molecules to the newly formed oil droplet surfaces and the reduction in the surface tension that follows. The newly formed oil droplets are protected from immediate coalescence into larger ones by the presence of adsorbed protein molecules at their surface. Hence, the emulsifying ability of a protein, that is its ability to aid in the preparation of an emulsion with droplets of a relatively small size, is bound to depend on protein molecular characteristics, such as size, solubility, surface hydrophobicity and structural flexibility, as these characteristics determine the protein adsorption properties. In spite of the fact that the proteins of the various pulses present many similarities, since they are mainly legumin- or vicilin-like globulins and albumins, the method applied for protein extraction may result in concentrates or isolates differing in composition, degree of protein denaturation and, hence, functionality.

The emulsifying ability of legume proteins is often expressed in terms of the emulsifying activity index (EAI), which is the maximum surface area created per unit protein. This index is usually determined by applying turbidimetry to highly diluted emulsion samples. During the first hours of emulsion life, the initially formed surface layer at the droplet surfaces is further strengthened as new protein molecules adsorb and rearrange to a state of minimum free energy and, at the same time, interact with other adsorbed molecules leading to the development of a mechanically strong protein layer. This membrane provides long-term stability to the system as it not only protects the droplets against coalescence, due to its inherent elasticity, but may also prevent their close approach and interaction leading to the appearance of undesirable phenomena, such as droplet aggregation and creaming upon long-term storage. One index often used for assessing the stability of legume protein-stabilized emulsions is the emulsifying stability index (ESI), which is determined by monitoring the decrease in turbidity with time of a diluted emulsion. Taking into account that the ESI is determined following short-term emulsion storage and the only change expected to take place within that period is emulsion serum separation resulting from the upward movement of the droplets, the ESI should in effect constitute an index of the emulsion stability against creaming.

Based on the determination of EAI, a number of studies reported that the emulsifying ability of pulse protein concentrates or isolates is affected by the type of pulse or the method applied for their preparation. For example, according to Fuhrmeister and Meuser (2003), a pea protein isolate prepared by isoelectric precipitation exhibited inferior emulsifying ability compared to the respective isolate obtained by ultrafiltration (EAI values of 10.1 or $14.0 \, m^2 \, g^{-1}$ and $27.4 \, m^2 \, g^{-1}$, respectively). Boye et al. (2010b) compared the EAI and ESI of a number of protein concentrates with a protein content ranging between 63.9 and 88.6%, obtained from pea, chickpea or lentil by applying isoelectric precipitation or ultrafiltration. These investigators concluded that the emulsifying properties of the proteins depended mainly on the type of pulse used for protein concentrate preparation, with the concentrates originating from chickpea exhibiting higher emulsifying ability compared to those of pea or lentil, while the method applied for the preparation of the concentrates had only a very limited impact.

The emulsifying properties of pulse proteins are also bound to depend on environmental parameters such as pH and ionic strength. Zhang et al. (2009) observed that a chickpea protein isolate obtained by isoelectric precipitation exhibited higher EAI at alkaline pH than at a pH close to the protein isoelectric point, where the emulsifying ability of protein declined dramatically. In addition, the EAI of chickpea protein decreased at relatively low ionic strength environments (around 0.1) and increased again as the ionic strength of the protein solution was increased. These differences in EAI were related to surface hydrophobicity and molecular secondary structure changes brought about by pH and ionic strength manipulation.

In spite of their relative usefulness in comparing the emulsifying potential of a number of proteins, it must be stressed that EAI and ESI are indices of a highly empirical nature since they do not provide any information about the emulsion droplet size distribution as well as the changes in droplet size resulting from aggregation/coalescence phenomena during long-term storage. Such information is only obtained when changes related to droplet flocculation and coalescence are continuously monitored over the emulsion storage time period. Monitoring of detailed droplet size changes with storage time also enables a more in-depth study of emulsion rheological properties and hence product textural characteristics.

Based on droplet size measurements, Papalamprou et al. (2005) concluded that lupin protein isolate obtained by isoelectric precipitation was inferior as an emulsifier compared to the respective isolate prepared from soybean. In addition, according to Chapleau and Lamballerie (2003), a high-pressure treated lupin protein isolate with modified surface adsorption characteristics exhibited improved emulsifying properties with respect to initial oil droplet size and stability against flocculation and creaming, compared to the control. In a very recent paper, Papalamprou et al. (2010) studied in a more systematic way the emulsifying properties of a number of chickpea protein isolates differing in composition, with respect to globulins and albumins. The isolates were recovered from alkaline chickpea protein extracts by applying either isoelectric precipitation or

Table 3.1 Effect of aging time on the mean particle size of emulsions prepared with chickpea protein isolate

	$d_{4,3}(\mu m)$		
	Storage time (days)		
Sample	0	60	120
TpI	0.60a (0.60a)	1.41a (0.72a)	12.78a (0.95a)
TUF	0.72b (0.72b)	1.76a (0.66a)	20.84b (1.66a)
TUFA	0.57a (0.57a)	16.73b (3.26b)	52.27c (33.40b)
TUFG	1.60 (0.87c)	5.09c (1.17c)	45.32c (25.51b)

Numbers in parentheses correspond to mean droplet size obtained after treatment with SDS/β-mercaptoethanol. Different letters within a column indicate significant differences ($P=0.05$).
From Papalamprou et al. (2010), with permission.

ultrafiltration. All the isolates exhibited very satisfactory emulsifying ability as they produced emulsions with a very small droplet size (Table 3.1). The method of protein recovery did not appear to influence to an appreciable extent the emulsifying ability of the protein, expressed in terms of the mean droplet size of fully dispersed emulsion samples. On the other hand, differences in protein composition among the isolates appeared to influence to a significant extent the tendency of the emulsion droplets to aggregate, thus affecting their stability against creaming or coalescence during long-term storage, with the droplets of emulsion stabilized with the proteins of the albumin fraction exhibiting a higher tendency to aggregate and coalesce upon long-term aging. The method of protein recovery also appeared to influence the long-term stability of the emulsions, especially with respect to creaming as well as to mean droplet size increase with prolonged emulsion heating at 95 °C (Fig. 3.2). More specifically, the isolate prepared by the method of isoelectric precipitation scored higher in this respect compared to most of the isolates obtained by ultrafiltration. The higher stabilizing ability of this isolate was attributed to the partial destabilization, during the precipitation step, of the subunit structure of proteins belonging to the legumin fraction, which enabled the molecules to anchor to the droplet surface in such a way as to minimize the possibility of any interactions with other protein molecules adsorbed to neighboring droplet surfaces.

Air entrapment in the form of bubbles by a protein solution is the basis for the preparation of a number of products, such as meringue, whipped toppings, mousses and cake batters. Although there are

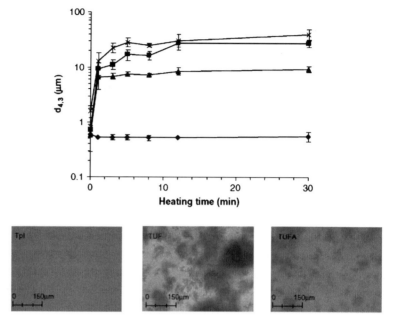

Figure 3.2 Development with heating time at 95 °C of the particle size of emulsions prepared with chickpea protein isolate. Key: TpI (◆), TUF (■), TUFA (▲), TUFG (×). Photographs at the bottom correspond to emulsions after 30 min of heating at 95 °C. (From Papalamprou et al., 2009, with permission.)

some striking differences between an emulsion system and a foam, the role of protein molecules in the formation of foam systems exhibits many similarities to that exhibited in the preparation of emulsions. This is to adsorb to the newly formed air–water interface and aid in the reduction of surface tension, followed by the formation of a rigid surface film around the air bubbles by the unfolded protein molecules that offers protection against bubble fusion, immediately after foam formation and, subsequently, during storage.

The foaming ability of proteins is usually expressed in terms of the foam expansion (FE) or foam capacity (FC) indices, which represent the relative volume of a protein solution increase, resulting from air incorporation. The foam stability (FS), on the other hand, is an index of the ability of the system to retain the air in the form of bubbles during aging. It is measured by either the foam volume decrease or the volume of serum separation from the system over a relatively short storage time period (usually, up to 30 min). Taking into account that these indices are very easily determined by employing common laboratory equipment, it does not come as a surprise that they have often been used to express the foaming properties of many proteins originating, for example, from chickpea (Paredes-Lopez et al., 1991), faba bean (Fernandez-Quintela et al., 1997) or lentil (Lee et al., 2007). Boye et al. (2010b) reported that, in general, the method of pea or lentil protein recovery does not influence to a significant extent the foaming capacity of the isolates, which ranges between 98 and 106%. On the other hand, the chickpea protein concentrate recovered by isoelectric precipitation exhibited significantly higher foaming capacity and lower foam stability compared to the rest of the concentrates. Makri and Doxastakis (2006a) studied the foaming properties of protein extracted from dry bean in the form of a concentrate and reported that they were fairly satisfactory and depended on the method of protein preparation and the pH of the system.

The foaming properties of pulse proteins may also be affected by the composition of the concentrates and isolates in terms of the presence of other pulse constituents (e.g. polysaccharides) and also by the extent of protein denaturation resulting during the process of its preparation. For example, Alamanou and Doxastakis (1997) attributed the higher FE and FS values of lupin protein isolate, obtained by ultrafiltration, compared to that prepared by isoelectric precipitation, to the presence in the former of seed polysaccharide molecules which may form a complex with the adsorbed protein molecules and stabilize the foam through the enhancement of steric repulsion interaction forces operating between neighboring bubbles. Pozani et al. (2002) reported that a relatively short heat treatment may result in an improvement of both foaming ability and foam liquid stability of lupin protein. These workers attributed this improvement to protein structure modification resulting from heating that affected the adsorption and interaction at the air–water interface properties of the protein.

In conclusion, pulse proteins in the form of concentrates or isolates may possess very satisfactory emulsifying and foaming ability, irrespective of the method applied for their recovery. On the other hand, the emulsions prepared with pulse proteins, as most protein-stabilized emulsions, may exhibit various degrees of instability, especially with respect to droplet aggregation and creaming, the extent of instability depending on protein composition and degree of denaturation related to the conditions prevailing during the process of protein recovery. In addition, the stability against bubble fusion and liquid separation of pulse protein foams is unsatisfactory. The problem of emulsion and foam destabilization during storage is often tackled by employing polysaccharides. When polysaccharides are present in the continuous phase of emulsions they tend to bring about an increase in viscosity that prevents the movement of the droplets and protects them against aggregation and/or coalescence with droplets in close proximity. In the case of foams, the increase in continuous phase viscosity may retard the thinning and subsequent rupture of the thin liquid films between the bubbles. Alternatively, adsorption of polysaccharide molecules to droplet or bubble surface, through their interaction with adsorbed protein molecules, may enhance emulsion or foam stability by preventing the close approach and interaction of emulsion droplets or foam bubbles.

Papalamprou et al. (2005) reported that incorporation of xanthan in emulsions based on lupin protein isolate enhanced their physical stability, an effect attributed to the strengthening of the droplet network structure, resulting from the presence of the polysaccharide molecules in the emulsion continuous phase and from the increased protein adsorption to droplet surfaces, due to depletion. According to Makri and Doxastakis (2006b), the stability against droplet coalescence and creaming of emulsions prepared with *Phaseolus* bean protein was found to be affected by the addition in the system of polysaccharides such as gum Arabic, locust bean gum or xanthan. The stabilizing effect of the biopolymer depended on the type of the polysaccharide, with the mixture of xanthan–locust bean gum being more effective, something that, according to the authors, was connected with the very high viscosity of the emulsion continuous phase. Liu et al. (2010) reported that the stability against creaming of emulsions stabilized by a pea protein isolate (PPI)–gum Arabic mixture was higher compared to that of an emulsion stabilized by PPI alone, with the effect being more pronounced in the pH range of 3.1–4.0. In addition, it was reported that gum Arabic incorporation did not influence the foaming capacity of the protein but only improved the stability of the resulting foam, the result depending on pH. The improvement in emulsion and foam stability resulting from the presence of gum Arabic was attributed to the higher continuous phase viscosity and to the formation of an electrostatic complex between the protein and the polysaccharide molecules at the droplet or bubble surfaces that increased the magnitude of the repulsive steric stabilizing forces. Gharsallaoui et al. (2010) reached similar conclusions when studying the effect of high methoxy pectin incorporation on the creaming stability of pea protein-stabilized emulsions. The increased stability of the system was connected with the formation of a protein–pectin complex at the oil droplet surface and the enhancement of the steric stabilization forces.

3.3.3 Protein gelation characteristics

As in the case of most pulse protein functional properties, the ability of pulse protein isolates and concentrates to form a self-supporting gel network structure could be affected by factors such as the pulse type and variety as well as by the method applied for the preparation of the protein material, its composition and storage conditions. Gel network formation by pulse proteins takes place following heating to a temperature higher than the protein denaturation point. The protein gelation ability is usually expressed by many investigators in terms of the least gelling concentration (LGC), which is the minimum concentration required for a protein dispersion to form a self-supporting network, determined either visually or, more accurately, by applying dynamic rheometry.

Fernandez-Quintela et al. (1997) studied the gelling properties of protein isolates obtained from legume seeds using isoelectric precipitation and reported that faba bean isolates could form gels at a lower concentration (14%) compared to pea (18%) or soybean isolates (16%). Similar LGC values, ranging between 14 and 18%, were also reported by Kaur and Singh (2007) for a number of various chickpea protein isolates.

Boye et al. (2010b) compared the gelling properties of a number of protein concentrates obtained from pea, chickpea and lentil varieties by applying the method of isoelectric precipitation, and reported that the lentil protein was more effective than the rest in producing a gel network (LGC 12% as opposed to 14%). These workers also observed that the gelling properties of the protein concentrates were improved when the method of ultrafiltration was applied for their preparation, with the LGC, for instance, of the lentil concentrate dropping to 8%. This latter finding emphasizes the importance of the method of protein manufacture on molecular interaction and gel network formation, something that was more systematically investigated for isolates obtained from lupin (Kiosseoglou et al., 1999) or chickpea (Papalamprou et al., 2008). These studies reported LGC values in the range of 4–8% for isolates obtained by ultrafiltration or dialysis. On the other hand, the LGC values for the isolates obtained by isoelectric precipitation were close to 12%. According to Papalamprou et al. (2008), the composition of chickpea protein isolate in terms of globulins and albumins influences only to a limited extent the LGC of the isolate manufactured by applying ultrafiltration (4.5% as opposed to 5.5% for the globulin- and the albumin-enriched isolate, respectively).

The method of protein preparation affects not only the LGC value of pulse proteins but also the mechanical properties of the resulting gels. At a relatively low protein concentration, but not above 10%, the gels of chickpea protein isolates prepared by ultrafiltration exhibited a more elastic character compared to that based on an isolate produced by isoelectric precipitation (Papalamprou et al., 2008). As was pointed out, the much gentler nature of ultrafiltration or the dialysis method, as opposed to isoelectric precipitation, was expected to result in less extensive protein structure changes and hence in more organized protein–protein interactions during the formation of the gel network. In the case of the isolate manufactured by precipitation, it was suggested that random protein subunit aggregation in the initial protein extract, during the precipitation step, probably resulted in the reduction of the effective protein concentration, leading to higher LGC and gels of a lower elastic character.

Apart from the method of protein isolate preparation, other parameters, such as the environmental conditions (pH, ionic strength), are expected strongly to affect the gelling properties of pulse proteins. As reported by Kiosseoglou et al. (1999), NaCl addition at levels of up to 0.5 M may bring about an increase in storage modulus and a reduction in tanδ values of lupin protein isolate gels. At a higher salt concentration, gel elasticity and strength appeared to decrease again, indicating that electrostatic interactions between the denatured protein molecules determined to some extent the properties of the gel network structure. Zhang et al. (2007) studied the effect of pH and NaCl or $CaCl_2$ content on the gelation behavior of chickpea protein isolate and reported that LGC reached a lower value at pH 7 than at pH 3, with LGC decreasing further in the presence of 0.1 M NaCl. Based on data obtained by applying oscillatory rheometry, these workers concluded that the gel elasticity and strength parameters depend on NaCl or $CaCl_2$ content. Zhang et al. (2010) investigated the thermal gelation at an acidic pH of a protein isolate obtained from kidney bean, as influenced by ionic strength. According to these investigators, the gelation behavior of kidney bean proteins is influenced to a marked extent by the ionic strength, with the extent of protein aggregation and storage moduli increasing and the LGC decreasing as the ionic strength is increased from 0 to 300 mM (Fig. 3.3).

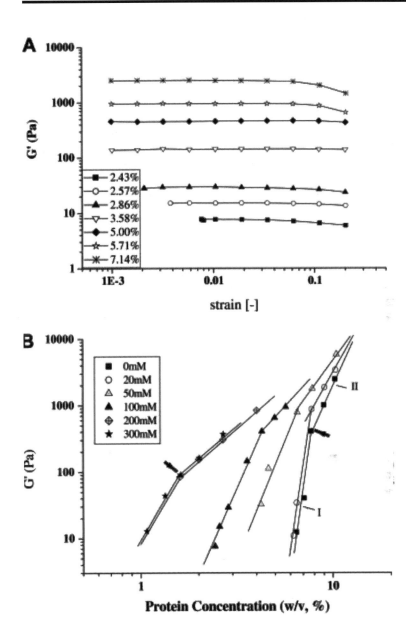

Figure 3.3 G' versus strain or protein concentration c for KPI gels (at pH 2.0) formed by heating at 85 °C for 1 h. (A) Typical profiles of G' versus strain for KPI gels, at ionic strength of 100 mM and different protein concentrations; (B) G' (in the linear regime) versus c, at varying ionic strengths. Symbols within figures: I and II, two distinct regimes, at low and high c values, respectively; arrows, the transition of these two regimes. (From Zhang et al., 2010, with permission.)

The main interactions responsible for gel network formation by pulse proteins appear to be hydrophobic in nature. Covalent disulfide bridges may also contribute to some extent to gel structure development (Kiosseoglou et al., 1999). Papalamprou et al. (2008) observed that 50% of protein in the structure of gels, based on chickpea protein isolate, was solubilized when treated with sodium dodecyl sulfate (SDS) or urea. This percentage, however, was much lower compared to the percentage of protein liberated from the structure of gels prepared with lupin protein isolate and led to the conclusion that the physical interactions play the dominant role in lupin protein gels. In the case of chickpea protein gels, on the other hand, the contribution of physical and disulfide bonds is more balanced, an indication that probably both the composition of the pulse proteins, in terms of sulfur amino acids, as well as the protein structure, may determine the gelation properties of pulse proteins. Additionally, the influence of parameters such as pH or ionic strength on the gelation characteristics of pulse proteins emphasizes the critical role that the protein charge plays in establishing a balance between protein intermolecular repulsive and attractive forces to form an elastic gel network structure.

3.4 Food applications of pulse proteins

Nowadays, the regular dietary intake of pulses or pulse ingredients as one of the ways to a healthy life is strongly supported by scientists. This is mainly due to their reported nutritional and health benefits, attributed to the presence of macro- and micronutrients. Among these nutrients, proteins play an important role in consideration of their amino acid composition which can easily be balanced in the diet. Besides, pulse proteins have gained increasing importance because they are also used to provide desired functional properties, including gelling and emulsifying properties. Pulse protein concentrates and isolates could, therefore, be proposed as a potential supplement in a great number of food applications offering opportunities for novel product development, some of which are presented below.

3.4.1 Baked products

Baked products (bread, biscuits, cakes, etc.) are an important source of nutrients, i.e. protein, iron, calcium and several vitamins. Wheat flour forms the main raw material of baked products. Wheat flour protein quality can be improved by the addition of supplements, which are high in lysine, to meet the specific needs of the target groups and vulnerable sections of the population who are undernourished and malnourished. Pulse proteins are a group of proteins of high quality since they provide almost all the essential amino acids. The higher lysine and lower methionine, cysteine and tryptophan contents in pulses complement well the wheat proteins, which are poor in lysine and relatively rich in the sulfur-containing amino acids (Duranti, 2006). Apart from being nutritious, pulse proteins are highly functional and exhibit properties like solubility, gelation and water binding that play a crucial role in structure formation and mouth feel of the finished products.

The inclusion of pulses in baked products has been reported in several studies. In most of these studies whole pulse flours were used and only in a few was the suitability of rich in protein materials, such as concentrates or isolates, evaluated. According to Mizrahi et al. (1967), isolated proteins often improve the appearance and taste compared with the original meal and therefore can be better utilized as nutritional and functional ingredients in food products. Patel and Johnson (1975) studied the effect of incorporation of horse bean flour or protein isolate to weak, medium and strong wheat flours on physical dough properties, bread quality and amino acid composition. Isolate addition produced satisfactory dough characteristics while the produced Moroccan-type breads exhibited superior quality features and were more desirable than those made with horse bean flour as protein supplement. Lorenz et al. (1979) used faba bean protein concentrate (68.2% protein), prepared by air classification, as a partial replacement for wheat flour in bread and cookie formulations. Bread and cookie quality were acceptable at up to 10% substitution level. Flavor and taste, although not disliked, were rated inferior to those of the control wheat flour. In another study (Sathe et al., 1981), up to 15% of great northern bean protein concentrate and isolate were used to replace wheat flour in a bread formulation. Addition of the protein materials at levels above 10% was unfavorable with respect to dough and bread quality.

Hsu et al. (1982) investigated the effect of germination on the physicochemical properties and bread-making performances of protein isolates from yellow pea, lentil and faba bean. Germination had an adverse effect on the baking properties of protein isolates from faba bean but not on those of pea or lentil. Generally, replacement of wheat flour with 5 or 8% pulse protein isolate had an adverse effect on loaf volume and crumb grain of bread.

Abdel-Aal et al. (1987) prepared a protein micellar mass (PMM) from chickpea and faba bean flours by applying the micellization method, which in turn was used at a 15% level to fortify bread. The protein content of the PMM ranged from 57.5% to 83.8%. Chickpea as well as faba bean PMM produced more acceptable fortified bread than did wheat flour.

Lupin protein concentrates and isolates have been utilized for bread fortification by a number of investigators (Mubarak, 2001; Güémes-Vera et al., 2004, 2008; Paraskevopoulou et al., 2010). Mubarak (2001) studied the effect of addition of products, such as lupin seed flour, two lupin protein isolates or lupin protein concentrate, on the rheological characteristics of dough and the physical, chemical, nutritional and sensory properties of bread. It was reported that by increasing the lupin protein level in wheat flour, water absorption, development time and dough weakening were increased. In addition, lupin protein isolates could be added to wheat flour at up to 9% level without any observed detrimental effect on loaf volume bread acceptability by the panelists. In another study (Güémes-Vera et al., 2008), the influence of reduction or elimination of non-nutritional compounds and yellow color of *Lupinus mutabilis* derivatives (flour, protein concentrate or protein isolate) on the quality properties of three types of bread products (loaf, bun and sweet) was evaluated. The detoxifying treatments resulted in the reduction of the non-nutritional and toxic compound level of the original lupin seed while the use of citric acid (1%) reduced yellow coloration in lupin flour and lupin protein concentrate. Addition of lupin products had variable effects on bread product firmness and volume, depending on the type of bread formulation as well as the lupin derivative added. These results are closely linked to results reported by Güémes-Vera et al. (2004), who performed structural analysis of dough originating from mixtures of wheat flour with lupin flour as well as with lupin protein concentrates and isolates and noted that the presence of the legume protein in the mixture led to the disruption of dough structure. Photomicrographs of the dough showed a progressive loss of interaction in the wheat gluten protein network with increasing lupin supplementation levels. Their study also revealed that lupin protein was present in the dough in the form of hydrated but not fully dispersed particles.

In a recent study, Paraskevopoulou et al. (2010) investigated the impact of addition of two lupin protein isolates, enriched in proteins belonging either to globulin (LPI G) or to albumin (LPI A) fraction, on wheat flour dough and bread characteristics. According to their results, LPI addition increased dough development time and stability plus the resistance to deformation and the extensibility of the dough. Besides, the water required for optimum bread making was increased in the case of LPI G, while this parameter remained unaffected when LPI A was added (Table 3.2). This was attributed to the fact that, during dough mixing, the proteins of LPI G require more water to become hydrated compared to the highly water-soluble albumins present in the LPI A. On the other hand, the elasticity of dough at 500 BU was decreased by LPI addition. Between the two isolates, the highest elasticity was observed at 5% (w/w) LPI A (87.5 BU) and the lowest at 5% (w/w) LPI G (70 BU) (see Table 3.2). These results suggest that the proteins of the LPI G do not readily bind with those of gluten to enhance dough elasticity. The presence of LPI proteins in dough affected bread quality in terms of volume, internal structure and texture, while extra gluten addition to the blends, to compensate for wheat gluten dilution resulting from

Table 3.2 Effect of 5% and 10% lupin protein isolate addition (LPI G and LPI A) on farinograph characteristics of wheat flour (WF)[a]

Dough formulation	Water absorption (%)	Dough development time (min)	Dough stability (min)	Elasticity (BU)
WF	53.6 ± 0.11a	1.53 ± 0.21a	1.40 ± 0.40a	97.5 ± 9.6c
WF and LPI G (5)	55.5 ± 0.16b	2.13 ± 0.25b,c	3.77 ± 0.85b	80.0 ± 14.1a,b
WF and LPI G (10)	56.0 ± 0.06b	2.07 ± 0.06b	13.07 ± 1.43e	70.0 ± 0.0a
WF and LPI G (10) and GI	57.0 ± 0.15b,c	2.53 ± 0.31c	4.23 ± 0.15b	83.3 ± 5.8a,b,c
WF and LPI A (5)	53.9 ± 0.08a	2.30 ± 0.24b,c	10.15 ± 1.66d	87.5 ± 9.6a,b
WF and LPI A (10)	54.0 ± 0.08a	2.15 ± 0.24b,c	12.03 ± 0.06e	82.5 ± 9.6b,c
WF and LPI A (10) and GI	55.7 ± 0.11b	2.227 ± 0.75b,c	5.93 ± 0.40c	93.3 ± 5.8b,c

[a]Mean ± standard deviation of three replicates; values in the column followed by the same letter are not significantly different ($P<0.05$).
From Paraskevopoulou et al. (2010), with permission.

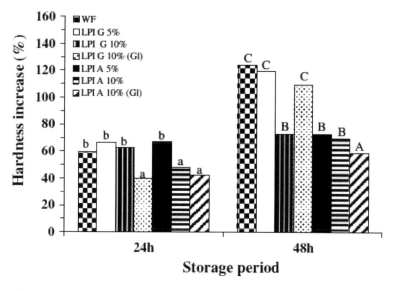

Figure 3.4 Increase (%) in hardness of control (WF) and LPI-enriched breadcrumbs after storage at room temperature. Values are means of three replicates; mean values for the first 24 h of storage with a different small letter are significantly different ($P<0.05$) and mean values for the 48 h of storage with a different letter are significantly different ($P<0.05$). (From Paraskevopoulou et al., 2010, with permission.)

LPI addition, led to an improvement of bread quality characteristics. In general, the incorporation of LP isolates in wheat flour delayed bread firming, the effect being more evident in the case of LPI A, when this was combined with supplementary gluten, indicating a kind of synergistic effect between the two additives (Fig. 3.4).

In another study (Mariotti et al., 2009), pea isolate, in combination with corn starch, amaranth flour and *Psyllium* fiber, was added to gluten-free dough. Generally, gluten replacement in cereal-based foods, aiming at producing foods for people suffering from celiac disease or other allergic reactions or intolerances to gluten consumption, impairs the dough capacity to develop properly during kneading, leavening and baking. As the absence of gluten results in bread with a crumbling texture and poor color, incorporation of ingredients that could simulate the viscoelastic properties of gluten in order to provide structure and retain gas, have been tested. Pea protein isolate ($\approx$87% protein) added at two levels (1 and 6%) induced an improvement of the gluten-free dough from both a physical (rheological and structural) and a nutritional point of view.

McWatters (1980) investigated the performance of pea protein (derived from cowpeas or field peas) as milk protein substitute in a baking powder biscuit (quick bread). Biscuits containing unheated and steamed (100 °C, 30 min) pea products were compared with reference biscuits made with whole milk. This study showed that replacement of milk protein with pea flour or protein concentrate in baking powder biscuits produced substantial changes in some sensory quality attributes, crust and crumb color and density. The "beany" aroma and flavor, which were the most objectionable sensory characteristics of biscuits made with unheated pea products, were lessened somewhat by preliminary steam treatment of the products.

Subagio and Morita (2008) studied the effect of hyacinth bean protein isolate on cake characteristics. The addition of protein isolate at concentrations up to 1% improved cake properties by increasing loaf and specific volume, softening the texture and improving overall preference. According to the authors, the protein isolate may have decreased the surface tension of the colloidal system, resulting in a reduction of the foam bubble size during cake mixing. This would improve the distribution of gas cells in the cake, resulting in the shortening of cake texture. When protein isolate of more than 1% was added, these properties tended to decrease while the color of the cake became more vivid and the staling rate lower.

3.4.2 Pasta products

Pasta products, traditionally manufactured using only durum wheat semolina, are regarded as a source of carbohydrates (74–77%) while their protein (11–15%) is deficient in lysine, in common with most cereal products. As a matter of fact, their nutritional properties can be improved and targeted to specific dietary requirements by the inclusion of pulse proteins, in the form of either flours or processed protein materials.

Several studies have examined various aspects of pulse protein incorporation into pasta. Faba bean protein concentrate (68.2% protein), produced by air classification, was used as a partial replacement for wheat flour in noodle formulation (Lorenz et al., 1979). The noodles had a desirable yellow color and were less brittle. Sensory analysis studies showed the 20% substituted noodles to be acceptable and with better flavor scores than the control. Nielsen et al. (1980) evaluated the fortification of hard red spring wheat flour with 33% pea flour or 20% air-classified pea protein concentrate in order to produce cost-effective noodles and spaghetti with improved protein and nutritional qualities. According to their observations, increasing the pea protein content induced a weakening of noodle dough and decreased cooking water absorption. The cooking time for noodles was reduced, but cooking losses were greater. Incorporation of raw pea protein concentrate produced spaghetti with similar texture characteristics to durum spaghetti, but its tolerance to overcooking and its flavor were inferior. In general, the color, flavor and texture of the fortified pasta compared favorably with those of 100% hard red spring wheat pasta. Precooking the pea products decreased noodle dough stickiness and improved noodle flavor, but it increased cooking loss and caused the deterioration of the fortified spaghetti's cooked texture. In another study (Yañez-Farias et al., 1999), chickpea protein concentrate (64% protein), obtained by ultrafiltration, was used to fortify pasta. Semolina was blended with 5, 10 and 15% (w/w) of the concentrate and the quality of the cooked pasta, evaluated by measuring water-absorption capacity, increased volume and resistance to disintegration as well as by a trained panel, was not greatly affected by increasing the amount of addition.

Seyam et al. (1983) used navy and pinto bean flour protein isolates, obtained by acid precipitation, as a source for protein enrichment of macaroni products and reported that the cooking quality of spaghetti made with semolina and navy bean protein isolate was better compared with that of spaghetti made with semolina and pinto bean protein isolate. Bahnassey et al. (1986) evaluated the nutritional properties of spaghetti fortified with non-roasted and roasted lentil, pinto bean and navy bean flours or their protein concentrates, and reported that the supplemented samples were nutritionally superior compared to spaghetti prepared from 100% semolina. The amino acid composition of spaghetti made with 10% or 15% (w/w) pulse flour or protein concentrate showed a better balance of lysine and sulfur amino acids than the control. Trypsin inhibitor activity in the fortified sample was, however, higher than in the control but was slightly reduced when the pulses were roasted. Additionally, semolina supplementation with the aforementioned pulse products caused an increase in farinograph water absorption, probably as a result of the higher protein content of the blends causing higher hydration capacity (Bahnassey and Kahn, 1986). Dough development time and stability were higher for blends containing navy or pinto bean protein products. However, fortified spaghetti shattered earlier

Table 3.3 Cooking characteristics[a] of spaghetti samples containing lupin protein isolate

Spaghetti composition S/LPI-E	Optimal cooking time (min)	Cooked weight (g)	Cooking loss (%)
100/0	3.9	33.20	7.50
95/5	3.8	32.04	8.71
90/10	4.1	28.10	14.47
85/15	3.8	25.10	15.31
80/20	4.1	23.61	16.93

[a]The values are the mean of three measurements.
From Doxastakis et al. (2007), with permission.

than control spaghetti. By increasing the level of supplementation, fortified spaghetti exhibited lower cooked weight and higher cooking loss and firmness values. Sensory analysis revealed that spaghetti supplemented with 10% pulse flour or protein concentrate was acceptable for all parameters tested.

Duszkiewicz-Reinhard et al. (1988) studied the shelf-life stability of spaghetti prepared with durum wheat semolina, containing 3% (w/w) gluten, and supplemented with 10% (w/w) pulse (navy and pinto bean) flour or their protein concentrate. Cooked weight of spaghetti was lower for fortified samples during storage while their firmness was higher compared to the control. The acidity of spaghetti increased upon storage, especially for spaghetti made with the pulse protein concentrate.

Lupin proteins are also an excellent choice for improving the nutritional value of pasta products. Doxastakis et al. (2007) investigated the rheological properties of dough and the cooking quality of spaghetti prepared by replacing semolina with different amounts of white lupin protein isolate. Dried spaghetti fortified with 5% lupin protein isolate exhibited color and rheological features comparable to those of the standard semolina sample. The behavior during cooking, in terms of cooked pasta weight, cooking loss and cooked product texture, was satisfactory. Incorporation of more than 5% lupin protein isolate resulted in an unacceptable cooking loss and weakening of the overall structure of spaghetti which was attributed to the dilution of semolina gluten network strength by the lupin proteins (Table 3.3). Moreover, as far as thermal damage is concerned, the furosine values (considered a useful marker of semolina quality and pasta processing conditions) of fortified spaghetti were found to differ only marginally from those of standard pasta while the available lysine content in the fortified samples was higher than in semolina spaghetti and the percentage lysine loss remained rather low.

The production of a lupin-based gluten-free pasta product for the needs of celiac individuals was the object of another study (Capraro et al., 2008). The protein pattern of the product, as well as any possible polypeptide covalent changes of the main polypeptide backbones appearing during the steps of production cycle, were monitored with an integrated modern protein analytical approach. For this purpose, lupin flour and its protein concentrate, some of the half-processed products (flour/protein concentrate mix), the gelatinized dough, the extruded dough and eventually the dry pasta were collected, extracted and analysed by two-dimensional gel electrophoresis. This analysis revealed that no major qualitative or quantitative changes of the overall protein patterns occurred during the process. In particular, there was no alteration of the covalent continuity of the main polypeptide backbones, while the disulfide pattern was not affected by the process either. In addition, the constancy of the glycosylation pattern was not apparently affected, thus suggesting a full stability of the covalently linked oligosaccharide side-chains during the process.

3.4.3 Meat products

Pulse proteins, when blended with meat proteins, may offer a promising alternative source for nutritional and functional proteins (Boye et al., 2010a). Their incorporation in products such as sausages, frankfurters, burgers, etc., apart from improving their nutritional quality, aims at a more acceptable texture and a better color and cooking performance.

Chickpea and faba bean flours, concentrates and PMM concentrates, prepared by applying a salting-out process, were used as sausage meat extenders at substitution levels ranging from 20% to 40% (Abdel-Aal et al., 1987). Protein-rich materials from lupin have also been used as extenders or as meat substitutes in comminuted meat products. Alamanou et al. (1996) conducted studies on the use of lupin protein isolate (LPI) from seeds of *Lupinus albus* ssp. *Graecus* in frankfurters at various levels. Incorporation of LPI increased ($P<0.05$) the pH and viscosity of batter and reduced jelly separation.

Lupin protein isolate addition at levels of up to 2% had no adverse effect on functional properties or consumer acceptance of the product. Moreover, the frankfurters exhibited a satisfactory processing yield and better water-holding capacity compared to the control. Previous LPI treatment (hydration or use as a stabiliser in pre-emulsified fat at 1% level) had a beneficial effect on the processing and sensory characteristics of frankfurters, probably due to better hydration of batter during these treatments.

In another study (Papavergou et al., 1999), fermented sausages were extended using lupin flours (LFs) and lupin protein isolates (LPIs) from a sweet variety of *Lupinus albus* and a bitter variety of *Lupinus albus* ssp. *Graecus*, at a 2% supplementation level. These workers found that LF addition from both varieties had detrimental effects on the odor and taste of the meat products, probably due to its high pro-oxidant character. Products containing LP isolates exhibited no difference in firmness, appearance and color compared to the control whereas their eating quality was affected by the type of isolate used, with that derived from the sweet variety being more susceptible to oxidation. The antioxidant effect exhibited by LPI, extracted from the bitter variety, was attributed to the occurrence of minor constituents, such as flavonoids, phenols and peptides, which do not promote the rancidity of the product.

Lupin proteins incorporated in comminuted meat products perform at the same time as fat emulsifiers, water binders and gelation agents, leading to the development of acceptable textural and sensory attributes. The way the proteins of lupin protein isolates, enriched in either albumins or globulins, interact in the meat gel network systems and affect the eating quality of products, where structure formation takes place upon cooking, was studied by Mavrakis et al. (2003). These researchers observed that LPI incorporation in a model comminuted meat system affected the compressive behavior of the gels, with the stress increasing more rapidly with strain for the gels containing lupin proteins. When 25% fat was incorporated into the system, to simulate the composition of a typical sausage product, the gel structure became weaker but again compressive stress reached higher values in the presence of lupin protein. This enhancement indicated that lupin proteins were involved in the development of the multicomponent comminuted gel structure during processing, probably through lupin protein participation in interactions, either at the fat particle surface or within isolated pockets of high lupin protein content (Mavrakis et al., 2003). According to Drakos et al.

(2007), the increase of the gel network resistance to compression upon incorporation of LPI in comminuted meat paste and heating at 90 °C depended on composition with respect to fat, water and salt. Although the lupin proteins tended to adsorb to the fat particle surfaces of the comminuted meat system, it was found that these surfaces were dominated by the salt-soluble proteins of meat.

3.4.4 Other food applications

The use of pulse proteins in the preparation of milk-based products, i.e. imitation milks, curds, dairy desserts, etc., is rather limited compared to the use of soy proteins. El-Sayed (1997) used chickpea, peanut or sesame protein isolates in order partially to replace skim milk powdered protein in processed cheese blends. The resulting products exhibited sensory characteristics quite similar to those of traditional ones. Swanson (1990) reported on the use of pea and lentil protein isolates in the development of imitation milk products which were of either equal or inferior quality in comparison to milk-based on soy protein isolate. Additionally, Cai et al. (2001) conducted studies on bean curd formation using protein materials originating from a number of pulses (chickpea, faba bean, lentil, mung bean, smooth pea, pea and winged bean), while Gebre-Egziabher and Summer (1983) prepared high-protein curd from pulses such as field pea. In another study (Nunes et al., 2003), the use of proteins from pulse seeds (lupin, pea) in combination with various polysaccharides (κ-carrageenan, gellan and xanthan gum) was studied as an alternative to milk proteins in dairy desserts. It was suggested that the ability of mixed systems to form gels could be an interesting alternative to milk puddings but a more detailed study is required.

Proteins, in general, have also been tested as food packaging materials in the form of edible films as they are effective lipid, oxygen and aroma barriers at low relative humidity conditions. However, limited information is available on the use of pulse proteins for this reason. Bamdad et al. (2006) produced an edible film for food packaging based on lentil protein and glycerine as plasticizer and reported that the film was found to be comparable to other protein edible films in terms of its mechanical, optical and barrier properties.

Finally, the development of gluten-free, pulse-based cracker snacks that take advantage of the anti-allergenic and health-enhancing nature of pulse ingredients was the objective of another study (Han et al.,

2010). These researchers focused on the formulation of gluten-free cracker snacks utilizing nine commercially available pulse flours and fractions (chickpea, green and red lentil, yellow pea, pinto and navy bean flours and pea protein, starch and fiber isolates). The produced crackers exhibited light color, good flavor and crisp texture, while those crackers containing pea protein isolate were in the "liking" range of the scale.

References

Abdel-Aal, E.M., Youssef, M.M., Shehata, A.A., El-Mahdy, A.R., 1987. Some legume proteins as bread fortifier and meat extender. Alexandria J. Agric. Res. 32, 179–189.

Alamanou, S., Doxastakis, G., 1997. Effect of wet extraction methods on the emulsifying and foaming properties of lupin seed protein isolates (*Lupinus albus* ssp. *Graecus*). Food Hydrocoll. 11, 409–413.

Alamanou, S., Bloukas, J.G., Paneras, E.D., Doxastakis, G., 1996. Influence of protein isolate from lupin seeds (*Lupinus albus* ssp. *Graecus*) on processing and quality characteristics of frankfurters. Meat Sci. 42, 79–93.

Bahnassey, Y., Kahn, K., 1986. Fortification of spaghetti with edible legumes. II. Rheological processing and quality evaluation studies. Cereal Chem. 63, 216–219.

Bahnassey, Y., Khan, K., Harrold, R., 1986. Fortification of spaghetti with edible legumes. I. Physicochemical, antinutritional, amino acid, and mineral composition. Cereal Chem. 63, 210–215.

Bamdad, F., Goli, A.H., Kadivar, M., 2006. Preparation and characterization of proteinous film from lentil (*Lens culinaris*). Edible film from lentil (*Lens culinaris*). Food Res. Int. 39, 106–111.

Boye, J., Zare, F., Pletch, A., 2010a. Pulse proteins: processing, characterization, functional properties and applications in food and feed. Food Res. Int. 43, 414–431.

Boye, J.I., Aksay, S., Roufik, S., 2010b. Comparison of the functional properties of pea, chickpea and lentil protein concentrates processed using ultrafiltration and isoelectric precipitation techniques. Food Res. Int. 43, 537–546.

Braudo, E.E., Plashchina, I.G., Schwenke, K.D., 2001. Plant protein interactions with polysaccharides and their influence on legume protein functionality. Nahrung/Food 45, 382–384.

Cai, R., Klamczynska, B., Baik, B.-K., 2001. Preparation of bean curds from protein fractions of six legumes. J. Agric. Food Chem. 49, 3068–3073.

Capraro, J., Magni, C., Fontanesi, M., Budelli, A., Duranti, M., 2008. Application of two-dimensional electrophoresis to industrial process analysis of proteins in lupin-based pasta. LWT – Food Sci. Technol. 41, 1011–1017.

Chakraborty, P., Sosulski, F., Bose, A., 1979. Ultracentrifugation of salt soluble proteins in ten legume species. J. Sci. Food Agric. 30, 766–771.

Chapleau, N., de Lamballerie-Anton, M., 2003. Improvement of emulsifying properties of lupin proteins by high pressure induced aggregation. Food Hydrocoll. 17, 273–280.

Doxastakis, G., Papageorgiou, M., Mandalou, D., 2007. Technological properties and non-enzymatic browning of white lupin protein enriched spaghetti. Food Chem. 101, 57–64.

Drakos, A., Doxastakis, G., Kiosseoglou, V., 2007. Functional effects of lupin proteins in comminuted meat and emulsion gels. Food Chem. 100, 650–655.

Duranti, M., 2006. Grain legume proteins and nutraceutical properties. Fitoterapia 77, 67–82.

Duszkiewicz-Reinhard, W., Khan, K., Dick, J.W., Holm, Y., 1988. Shelf life stability of spaghetti fortified with legume flours and protein concentrates. Cereal Chem. 65, 278–281.

Elkowicz, K., Sosulski, F.W., 1982. Antinutritive factors in eleven legumes and their air-classified protein and starch fractions. J. Food Sci. 47, 1301–1304.

El-Sayed, M.M., 1997. Use of plant protein isolates in processed cheese. Nahrung/Food 41, 91–95.

Fan, T.Y., Sosulski, F.W., 1974. Dispersability and isolation of protein from legume flours. Can. Inst. Food Sci. Technol. J. 7, 256–259.

Fernandez-Quintela, A., Macarulla, M.T., Del Barrio, A.S., Martinez, J.A., 1997. Composition and functional properties of protein isolates obtained from commercial legumes grown in northern Spain. Plant Foods Hum. Nutr. 51, 331–342.

Fuhrmeister, H., Meuser, F., 2003. Impact of processing on functional properties of protein products from wrinkled peas. J. Food Eng. 56, 119–129.

Gebre-Egziabher, A., Summer, A.K., 1983. Preparation of high protein curd from field peas. J. Food Sci. 48, 375–377.

Gharsallaoui, A., Yamauci, K., Chambin, O., Cases, E., Saurel, R., 2010. Effect of high methoxyl pectin in aqueous solution and at oil/water interface. Carbohydr. Polym. 80, 817–827.

Gueguen, J., Cerleti, P., 1994. Proteins of some legume seeds: soybean, pea, fababean and lupin. In: Hudson, B.J.F. (Ed.), New and developing sources of food proteins. Chapman and Hill, pp. 145–193.

Güémes-Vera, N., Arciniega-Ruiz Esperza, O., Dávila-Ortiz, G., 2004. Structural analysis of the *Lupinus mutabilis* seed, its flour, concentrate, and isolate as well as their behavior when mixed with wheat flour. LWT – Food Sci. Technol. 37, 283–290.

Güémes-Vera, N., Peña-Bautista, R.J., Jiménez-Martínez, C., Dávila-Ortiz, G., Calderón-Domínguez, G., 2008. Effective detoxification and decoloration of *Lupinus mutabilis* seed derivatives, and effect of these derivatives on bread quality and acceptance. J. Sci. Food Agric. 88, 1135–1143.

Gupta, R., Dhillon, S., 1993. Characterization of seed storage proteins of lentil (*Lens culinaris* M.). Ann. Biol. 9, 71–78.

Han, J., Janz, J.A.M., Gerlat, M., 2010. Development of gluten-free cracker snacks using pulse flours and fractions. Food Res. Int. 43, 627–633.

Hsu, D.L., Leung, H.K., Morad, M.M., Finney, P.L., Leung, C.T., 1982. Effect of germination on electrophoretic, functional, and bread-baking properties of yellow pea, lentil, and faba bean protein isolates. Cereal Chem. 59, 344–350.

Kaur, M., Singh, N., 2007. Characterization of protein isolates from different Indian chickpea (*Cicer arietinum* L.) cultivars. Food Chem. 102, 366–374.

Kiosseoglou, A., Doxastakis, G., Alevisopoulos, S., Kasapis, S., 1999. Physical characterization of thermally induced networks of lupin protein isolates prepared by isoelectric precipitation and dialysis. Int. J. Food Sci. Technol. 34, 253–263.

Kohnhorst, A.L., Uebersax, M.A., Zabik, M.E., 1990. Production and functional characteristics of protein concentrates. J. Am. Oil Chem. Soc. 67, 285–292.

Lee, H.C., Htoon, A.K., Uthayakumaran, S., Paterson, J.L., 2007. Chemical and functional quality of protein isolated from alkaline extraction of Australian lentil cultivars: Matilda and Digger. Food Chem. 102, 1199–1207.

Liu, S., Elmer, C., Low, N.H., Nickerson, M.T., 2010. Effect of pH on the functional behavior of pea protein isolate–gum Arabic complexes. Food Res. Int. 43, 489–495.

Lorenz, K., Dilsaver, W., Wolt, M., 1979. Fababean flour and protein concentrate in baked goods and in pasta products. Bakers Digest 39, 45–51.

Makri, E., Doxastakis, G., 2006a. Emulsifying and foaming properties of *Phaseolus vulgaris* and *coccineus* proteins. Food Chem. 98, 558–568.

Makri, E., Doxastakis, G., 2006b. Study of emulsions stabilized with *Phaseolus vulgaris* or *Phaseolus coccineus* with the addition of Arabic gum, locust bean gum and xanthan gum. Food Hydrocoll. 20, 1141–1152.

Mariotti, M., Lucisano, M., Ambrogina Pagani, M., Ng, P.K.W., 2009. The role of corn starch, amaranth flour, pea isolate, and *Psyllium* flour on the rheological properties and the ultrastructure of gluten-free doughs. Food Res. Int. 42, 963–975.

Mavrakis, C., Doxastakis, G., Kiosseoglou, V., 2003. Large deformation properties of gels and model comminuted meat products containing lupin protein. J. Food Sci. 68, 1371–1376.

McWatters, K.H., 1980. Replacement of milk protein with protein from cowpea and field pea flours in baking powder biscuits. Cereal Chem. 57, 223–226.

Mizrahi, S., Zimmerman, G., Berk, Z., Cogan, U., 1967. The use of isolated proteins in bread. Cereal Chem. 44, 193–203.

Mubarak, A.E., 2001. Chemical, nutritional and sensory properties of bread supplemented with lupin seed (*Lupinus albus*) products. Nahrung/Food 45, 241–245.

Nielsen, M.A., Sumner, A.K., Whalley, L.L., 1980. Fortification of pasta with pea flour and air-classified pea protein concentrate. Cereal Chem. 57, 203–206.

Nunes, M.C., Batista, P., Raymundo, A., Alves, M.M., Sousa, I., 2003. Vegetable proteins and milk puddings. Colloid Surf. B: Biointerfaces 31, 21–29.

Papalamprou, E.M., Makri, E.A., Kiosseoglou, V.D., Doxastakis, G.I., 2005. Effect of medium molecular weight xanthan gum in rheology and stability of oil-in-water emulsion stabilized with legume proteins. J. Sci. Food Agric. 85, 1967–1973.

Papalamprou, E., Doxastakis, G., Biliaderis, C., Kiosseoglou, V., 2008. Influence of preparation methods on physicochemical and gelation properties of chickpea protein isolates. Food Hydrocoll. 23, 337–343.

Papalamprou, E., Doxastakis, G., Kiosseoglou, V., 2010. Chickpea protein isolates obtained by wet extraction as emulsifying agents. J. Sci. Food Agric. 90, 304–313.

Papavergou, E.J., Bloukas, J.G., Doxastakis, G., 1999. Effect of lupin seed proteins on quality characteristics of fermented sausages. Meat Sci. 52, 421–427.

Paraskevopoulou, A., Provatidou, E., Tsotsiou, D., Kiosseoglou, V., 2010. Dough rheology and baking performance of wheat flour–lupin protein isolate blends. Food Res. Int. 43, 1009–1016.

Paredes-Lopez, O., Ordorica-Falomir, C., Olivares-Vazquez, M.R., 1991. Chickpea protein isolates: physicochemical, functional and nutritional characterization. J. Food Sci. 56, 726–729.

Patel, K.M., Johnson, J.A., 1975. Horsebean protein supplements in breadmaking. II. Effect on physical dough properties, baking quality and amino acid composition. Cereal Chem. 52, 791–800.

Peterson, D.S., 1998. Composition and food uses of lupins. In: Gladstones, J.S., Atkins, C., Hamblin, J. (Eds.), Lupins as crop plants: biology, production and utilization. CAB International, Oxford, pp. 353–384.

Pozani, S., Doxastakis, G., Kiosseoglou, V., 2002. Functionality of lupin seed protein isolate in relation to its interfacial behaviour. Food Hydrocoll. 16, 241–247.

Roy, F., Boye, J.I., Simpson, B.K., 2010. Bioactive proteins and peptides in pulse crops: pea, chickpea and lentil. Food Res. Int. 43, 432–442.

Sanchez-Vioque, R., Clemente, A., Vioque, J., Bautista, J., Millan, F., 1999. Protein isolates from chickpea (*Cicer arietinum* L.): chemical composition, functional properties and protein characterization. Food Chem. 64, 237–243.

Sathe, S.K., Salunkhe, D.K., 1981. Functional properties of great northern bean proteins: emulsion, foaming, viscosity and gelation properties. J. Food Sci. 46, 71–81.

Sathe, S.K., Ponte, J.G., Rangnekar, P.D., Salunkhe, D.K., 1981. Effects of addition of great northern bean flour and protein concentrates on rheological properties of dough and baking quality of bread. Cereal Chem. 58, 97–100.

Satterlee, L.D., Bembers, M., Kendrick, G.J., 1975. Functional properties of the great northern bean (*Phaseolus vulgaris*) protein isolate. J. Food Sci. 40, 81–84.

Seyam, A.A., Banasik, O.J., Breen, M.D., 1983. Protein isolates from navy and pinto beans: their uses in macaroni products. J. Agric. Food Chem. 31, 499–502.

Subagio, A., Morita, N., 2008. Effects of protein isolate from hyacinth beans (*Lablab purpureus* (L.) Sweet) seeds on cake characteristics. Food Sci. Technol. Res. 14, 12–17.

Swanson, B.G., 1990. Pea and lentil protein extraction and functionality. J. Am. Oil Chem. Soc. 67, 276–280.

Tomoskozi, S., Lasztiti, R., Dudek, S., 2001. The functional potential of legume proteins for production of food ingredients. Grain Legumes 33, 16–18.

Tyler, R.T., 1984. Impact milling quality of grain legumes. J. Food Sci. 49, 925–930.

Tyler, R.T., Panchuk, B.D., 1982. Effect of seed moisture content on the air classification of field peas and faba beans. Cereal Chem. 59, 31–33.

Tyler, R.T., Youngs, C.G., Sosulski, F.W., 1981. Air classification of legumes I – Separation efficiency, yield, and composition of the starch and protein fractions. Cereal Chem. 58, 144–148.

Vose, J.R., 1980. Production and functionality of starches and protein isolates from legume seeds: field peas (*Pisum sativum* cultivar trapper) and horse beans (*Vicia faba-equina* cultivar Diana). Cereal Chem. 57, 406–410.

Yañez-Farias, G.A., Bernal-Aguilar, V., Ramírez-Rodríguez, L., Barrón-Hoyos, J.M., 1999. Note. Fortification of some cereal foods with a chickpea protein concentrate. Food Sci. Technol. Int. 5, 89–93.

Zhang, T., Jiang, B., Wang, Z., 2007. Gelation properties of chickpea protein isolates. Food Hydrocoll. 21, 280–286.

Zhang, T., Jiang, B., Mu, W., Wang, Z., 2009. Emulsifying properties of chickpea protein isolates: influence of pH and NaCl. Food Hydrocoll. 23, 146–152.

Zhang, Y.-H., Tang, C.-H., Wen, Q.-B., Yang, X.-Q., Li, L., Deng, W.-L., 2010. Thermal aggregation and gelation of kidney bean protein isolate at pH 2: influence of ionic strength. Food Hydrocoll. 24, 266–274.

Functional and physicochemical properties of pulse starch

4

Narpinder Singh
Department of Food Science and Technology, Guru Nanak Dev University,
Amritsar, India

4.1 Introduction

Chickpea (*Cicer arietinum* L.), black gram (*Phaseolus mungo* L.), pigeon pea (*Cajanus cajan* L.), field pea (*Pisum sativum* L.), kidney bean (*Phaseolus vulgaris* L.) and lentil (*Lens culinaris*) are some of the important pulses grown throughout the world. Chickpea (*Cicer arietinum* L.) is one of the most important pulse crops in the Indian subcontinent. Black gram (*Phaseolus mungo* L.) is an important pulse crop throughout a large part of the tropics and is widely grown in India, Iran, South-East Asia, Greece and East Africa (Sathe et al., 1982). Field pea (*Pisum sativum* L.) is extensively grown in northern Europe, the USA, Canada, Russia and China but is grown in a limited area in India (Singh, 1999). India, Canada and Turkey are the main lentil-producing countries. India alone produces one-third of the world production of lentils. Pulses are the major source of

Pulse Foods: Processing, Quality and Nutraceutical Applications. DOI: 10.1016/B978-0-1238-2018-1.00007-0

proteins and other nutrients in the diets of malnourished areas of the world. They are also good sources of slowly digestible starch, the most desirable form of dietary starch that is completely, but more slowly, digested in the small intestine, and attenuates postprandial plasma glucose and insulin levels (Jenkins et al., 1981).

Extensive research has been conducted on the structure and functional properties of the main starches of commerce such as corn, wheat, potato and rice due to their ready availability, low price compared to pulses, and extensive utilization in food and non-food applications (Singh et al., 2003). Pulse starches, on the other hand, are not commercially produced, to a large extent due to their higher price, and consequently have remained somewhat neglected. Pulses have been primarily looked upon as a protein source rather than as a carbohydrate source, although starch is their major component, ranging between 35 and 60% (Deshpande et al., 1982). The protein content of pulses varies between 14.9 and 39.4% (Salunkhe et al., 1985). The separation of starch of high purity from certain pulses is difficult because of the presence of highly hydrated fine fiber fractions and insoluble protein (Schoch and Maywald, 1968). Pulse starches are isolated using aqueous techniques (wet) and dry milling (Schoch and Maywald, 1968; Reichert and Youngs, 1978). The information on the characteristics of pulse starches in comparison to cereals and tuber starches is limited. The majority of pulse starches have greater stability against heat and mechanical shear; they represent a good source of resistant starch, which can make these starches an exciting food ingredient for a number of food applications. Pulse starches can be used in processed meat products, particularly where extensive heating and mechanical stirring is required. Canned foods, cooked sausages, soups, sauces, noodles and vermicelli prepared using pulse starches have better sensory attributes than those made from starches from traditional sources. However, the high cost of pulses in comparison to the traditional source of starch (corn) limits their use as a commercial source of starch. In this chapter, the isolation, physicochemical and functional properties of pulse starches are discussed.

4.2 Starch isolation

The starch separation method from pulses, which was originally reported by Kawamura et al. (1955), involves treatment with 0.2% NaOH solution, washing with water, and dehydration with ethanol and water. Later, three methods for the separation of starch from pulses were suggested by Schoch and Maywald (1968). The first method, suggested for separation of starch from mung beans, garbanzo beans and dehulled split yellow peas, involved steeping in warm water containing toluene to prevent fermentation, followed by wet grinding and repeated screening. These authors suggested a second method for separation of starch from lentils, lima beans and white navy beans, which were difficult to process. This method involved steeping in warm water in the presence of toluene followed by resuspension in 0.2% NaOH solution (to dissolve most of the protein). The alkaline suspension was then screened through 220-mesh nylon to remove a portion of fine fiber and then slowly flowed down an inclined "table", which was a flat, shallow, trough of heavy-gauge stainless-steel with a total slope of 0.5 inch (1.27 cm). A third method was suggested for wrinkled seeded peas that required exhaustive alkaline steeping and washings of the isolated starch. Pulse grains contain predominantly soluble proteins, which can be easily separated from starch. However, the fine fiber along with some of the flocculent proteins present in pulse grains co-settle with starch during the separation operation and make the isolation of starch difficult (Schoch and Maywald, 1968). The separation of starch from cereals and tubers is easier as compared to pulse grain. The starch from pulses is separated by dry milling and wet milling; the former method is used in commercial-scale production. The dry milling process involves extensive particle size reduction of grain, generally done by pin mills, followed by air classification. Air classification separates the low-protein starch fraction from fine fractions high in protein. The repeated milling followed by air classification is done to reduce the protein content in starch. However, washing with water and dilute alkali is necessary to produce starch of high purity having a protein content less than 0.2%.

The starch obtained by wet milling has higher purity than that from dry milling. A dilute alkali with pH between 8.5 and 10.0 was employed to extract the protein (Hoover and Sosulski, 1985a,b; Bogracheva et al., 1995). The starch produced by wet milling has a protein content between 0.01 and 0.50% (Hoover and Sosulski, 1991). The wet milling process involves repeated filtration through screens and alkaline washing (0.02% NaOH), which causes a substantial decrease in the protein content in pea starch (Hoover and Sosulski, 1985a). The starch of higher extraction rates (93.8–96.7%) from smooth pea flour could be produced after protein extraction at pH 9 using different sieves and washing conditions (Colonna et al., 1981). Meuser et al. (1995) developed a process to extract starch from wrinkled peas. This process involved the steeping of peas in warm water followed by dehulling with rubber rollers (to separate the hulls from cotyledons), gentle reduction of cotyledons and high-pressure disintegration of the screened out protein/starch fraction suspended in the process water. This process gave a starch yield of 89% with a protein content of around 1%. Singh et al. (2004a) isolated starches from different chickpea cultivars by steeping seeds in water containing 0.16% sodium hydrogen sulfite for 12 h at 50 °C, followed by grinding, sieving and centrifugation. The repeated washings could yield starches with a protein content of less than 0.8%. The avoidance of amylolytic or mechanical damage to the starch granules during the initial isolation steps, effective deproteinization of the starch, and minimizing the loss of small granules are some of the important considerations that must be kept in mind during starch separation (Schulman and Kammiovirta, 1991).

4.3   Physicochemical properties

4.3.1   Morphology

The granule size of pulse starches ranges from 0.4 to 103 μm depending on the botanical source. Starch granules with different shapes such as oval, spherical, round, elliptical, disks and irregular shapes have been reported in various studies (Fig. 4.1). The surface of the granules is generally smooth without any fissures, cracks or pores. The majority of pulses contain simple starch granules; however, a mixture of simple and compound (consisting of 3–10 individual units) granules in wrinkled and smooth pea starch has been reported (Colonna et al., 1982; Bertoft et al., 1993). A granule size between 2.72 and 31.8 μm was observed for pea starches (Aggarwal et al., 2004). The large oval to small spherical shaped granules were reported to be present in starch from different chickpea cultivars (Singh et al., 2004a). Chickpea starch had a granule length and width ranging between 17.0 and 20μm and 11.0 and 14.2 μm, compared to 24.5 and 19.5 μm for pigeon pea starch (Singh et al., 2008). Kidney bean starch has predominantly granules of size 10–30 μm (Singh, 2010, unpublished data). Black gram starches are relatively small compared to other pulse starches, with characteristic granule

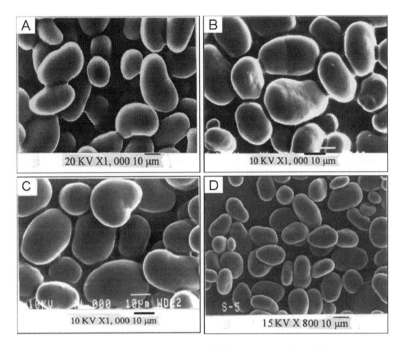

Figure 4.1 Scanning electron micrographs (SEM) of starches from different pulses: (A) mung bean, (B) pigeon pea, (C) kidney bean, (D) chickpea. (Source: Singh et al., 2004a.)

diameter in the range between 12.8 and 14.3 μm and shape varying from oval to elliptical (Singh et al., 2004b). The average granule size among the starches from different plants has been related to their physicochemical properties such as light transmittance, amylose content, swelling power and water-binding capacity (Zhou et al., 1998; Singh and Singh, 2001; Kaur et al., 2002).

4.3.2 Composition

Pulse starch, like cereal and tuber starch, consists of amylose and amylopectin. The amylose content of the starches varies with the botanical source and is affected by the climatic conditions and soil type during growth (Inatsu et al., 1974; Morrison et al., 1984; Asaoka et al., 1985; Morrison and Azudin, 1987). Pulse starches are characterized by high amylose content (24–65%). Singh et al. (2004b) and Sathe et al. (1982) reported the amylose content of black gram starches in the range of 30.2–34.6%. Amylose content in the range of 28.6–34.3% for chickpea starches has been reported (Lineback and Ke, 1975; El Faki et al., 1983; Singh et al., 2004a). Hoover and Sosulski (1986) reported an amylose content of 38.5%, 32%, and 34.2% for lentil, faba bean and field pea starches, respectively. A wide range between 21.4 and 58.3% of amylose content in starch from field pea lines has been reported (Singh et al., 2010a). The difference in amylose content among pulse starches may be due to genetic differences, physiological state of the seed or the amount of amylose complexed within lipids (Morrison and Laignelet, 1983). Differences in amylopectin chain length distribution among different pulse starches have also been reported. The distribution of amylopectin chain DP after hydrolysis with isoamylase was determined by high-performance anion exchange chromatography equipped with a pulsed amperometric detector (HPAEC-PAD), capillary electrophoresis and mass spectrometry (Nakamura et al., 2002). The amylopectin chains with DP6–12, 13–18, 19–24 and 25–30 ranged from 36.2 to 43.25%, 36.44 to 38.68%, 14.86 to 18.22% and 4.95 to 6.9%, respectively, among starch separated from 20 chickpea cultivars (Singh et al., 2010b). Chung et al. (2008a) reported DP6–12, DP13–24 and DP25–36 ranged from 16.9 to 17.2%, 28.7 to 30.6% and 54.8 to 56.0%, respectively, in chickpea starch. Asaoka et al. (1985) reported measurement of fine structure of amylopectin by disbranching of starches followed by fractionation on gel permeation chromatography according to the wavelength range at the maximum absorption (λ_{max}) in the absorption spectra of the glucan–iodine complexes. They fractionated starch into apparent amylose content (Fr. I), intermediate fraction, i.e. mixture of relatively short amylose and long side-chains of amylopectin, long side-chains of amylopectin (Fr. II) and short side-chains of amylopectin (Fr. III). Singh et al. (2008) compared the fine structure of pulse starches using this method. Chickpea and black gram starches were reported to contain Fr. I in the range of 34.4–35.5% and 32.9–35.6%, respectively. Pigeon pea and kidney bean starches showed Fr. I of 31.8 and 35.9%, respectively. The intermediate fraction, Fr. II and Fr. III ranged between 6.0 and 8.5%, 14.7 and 15.4%, and 41.7 and 43.8%, respectively, among the chickpea starches compared to 2.7–3.0%, 16.7–18.5%, and 44.0–46.5% of similar fractions in black gram starches. Field pea and kidney bean starches showed a lower proportion of fraction II compared to other pulse starches. Black gram and pigeon pea starch showed higher proportions of Fr. II than other pulse starches. Black gram starch had the lowest content of intermediate fraction between 2.5 and 3.0%, compared to the highest amount of the same fraction (6–8.5%) in chickpea starches. The distribution of α-1,4-chains of amylopectin (Fr. III/Fr. II) ranged between 2.5 and 3.3, lower for black gram and pigeon pea starch. Black gram and pigeon pea starches showed the highest proportion of Fr. III, followed by chickpea, kidney bean and field pea starch. Pulse starch was observed to have lower Fr. II and Fr. III compared to rice starches. Fr. I, intermediate fraction, Fr. II and Fr. III among rice starches ranged between 9.7 and 28.3%, 3.7 and 5.0%, 20.6 and 26.6%, and 45.8 and 59.4%, respectively (Singh et al., 2007). Pulse starches have a lipid content between 0.02 and 1.40, which varies with type and variety (Hoover et al., 2010).

4.3.3 Structure

The packing of amylose and amylopectin within the granules varies among starches from different species. X-ray diffractometry is used to reveal the presence and characteristics of the crystalline structure of the starch granules. The "A", "B" and "C" pattern are the different polymeric forms present in starches of different botanical sources that differ in the packing of the amylopectin double helices. The packing of double helices within the A-type polymorphic (crystalline) structure is compact with low water content, while the B-type polymorph has a more open structure containing a hydrated helical core (Wu and Sarko, 1978b; Tester et al., 2004). The cereal starches exhibit A-type, where double helices comprising the crystallites are densely packed. Tuber starches show the B-form, wherein crystallites are less densely packed. Pulse starches have a mixed A- and B-type pattern known as C-type (Singh et al., 2008). X-ray diffractograms of pulse starches, maize starch and potato starch are illustrated in Figure 4.2. Pulses starches are characterized by strong intensity peaks corresponding to approximately $2\theta=15°$, $17°$ and $23°$. Pulse starches also showed a small peak at $2\theta=5.6°$ as observed for potato starch; however, the intensity of peak was very low (Singh et al., 2008). The intensity of peak at $2\theta=5.6°$ for field pea starch was observed to be higher than that from chickpea, pigeon pea and kidney bean starches. It has been shown that certain pea mutants have different proportions of A- and B-type polymorphs with a weak peak at $2\theta=5.6°$ and a strong peak at $2\theta=17.9°$ within the starch granule, that can serve as models to understand the development of different polymorphs (Bogracheva et al., 1999; Hedley et

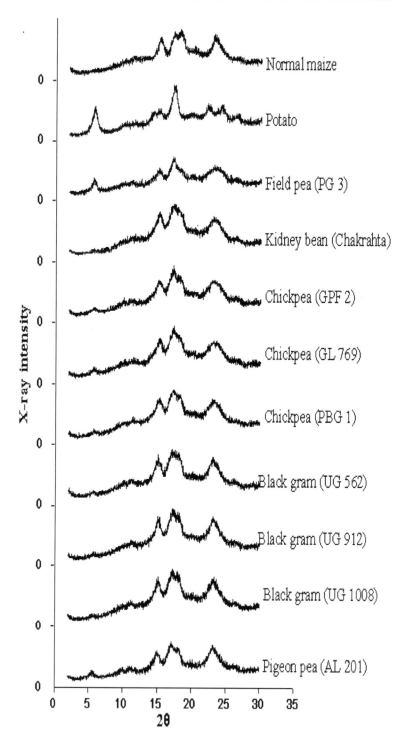

Figure 4.2 X-ray diffractograms of pulse starches. (Source: Singh et al., 2008.)

al., 2002). A strong intensity peak at $2\theta=7.9°$ and a weak intensity peak at $2\theta=5.6°$ are characteristic of A- and B-type, respectively. A-type starches have strong intensity peaks at $2\theta=15.18°$, $17.13°$, $18°$ and $22.86°$ while B-type have strong intensity peaks at $2\theta=5.6°$ and $2\theta=17.16°$. The central and peripheral starch granules are rich in B- and A-type polymorphs, respectively (Bogracheva et al., 1998). A weak peak at around $2\theta=20°$ present in wheat and rice related to an amylose–lipids complex has also been reported and is generally absent in pulse starches (Singh et al., 2008). Crystallinity among pulse starches ranges between 17 and 34%, which could be influenced by moisture content, size and number of crystallites arranged in a crystalline array and polymorphic content (Hoover et al., 2010). Crystallinity of 33.4%, 29.1%, 28.1%, 27.7–30.3% and 17.7%, respectively, for pigeon pea, mung bean, black gram, kidney bean and wrinkled pea starches has been reported (Zhou et al., 2004; Chung et al., 2008a,b; Sandhu and Lim, 2008). Pigeon pea, mung bean and black gram starches showed greater crystallinity as compared to kidney beans and chickpea starches. It was indicated that the proportion of longer amylopectin chains was responsible for the variation in crystallinity among starches from different pulses (Singh et al., 2008). It has also been reported that maize starches with higher amounts of long side-chains of amylopectin are more crystalline than those with lesser amounts of the similar fractions (Singh et al., 2006). Pulses starches showed a single peak at $2\theta=15.2°$ and $23.2°$, and dual peaks at $2\theta=17–18.1°$ as observed in maize starch. The intensity of other peaks in field pea starch was much lower than that for other pulse starches. A peak at $2\theta=5.6°$, which is the characteristics of the "B" polymorphic form, has been observed earlier in wrinkled pea starch (Zhou et al., 2004). Potato starch showed a dual peak at $2\theta=22.4–24.1°$ which appeared as a single peak in pulse starch. Maize starch showed a single peak at $2\theta=23°$, characteristic of typical A-type starch. Black gram and pigeon pea starches showed greater crystallinity as compared to kidney bean and chickpea starches as indicated by the greater intensity of peaks at $2\theta=17°$, $18.1°$ and $23.2°$.

4.3.4 Swelling and solubility

When starch molecules are heated in excess water, the crystalline structure is disrupted and water molecules become linked by hydrogen bonding to the exposed hydroxyl groups of amylose and amylopectin, which causes an increase in granule swelling and solubility. Swelling of starch is measured as swelling factor or as swelling power. Swelling factor measures only intergranular water (Tester and Morrison, 1990a) while swelling power measures both inter- and intragranular water (Leach et al., 1959). Swelling power of chickpea and black gram starches in the range of 11.4–13.6 g g^{-1} (Singh et al., 2004a) and 11–26 g g^{-1}, respectively (Hoover and Sosulski, 1991; Singh et al., 2004b) has been reported. Schoch and Maywald (1968) reported swelling power in the range of 16–20 g g^{-1} for yellow pea, navy bean, lentil and garbanzo bean starches. Pulse starches have much lower swelling power and solubility than potato starch. The higher swelling power and solubility of potato starch has been linked to the presence of a higher content of phosphate groups on amylopectin (repulsion between phosphate groups on adjacent chains will increase hydration by weakening the extent of bonding within the crystalline domain) (Galliard and Bowler, 1987). The presence of lipids in starch has a reducing effect on the swelling of the individual granules (Galliard and Bowler, 1987). Cereal starch granules contain more lipids as compared to potato starch granules, which may possibly explain the difference in the swelling power of these starches. Pulse starches have been reported to exhibit a single stage restricted swelling and low solubility patterns (Schoch and Maywald, 1968; Wankhede and Ramteke, 1982; El-Faki et al., 1983; Hoover and Sosulski, 1985a). This is indicative of the strong bonding forces between starch components that relax over one temperature (Schoch and Maywald, 1968). The high intermolecular attraction between starch components may reflect an orderly arrangement of polymer chains within the starch granule, permitting a close parallel alignment, thus favoring maximum interaction via hydrogen bonding (Hoover et al., 1991).

The swelling power and solubility provide evidence of the magnitude of interaction between starch chains within the amorphous and crystalline domains. The extent of this interaction is influenced by the amylose to amylopectin ratio, and by the characteristics of amylose and amylopectin in terms of molecular weight/distribution, degree and length of branching and conformation (Hoover, 2001). The differences in swelling power and solubility of starches from different sources may also be due to the difference in morphological structure of starch granules. Higher swelling power and lower solubility for potato starches having large and irregular or cuboidal granules have also been reported (Kaur et al., 2002). The large and irregular or cuboidal granules may be helpful in immobilizing the starch substance within the granule even at very high levels of swelling, which results in lower solubility levels. Granules continue to swell as the temperature of the suspension is increased above the gelatinization range. According to deWilligen (1976a,b), corn and wheat granules may swell up to 30 times their original volume and potato starch granules up to 100 times their original volume, without disintegration. It has been suggested that amylose plays a role in restricting initial swelling because this form of swelling proceeds more rapidly after amylose has been exuded. The lower swelling of pulse starches is mainly due to their higher amylose content. The increase in starch solubility, with the concomitant increase in suspension clarity, is seen mainly as the result of the granule swelling permitting the exudation of amylose. The extent of leaching of solubles mainly depends on the lipid content of the starch and the ability of the starch to form amylose–lipid complexes. The amylose–lipid complexes are insoluble in water and require higher temperatures to dissociate (Morrison, 1988; Raphaelides and Karkalas, 1988). The amylose involved in complex formation with lipids is prevented from leaching out (Tester and Morrison, 1990a).

4.4 Thermal properties

Differential scanning calorimetry (DSC) is the most commonly used technique to measure the gelatinization transition temperatures (T_o, onset; T_p, peak; T_c) and enthalpy of gelatinization (ΔH_{gel}) of starches. T_o, T_p, T_c and ΔH_{gel} have been reported to be influenced by the molecular architecture of the crystalline region, which is related to the distribution of amylopectin short chains (DP6–11) and not by the proportion of crystalline region, which corresponds to the amylose to amylopectin ratio (Noda et al., 1996). ΔH_{gel} gives an overall measure of crystallinity (quality and quantity) and is an indicator of the loss of molecular order within the granule (Cooke and Gidley, 1992). T_o, T_p and T_c of chickpea starches have been reported to range between 59.3 and 60.25 °C, 66.6 and 68.6 °C, and 76.1 and 77.3 °C, respectively, compared to 66.8–79.6 °C, 71.4–74.6 °C and 77–79.6 °C for black gram starches (Table 4.1). Kidney bean and pigeon pea starch showed values of 63.5 and 68.4 °C for T_o, 67.8 and 71.56 °C for T_p, and 73.2 and 76.3 °C for T_c, respectively. Field

Table 4.1 Thermal properties of starches separated from different pulses

Pulse	T_o (°C)	T_p (°C)	T_c (°C)	ΔH_{gel} (J g^{-1})
Field pea[a]	55.4	60.8	67.6	3.6
Kidney bean[a]	68.3	73.4	79.1	3.0
Chickpea[a]	59.3–60.2	66.6–68.6	76.1–77.3	2.6–4.2
Black gram[a]	66.8–70.6	71.4–74.6	77.0–79.6	1.6–1.7
Pigeon pea[a]	72.5	77.7	83.4	2.6
Lentil[b]	57.8–68.4	66.0–76.1	71.0–82.0	3.0–13.0
Mung bean[c]	58.0–62.2	67.0–67.4	72.1–82.0	7.9–18.5
Navy bean[d]	61.0–66.8	68.6–75.1	71.0–91.0	8.8–15.3
Azuki bean[e]	69.9	73.4	76.6	11.7

Source: [a]Singh et al. (2008); [b]Chung et al. (2008a), Sandhu and Lim (2008), Zhou et al. (2004), Hoover and Ratnayake (2002); [c]Sandhu and Lim (2008), Hoover et al. (1997); [d]Chung et al. (2008b), Hoover and Ratnayake (2002), Su et al. (1997); [e]Su et al. (1997).

pea starch showed lower transition temperatures while pigeon starch showed the highest. Lower X-ray diffraction intensities in field pea starch, indicative of the presence of a more amorphous region, have been attributed to a lower transition temperature in such starches (Singh et al., 2008). Pigeon pea starches with higher crystallinity have been shown to have higher transition temperatures as well as ΔH_{gel} (Singh et al., 2008). The ΔH_{gel} has been attributed to the disruption of the double helices rather than long-range disruption of crystallinity (Cook and Gidley, 1992). The lower transition temperatures and higher ΔH_{gel} of field pea starch suggested that disruption of double helices (in amorphous and crystalline regions) during gelatinization was more pronounced in this starch than in other starches. Pigeon pea and black gram starches showed a higher proportion of long side-chains of amylopectin and a lower amount of amylose plus intermediate fractions as compared to other starches, which is responsible for the difference in transition temperature. Barichello et al. (1990) also reported that higher transition temperatures resulted from a higher degree of crystallinity, which provides structural stability and makes the granules more resistant to gelatinization. The starches with long-branch chain length amylopectin generally display higher ΔH_{gel}, indicating that more energy is required to gelatinize the crystallites of long chain length in such starches. An inverse relationship of DP6–12 and positive of DP13–30 with T_p and T_c was observed (Singh et al., 2010b). However, the correlation of DP6–10 with T_p ($r=-0.483$ to -0.650, $P \leq 0.05$) and T_c ($r=-0.426$ to

−0.585) was significant. DP14, 15, 16, 17 and 18 showed significant correlation with T_p (r=0.580, 0.630, 0.551, 0.506 and 0.445, respectively, $P \leq 0.05$) and DP14, 15 and 16 showed significant correlation with T_c (r=0.456, 0.507 and 0.465, respectively, $P \leq 0.05$). An inverse relationship of short chains DP6–12 with T_o, T_p and T_c among wheat starches has also been observed (Singh et al., 2009). Noda et al. (2003) also observed a negative correlation between the amount of amylopectin short chains of DP6–12 and T_o and T_p. Shorts chains are known to locate on the external part of the crystalline structure (Hizukuri, 1986) and seem to form a less stable double helical structure and consequently get disrupted by heat at lower temperatures. Amylopectin chains with long DP showed significant positive correlations with T_p and T_c. These results reflect a more stable crystalline network formation by these chains in the starch granule.

4.5 Dynamic rheological properties

Small strain dynamic rheometry is used to study the viscoelastic properties of starches. In a number of studies, dynamic rhemetry has been used to evaluate the rheological parameters (G', G'' and tan δ) of starches from cereals and pulses (Singh et al., 2006, 2007, 2008). G' (storage modulus) is a measure of the energy that is stored in the material or recoverable per cycle of deformation while G'' (loss modulus) is a measure of the energy that is lost as viscous dissipation per cycle of deformation. The ratio of the energy lost to the energy stored for each cycle can be defined by tan δ. G' and G'' of pulse starch suspensions (20%) measured during heating from 50 to 90 °C at a rate of 0.5 °C min^{-1} and holding at 90 °C for 30 min, using a Fluids Spectrometer RFS II (Rheometrics Co., New Jersey, USA) equipped with parallel plate (25 mm diameter) geometry were reported (Singh et al., 2008). Both the moduli increased during heating to a maximum followed by a drop. Among the different pulse starches studied, chickpea starch had the highest peak G' and pigeonpea starch showed the least (Fig. 4.3). Peak G' of starch from different chickpea cultivars ranged between 2.0×10^3 and 2.5×10^3 Pa while field pea and pigeon pea starch showed peak G' at 1.8×10^3 and 1.5×10^3 Pa, respectively (see Fig. 4.3). Heating caused swelling of the starch granules, followed by dissolving of amylose molecules resulting in an increase

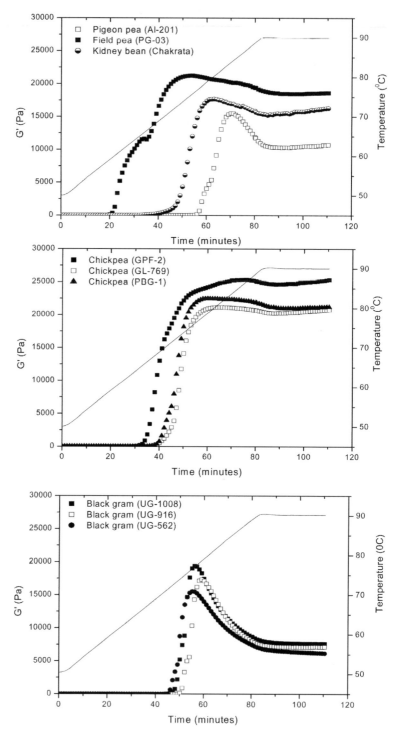

Figure 4.3 Changes in pulse starches during heating measured using a fluids spectrometer RFS II (Rheometrics Co. Ltd, New Jersey, USA). (Source: Singh et al., 2008.)

in G' and G'' to a maximum value. These changes have been attributed to the formation of a network of swollen starch granules (Hsu et al., 2000). The moduli started decreasing after reaching a maximum value that indicates destruction of the gel structure (Tsai et al., 1997). Starch gels have been defined as composites consisting of swollen granules filling an interpenetrating polymer network (Liu and Lelievere, 1992), and the major polymer in this network is amylose (Brownsey et al., 1987). Lindqvist (1979) indicated that both amylose and amylopectin of starch played important roles in the formation of starch gel. A strong interaction between the amylose matrix and rigid filler is required to produce gels with higher strength (Liu and Lelievere, 1992). Chickpea, kidney bean and field pea starches showed flat peaks while black gram and pigeon pea starches showed sharp peaks (see Fig. 4.3). The results revealed that the destruction in gel is slower in chickpea, kidney bean and field pea than black gram and pigeon pea starches. High crystallinity in black gram and pigeon pea starch has accounted for their lower peak viscosity, sharper peak and faster breakdown of gels than other pulse starches. Black gram and pigeon pea starches showed similar behavior to that of waxy maize starches (Singh et al., 2006). The breakdown in G' has been defined as the difference between peak G' and minimum G' at 90 °C after heating for 30 min (Singh et al., 2006). Higher breakdown was observed in starches with higher crystallinity. Black gram and pigeon pea starch exhibited higher breakdown in G' than chickpea, field pea and kidney bean. Among the chickpea, kidney bean and field pea, chickpea starch showed a higher breakdown in G'. Higher breakdown in black gram and pigeon pea starches has been attributed to more disintegration of swollen starch granules in the presence of lower amylose and intermediate fraction (Singh et al., 2008). Amylose and lipids have been reported to assist in maintaining granule integrity during heating (Morrison et al., 1993). The difference in rheological parameters has also been attributed to the difference in amylose/amylopectin ratio (Doublier and Llamas, 1993).

4.5.1 Pasting properties

When starch suspensions are heated in excess water, starch granules absorb water and swell to several times their original size. The swelling results from disruption of hydrogen bonds in amorphous regions. As the temperature is increased, granules swell more and this involves the disruption of both amorphous and crystalline structure. The Brabender viscoamylograh (BVA) and rapid visco-analyser (RVA) are used to study the pasting behavior of starches. These instruments measure the pasting behavior of starches under controlled increase in temperature and constant shearing conditions. A paste is formed during heating, which consists of a continuous phase of solubilized amylose and/or amylopectin and a discontinuous phase of granule ghosts and fragments (Hoover et al., 2010). Pasting refers specifically to changes in the starch upon further heating after gelatinization has occurred, including further swelling and leaching of polysaccharides from the starch granule, and increased viscosity which occurs on application of shear forces (Atwell et al., 1988; Tester and Morrison, 1990a,b). At the initial stage, granules absorb water, followed by a rapid increase in viscosity with the increase in temperature. After reaching the maximum, viscosity starts decreasing as a result of rupturing and fragmentation of granules during continuous stirring. This is known as breakdown, which is a measure of the ease with which the swollen granules become disintegrated upon shearing. A breakdown value of between 32 and 123 cP has been observed in chickpea flours, which indicated their higher stability (Singh et al., 2010). The lower breakdown may be attributed to lower disintegration of the swollen starch granules in the presence of a higher amylose content (Singh et al., 2008). Amylose and lipids have been reported to assist in maintaining granule integrity during heating (Morrision et al., 1993).

4.5.2 Retrogradation properties

Starch granules, when heated in excess water, undergo an order–disorder phase transition called gelatinization. On cooling, the starch chains (amylose and amylopectin) in the gelatinized paste interact, leading to the formation of a more ordered structure termed "retrogradation". Retrogradation is accompanied by increases in the degree of crystallinity and gel firmness, exudation of water (syneresis), and the appearance of a "B"-type X-ray pattern (Miles et al., 1985; Hoover, 1995). Many techniques such as DSC, turbidimetry, syneresis, X-ray diffraction, nuclear magnetic resonance (NMR), Raman spectroscopy, Fourier transform infrared spectroscopy and dynamic rheometry have been used to study the retrogradation behavior of starches (Karim et al., 2000; Singh et al., 2008). DSC provides information on amylopectin crystallization, whereas NMR elucidates changes in water mobility during retrogradation. Retrogradation of pulse starches has been determined mainly by measuring the amount of water exuded (syneresis) after thawing frozen gelatinized starch gels to room temperature. The syneresis results reported in various studies for different pulse starches may not facilitate comparison because of differences in storage temperature, number of freeze–thaw cycles, thawing time, centrifugal speed, centrifugation time and method of measuring the exuded water used (Hoover et al., 2010). Furthermore, in most studies, only one or a few cultivars of pulses have been evaluated. Consequently, it is difficult to ascertain whether the data reported truly represent a particular pulse because a wide range of amylose content among lines of field pea and chickpea has been reported (Singh et al., 2010a,b).

The changes in moduli (G', G'') at lower temperature have also been used to determine the retrogradation tendency of cooked pulse starch pastes (Singh et al., 2008). Field pea, chickpea and kidney bean starches have been shown to have greater tendency towards retrogradation as compared to black gram and pigeon starches on the basis of increase in G' at 10 °C over 10 h. The changes in moduli of cooked pulse starch pastes during cooling from 80 to 10 °C, holding at 10 °C for 10 h followed by reheating to 80 °C were reported. Both the moduli increased with decreasing temperature; however, the increase was highest in kidney bean and lowest in black gram. Starches from different varieties from the same pulse also showed a variation in retrogradation tendency. The retrogradation in order of: field pea > kidney bean > chickpea > black gram > pigeon pea starch was revealed from the changes in G' measured during 10 h at 10 °C. Black gram and pigeon pea starches showed a lower increase in moduli than other pulse starches. Chickpea, field pea and kidney bean starches with higher amounts of amylose content and intermediate fraction (mixture of short amylose and long side-chains of amylopectin) and lower amounts of short side-chain amylopectin fractions showed a greater change in moduli at 10 °C during a holding period of 10 h. Black gram and pigeon pea starches with lower amounts of amylose and intermediate fraction and higher amounts of short side-chains of amylopectin showed less change in moduli (Singh et al., 2008). Both the amylose and long branch chain amylopectin probably contribute more to the increase in moduli during retrogradation.

4.6 Digestibility

The retrogradation rate and its extent vary with starch properties (molecular and crystalline structure) and storage conditions (temperature, duration, water content, etc.). Pulse starches in general have higher a retrogradation rate due to a higher amylose content. The retrogradation makes the starch resistant towards breakdown by digestive enzymes and consequently reduces the glycemic index (GI) (Fredriksson et al., 2000). GI characterizes the carbohydrates consumed in the form of different foods on the basis of postprandial level of blood glucose (Jenkins, 2007). Carbohydrates that break down quickly during digestion and release glucose rapidly into the bloodstream are considered to have a high GI. Pulse starches have a low GI, attributed to their higher amylose content and resistant starch (RS) content along with a lesser tendency to lose completely crystalline and granular starch structure during heating. The GI values of pulses, pulse starches, pulse-based products, cereal and cereal-based products are reported in Table 4.2.

Starch is classified into three groups according to the rate of glucose release and its absorption in the gastrointestinal tract: rapidly digestible starch (RDS), slowly digestible starch (SDS) and RS. RDS is the group of starches that can be rapidly hydrolyzed by digestive enzymes; SDS is the group that is digested at a relatively slow rate (Englyst et al., 1992); and RS is not digested by digestive enzymes and is consequently transferred into the colon. RDS and SDS are measured as the glucose released after 20 and 100 min, respectively, of incubation. The starch that is not hydrolyzed after 120 min incubation is termed RS. RS is more commonly defined as the sum of starch and products of starch degradation not absorbed in the small intestine of healthy individuals. RS escapes digestion in the small intestine and is digested/fermented by bacteria in the colon (Asp, 1992). Pulse starches are less digestible than cereal starches (Chung et al., 2009). The lower digestibility of pulse starches has been attributed to their higher amylose content, C-type crystalline structure and absence of pores on the granule surface. Contrarily, normal corn starch is more susceptible to amylolysis compared to

Table 4.2 Glycemic index (GI) of pulses, pulse starches, pulse based products, cereal and cereal products

Food	GI
Amaranth grain (roasted)[a]	106
Amaranth grain (flaked)[a]	106
Amaranth grain (popped)[a]	101
Amaranth grain (popped)[b]	97
White bread[a]	94
Amaranth grain (extruded)[a]	91
Amaranth grain (raw)[a]	87
Instant mashed potatoes[b]	85
Cornflakes[b]	84
Puffed wheat[b]	74
Wheat bread[b]	70
Wholemeal bread[b]	69
Sweet corn[b]	53
White rice[b]	64
Brown rice[b]	55
Mung bean, fried	53
Kidney bean, canned[b]	52
Mung bean starch	50.7
Chickpea starch[c]	49.8
Baked beans (canned)[b]	48
Black gram starch[c]	48.7
Pigeon pea starch[c]	44.2
Mung bean, pressure cooked[b]	42
Chickpea, curry, canned[b]	41
Mung bean noodles, dried and boiled	39
Kidney bean, autoclaved[b]	34
Dhokla (leavened, fermented, steamed cake made from dehusked chickpea and wheat semolina)	33
Lentils, red, boiled[b]	32
Chickpea, dried, boiled[b]	31
Lentils, red[b]	28
Chapati, baisen (chickpea)[b]	27
Mung bean, germinated[b]	25
Black bean, soaked, cooked 45 min[b]	20
Peas, dried and boiled[b]	22
Pigeon pea, soaked and boiled 45 min	22
Kidney bean[b] (India)[b]	19
Soybean[b]	18
Bengal gram (chickpea) dahl[b]	11

Source: [a]Capriles et al. (2008) [Glycemic index (predicted), determined using equation = 39.71 + 0.549 (hydrolysis index) of Goni et al., 1997]; [b]Foster-Powell et al. (2002), reference food is glucose; [c]Sandhu and Lim (2008) [GI determined using equation = 39.71 + 0.549 (hydrolysis index) of Goni et al., 1997].

high amylose corn starch, which is attributed to the presence of surface pores and channels that facilitate enzymatic diffusion (Zhang et al., 2006). The association between amylose chains and its potential towards amylose–lipid complex formation (Morita et al., 2007), higher crystalline lamella thickness and a thicker peripheral layer (Jenkins and Donald, 1995) are factors that make the high amylose starch granules resistant to amylolysis. RS, RDS and SDS of 60–65%, 5.2–8.1% and 29.6–31.0%, respectively, for pigeon pea starches have been reported (Kaur and Sandhu, 2010). Waxy starches are more rapidly digested than high amylose starch; this may be due to more surface area per molecule of the amylopectin than amylose. RS has been associated with health benefits such as improved cholesterol metabolism, reduced risk of type 2 diabetes and colon cancer (Hoebler et al., 1999). RS has also been linked to other health benefits; for example, it may act as a substrate for probiotic organisms and facilitate increased absorption of minerals (Hoover and Zhou, 2003; Sajilata et al., 2006). RS rich foods cause slow and steady insulin release that enables a greater part of the stored fat to be transported to the liver and eventually used for energy. Therefore, RS plays an important role in the insulin response moderation and regulating the glucose metabolism. SDS is linked to stable glucose metabolism and satiety (Lehmann and Robin, 2007).

Physical, enzymatic, thermal, and chemical and food processing treatments change the structure and digestibility of starch and starch-based products. Waxy starches and low amylose corn starches are readily damaged by milling, rendering them more vulnerable to amylolysis (Tester and Morrison, 1994). Thermal treatment, such as autoclaving, baking, steam cooking and parboiling, affects the gelatinization and retrogradation processes and, consequently, the formation of RS in foods. Pulses are processed and consumed in different forms all over the world. The most common home-scale processing methods of pulses include dehulling (splitting), soaking, boiling, pressure cooking and germination. These processing treatments of pulses affect the *in vitro* digestibility of the pulse starch differently (Bravo et al., 1998; Rehman and Salariya, 2005; Costa et al., 2006). Eyaru et al. (2009) reported that both ordinary cooking and pressure cooking of soaked as well as unsoaked pulse grains led to a significant decrease in RS and increase in RDS; however, the effect of pressure cooking was greater. A number of fermented products are made from pulse and mixtures of cereals and pulses. *Idli*, *wadies* and *dokhla* are examples of pulse-based fermented products prepared in North and South India. Germination and fermentation reduced the RS content which is attributed to the loss in structural integrity of the starch granules and breakdown of starch into simple sugars by enzymes of the fermenting microflora. The chemical modifications of starches are done to change the functional properties that also change the susceptibility toward the action of enzymes. Esterification, etherification and cross-linking of starch make it resistant to α-amylase (Hood and Arneson, 1976). Chung et al. (2009) reported that substitution and oxidation of corn starch increase RS whereas cross-linking does not affect starch digestibility considerably. However, annealing and heat moisture treatment (HMT) decreased the RS content of pea, lentil and navy bean starches (Chung et al., 2010). These authors reported a greater effect of HMT as compared to annealing on RS. Chung et al. (2009) reported that the oxidized starch had higher amounts of RDS (33.5%) compared to hydroxypropylated starch (27.9%), cross-linked and acetylated starches ($\approx$24%) and unmodified starch (25.6%). They observed that the oxidized starch was rapidly hydrolyzed during the early stage of digestion (up to 60 min) because the chain degradation of starch occurred during the oxidation. Therefore, the fast rate of hydrolysis resulted in the greatest amount of RDS. The higher swelling ability of hydroxypropylated starch enhanced the access of digestive enzymes inside the granules, thereby increasing the RDS content. On the other hand, acetylated starch swelled more readily than unmodified starch, but the acetyl groups hinder the enzymatic action during the hydrolysis.

4.7 Conclusions

The separation of starch from pulses is difficult and generally gives lower yield, particularly when starch with high purity is required. A wide variation in granule size is reported in pulse starch. The role of granule size and surface characteristics in controlling the functional and digestibility properties is not clear. Most of the studies have been conducted using limited varieties of pulses. There is a need to conduct studies on pulse germplasms available in different countries to gain an improved understanding of the functionality and digestibility of their starches. Pulse starches have higher stability towards heat and mechanical shearing compared to starches from traditional sources, which can be exploited in the development of new products. Pulse starches have low GI, attributed to their higher amylose and RS content, which play an important role in the insulin response moderation and in regulating glucose metabolism.

References

Aggarwal, V., Singh, N., Kamboj, S.S., Brar, P.S., 2004. Some properties of seeds and starches separated from different Indian pea cultivars. Food Chem. 85, 585–590.

Asaoka, M., Okuna, K., Fuwa, H., 1985. Effect of environmental temperature at the milky stage on the amylose content and fine structure of amylopectin of waxy and nonwaxy endosperm starches of rice. Agric. Biol. Chem. 49, 373–379.

Asp, N.G., 1992. Resistant starch. Proceedings from the second plenary meeting of EURESTA: European FLAIR Concerted Action No. 11 on physiological implications of the consumption of resistant starch in man. Eur. J. Clin. Nutr. 46(Suppl. 2).

Atwell, W.A., Hood, L.F., Lineback, D.R., Varriano-Marston, E., Zobel, H.F., 1988. The terminology and methodology associated with basic starch phenomena. Cereal Foods World 33, 306–311.

Barichello, V., Yada, R.Y., Coffin, R.H., Stanley, D.W., 1990. Low temperature sweetening in susceptible and resistant potatoes: starch structure and composition. J. Food Sci. 55, 1054–1059.

Bertoft, E., Manelius, R., Qin, Z., 1993. Studies on the structure of pea starches – Part I. Initial stages in a-amylolysis of granular smooth pea starch. Starch 45, 215–220.

Bogracheva, T.Y., Davydova, N.I., Genin, Y.V., Hedley, C.L., 1995. Mutant genes at the r and rb loci affect the structure and physicochemical properties of pea seed starches. J. Exp. Bot. 46, 1905–1913.

Bogracheva, T.Y., Morris, V.J., Ring, S.G., Hedley, C.L., 1998. The granular structure of C-type pea starch and its role in gelatinization. Biopolymers 45, 323–332.

Bogracheva, T.Y., Cairns, P., Noel, T.R., Hulleman, S., Wang, T.L., Morris, V.J., 1999. The effect of mutant genes at the r, rb, rug3, rug4, rug5 and lam loci on the granular structure and physicochemical properties of pea seed starch. Carbohydr. Polym. 39, 303–314.

Bravo, L., Siddhuraju, P., Saura-Calixto, F., 1998. Effect of various processing methods on the in vitro starch digestibility and resistant starch content of Indian tribal pulses. J. Agric. Food Chem. 46, 4667–4674.

Brownsey, G.J., Ellis, H.S., Ridout, M.J., Ring, S.G., 1987. Elasticity and failure in composite gels. J. Rheol. 31, 635–649.

Capriles, V.D., Coelho, K.D., Guerra-Matias, A.C., Areas, J.A., 2008. Effects of processing methods on amaranth starch digestibility and predicted glycemic index. J. Food Sci. 73, H160–H164.

Chung, H.J., Liu, Q., Donner, E., Hoover, R., Warkentin, T.D., Vandenberg, B., 2008a. Composition, molecular structure, properties and *in vitro* digestibility of starches from newly released Canadian pulse cultivars. Cereal Chem. 85, 471–479.

Chung, H.J., Liu, Q., Hoover, R., Warkentin, T.D., Vandenberg, B., 2008b. *In vitro* starch digestibility, expected glycemic index, and thermal and pasting properties of flours from pea, lentil and chickpea cultivars. Food Chem. 111, 316–321.

Chung, H.J., Liu, Q., Hoover, R., 2009. The impact of annealing and heat–moisture treatments on rapidly digestible, slowly digestible and resistant starch levels in native and gelatinized corn, pea and lentil starches. Carbohydr. Polym. 75, 436–447.

Chung, H.J., Liu, Q., Hoover, R., 2010. Effect of single and dual hydrothermal treatments on the crystalline structure, thermal properties, and nutritional fractions of pea, lentil, and navy bean starches. Food Res. Int. 43, 501–508.

Colonna, P., Buleon, A., Mercier, C., 1981. *Pisum sativum* and *Vicia faba* carbohydrates: structural studies of starches. J. Food Sci. 46, 88–93.

Colonna, P., Buléon, A., LeMaguer, M., Mercier, C., 1982. *Pisum sativum* and *Vicia faba* carbohydrates. Part IV. Granular starches of wrinkled pea starch. Carbohydr. Polym. 2, 43–59.

Cooke, D., Gidley, M.J., 1992. Loss of crystalline and molecular order during starch gelatinization. Origin of the enthalpic transition. Carbohydr. Res. 227, 103–112.

Costa, G.E.A., Queiroz-Monici, K.S., Reis, S.M.P.M., Oliveira, A.C., 2006. Chemical composition, dietary fibre and resistant starch contents of raw and cooked pea, common bean, chickpea and lentil legumes. Food Chem. 94, 327–330.

Deshpande, S.S., Sathe, S.K., Rangnekar, P.D., Salunkhe, D.K., 1982. Functional properties of modified black gram (*Phaseolus mungo* L.) starch. J. Food Sci. 47, 1528–1533.

deWilligen, A.H.A., 1976a. The rheology of starch. In: Radley, J.A. (Ed.), Examination and analysis of starch and starch products. Applied Science Publishers Ltd, London, pp. 61–90.

deWilligen, A.H.A., 1976b. The manufacture of potato starch. In: Radley, J.A. (Ed.), Starch production technology. Applied Science Publishers, London, pp. 135–154.

Doublier, J.L., Llamas, G.A., 1993. Rheological description of amylose–amylopectin mixtures in food colloids and polymers: stability and mechanical properties. In: Dickinson, E., Walstra, P. (Eds.), Food colloids and polymers: stability and mechanical properties. Royal Society of Chemistry, London, pp. 139–146.

El-Faki, H.A., Desikachar, H.S.K., Paramahans, S.V., Tharanathan, R.N., 1983. Physicochemical characteristics of starches from chickpea, cowpea and horsegram. Starch 38, 118–122.

Englyst, H.N., Kingman, S.M., Cummings, J.H., 1992. Classification and measurement of nutritionally important starch fractions. Eur. J. Clin. Nutr. 46, S33–S50.

Eyaru, R., Shrestha, A.K., Arcot, J., 2009. Effect of various processing techniques on digestibility of starch in red kidney bean (*Phaseolus vulgaris*) and two varieties of peas (*Pisum sativum*). Food Res. Int. 42, 956–962.

Foster-Powell, K., Holts, S., Brand-Miller, J.C., 2002. International table of glycaemic index and glycemic load values. Am. J. Clin. Nutr. 76, 5–56.

Fredriksson, H., Bjorck, I., Andersson, R., et al. 2000. Studies on α-amylase degradation of retrograded starch gels from waxy maize and high-amylopectin potato. Carbohydr. Polym. 43, 81–87.

Galliard, T., Bowler, P., 1987. Morphology and composition of starch. In: Galliard, T. (Ed.), Starch properties and potential. Wiley, Chichester, pp. 57–78.

Goni, I., Garcia-Alonso, A., Saura-Calixto, F., 1997. A starch hydrolysis procedure to estimate glycemic index. Nutr. Res. 17, 427–437.

Hedley, C.L., Bogracheva, T.Y., Wang, T.L., 2002. A genetic approach to studying the morphology, structure and function of starch granules using pea as a model. Starch 54, 235–242.

Hizukuri, S., 1986. Polymodal distribution of the chain lengths of amylopectins, and its significance. Carbohydr. Res. 147, 342–347.

Hoebler, C., Karinthi, A., Chiron, H., Champ, M., Barry, J.L., 1999. Bioavailability of starch in bread rich in amylose: metabolic responses in healthy subjects and starch structure. Eur. J. Clin. Nutr. 53, 360–366.

Hood, L.F., Arneson, V.G., 1976. *In vitro* digestibility of hydroxypropyl distarch phosphate and unmodified tapioca starch. Cereal Chem. 53, 282–290.

Hoover, R., 1995. Starch retrogradation. Food Res. Int. 11, 331–346.

Hoover, R., 2001. Composition, molecular structure and physicochemical properties of tuber and root starches: a review. Carbohydr. Polym. 4, 253–267.

Hoover, R., Ratnayake, W.S., 2002. Starch characteristics of black bean, chick pea, lentil navy bean and pinto bean cultivars grown in Canada. Food Chem. 78, 489–498.

Hoover, R., Sosulski, F.W., 1985a. Studies on the functional characteristics and digestibility of starches from *Phaseolus vulgaris* biotypes. Starch 37, 181–191.

Hoover, R., Sosulski, F.W., 1985b. A comparative study of the effect of acetylation on starches of *Phaseolus vulgaris* biotypes. Starch 37, 397–403.

Hoover, R., Sosulski, F.W., 1986. Effect of cross-linking on functional properties of legume starches. Starch 38, 149–155.

Hoover, R., Sosulski, F.W., 1991. Composition, structure, functionality and chemical modification of legume starches: a review. Can. J. Physiol. Pharmacol. 69, 79–92.

Hoover, R., Zhou, Y., 2003. *In vitro* and *in vivo* hydrolysis of legume starch α-amylase and resistant starch formation in legumes – a review. Carbohydr. Polym. 54, 401–417.

Hoover, R., Hughes, T., Chung, H.J., Liu, Q., 2010. Composition, molecular structure, properties, and modification of pulse starches: a review. Food Res. Int. 43, 399–413.

Hoover, R., Li, Y., Hynes, G., Senanayake, N., 1997. Physicochemical characterization of mung bean Starch. Food Hydrocolloids 11, 401–408.

Hsu, S., Lu, S., Huang, C., 2000. Viscoelastic changes of rice starch suspensions during gelatinization. J. Food Sci. 65, 215–220.

Inatsu, O., Watanabe, K., Maida, I., Ito, K., Osani, S.J., 1974. Studies to improve the quality of rice grown in Hokkaido. I. Amylose contents of different rice starches. J. Jpn. Soc. Starch Sci. 21, 115–117.

Jenkins, A.L., 2007. The glycemic index: looking back 25 years. Cereal Foods World 52, 50–53.

Jenkins, D.J.A., Wolever, T.M.S., Taylor, R.H., et al. 1981. Glycemic index of foods: a physiological basis for carbohydrate exchange. Am. J. Clin. Nutr. 34, 362–366.

Jenkins, P.J., Donald, A.M., 1995. The influence of amylose on starch granule structure. Int. J. Biol. Macromol. 17, 315–321.

Karim, A.A., Norziah, M.H., Seow, C.C., 2000. Methods for the study of starch retrogradation. Food Chem. 71, 9–36.

Kaur, L., Singh, N., Sodhi, N.S., 2002. Some properties of potatoes and their starches. II. Morphological, thermal and rheological properties of starches. Food Chem. 79, 183–192.

Kaur, M., Sandhu, K.S., 2010. *In vitro* digestibility, structural and functional properties of starch from pigeon pea (*Cajanus cajan*) cultivars grown in India. Food Res. Int. 43, 263–268.

Kawamura, S., Tuboi, Y., Huzii, T., 1955. Studies on legume starches. I. Tech. Bull. Coll. 7, 87–90.

Leach, H.W., McCowen, L.D., Schoch, T.J., 1959. Structure of the starch granule. I Swelling and solubility patterns of various starches. Cereal. Chem. 36, 534–537.

Lehmann, U., Robin, F., 2007. Slowly digestible starch – its structure and health implications: a review. Trends Food Sci. Technol. 18, 346–355.

Lindqvist, I., 1979. Cold gelatinization of starch. Starch 31, 195–200.

Lineback, D.R., Ke, C.H., 1975. Starches and low-molecular weight carbohydrates from chick pea and horse bean flours. Cereal Chem. 52, 334–347.

Liu, H., Lelievre, J., 1992. Differential scanning calorimetric and rheological study of the gelatinization of starch granules embedded in gel matrix. Cereal Chem. 70, 385–391.

Meuser, F., Pahne, N., Möller, M., 1995. Extraction of high amylose starch from wrinkled peas. Starch 47, 56–61.

Miles, M.J., Morris, V.J., Orford, P.D., Ring, S.G., 1985. The roles of amylose and amylopectin in the gelation and retrogradation of starch. Carbohydr. Res. 35, 271–278.

Morita, T., Ito, Y., Brown, I.L., Ando, R., Kiriyama, S., 2007. *In vitro* and *in vivo* digestibility of native maize starch granules varying in amylose contents. J. AOAC Int. 90, 1628–1634.

Morrison, W.R., 1988. Lipids in cereal starches: a review. J. Cereal Sci. 8, 1–15.

Morrison, W.R., Azudin, M.N., 1987. Variation in the amylose and lipid contents and physical properties of rice starches. J. Cereal Sci. 5, 35–37.

Morrison, W.R., Laignelet, B., 1983. An improved colorimetric procedure for determining apparent and total amylose in cereal and other starches. J. Cereal Sci. 1, 9–20.

Morrison, W.R., Milligan, T.P., Azudin, M.N., 1984. A relationship between the amylose and lipids contents of starches from diploid cereals. J. Cereal Sci. 2, 257–260.

Morrison, W.R., Tester, R.F., Snape, C.E., Law, R., Gidley, M.J., 1993. Swelling and gelatinisation of cereal starches. IV. Some effects of lipids complexed amylose and free amylose in waxy and normal barley starches. Cereal Chem. 70, 385–391.

Nakamura, Y., Sakurai, A., Inaba, Y., Kimura, K., Iwasawa, N., Nagamine, T., 2002. The fine structure of amylopectin in endosperm from Asian cultivated rice can be largely classified into two classes. Starch 54, 117–131.

Noda, T., Takahata, Y., Sato, T., Ikoma, H., Mochida, H., 1996. Physicochemical properties of starches from purple and orange fleshed sweet potato roots at two levels of fertilizer. Starch 48, 395–399.

Noda, T., Nishiba, Y., Sato, T., Suda, I., 2003. Properties of starches from several low-amylose rice cultivara. Cereal Chem. 80, 193–197.

Raphaelides, S., Karkalas, J., 1988. Thermal dissociation of amylose–fatty acid complexes. Carbohydr. Res. 172, 65–82.

Rehman, Z., Salariya, A.M., 2005. The effects of hydrothermal processing on antinutrients, protein and starch digestibility of food legumes. Int. J. Food Sci. Technol. 40, 695–700.

Reichert, R.D., Youngs, C.G., 1978. Nature of the residual protein associated with starch fractions from air classified pea. Cereal Chem. 55, 469–480.

Sajilata, M.G., Singhal, R.S., Kulkarni, P.R., 2006. Resistant starch – a review. Comp. Rev. Food Sci. Food Saf. 5, 1–17.

Salunkhe, D.K., Kadam, S.S., Chavan, J.K., 1985. Processing and utilization. In: Postharvest biotechnology of food legumes. CRC Press, Boca Raton, pp. 109–140.

Sandhu, K.S., Lim, S., 2008. Digestibility of legume starches as influenced by their physical and structural properties. Carbohydr. Polym. 71, 245–252.

Sathe, S.K., Rangnekar, P.D., Deshpande, S.S., Salunkhe, D.K., 1982. Isolation and partial characterization of black gram (*Phaseolus mungo* L.) starch. J. Food Sci. 47, 1524–1527.

Schoch, T.J., Maywald, E.C., 1968. Preparation and properties of various legume starches. Cereal Chem. 45, 564–571.

Schulman, A.H., Kammiovirta, K., 1991. Purification of barley starch by protein extraction. Starch 43, 387–389.

Singh, J., Singh, N., 2001. Studies on the morphological, thermal and rheological properties of starch separated from some Indian potato cultivars. Food Chem. 75, 67–77.

Singh, N., Singh, J., Kaur, L., Sodhi, N.S., Gill, B.S., 2003. Morphological, thermal and rheological properties of starches from different botanical sources – a review. Food Chem. 81, 219–231.

Singh, N., Sandhu, K.S., Kaur, M., 2004a. Characteristics of starches from Indian chickpea cultivars. J. Food Eng. 63, 441–449.

Singh, N., Kaur, M., Sandhu, K.S., Guraya, H.S., 2004b. Physicochemical, thermal, morphological and pasting properties of starches from some Indian black gram varieties (*Phaseolus mungo*. L). Starch 56, 535–544.

Singh, N., Inouchi, N., Nishinari, K., 2006. Structural, thermal and viscoelastic characteristics of starches separated from normal, sugary and waxy maize. Food Hydrocoll. 20, 923–935.

Singh, N., Nakaura, Y., Inouchi, N., Nishinari, K., 2007. Fine structure, thermal and viscoelastic properties of starches separated from Indica rice cultivars. Starch 60, 349–357.

Singh, N., Nakaura, Y., Inouchi, N., Nishinari, K., 2008. Structure and viscoelastic properties of starches separated from different legumes. Starch 60, 349–357.

Singh, N., Singh, S., Isono, N., Noda, T., Singh, A.M., 2009. Diversity in amylopectin structure, thermal and pasting properties of starches from wheat varieties/lines. Int. J. Biol. Macromol. 45, 298–304.

Singh, N., Kaur, N., Rana, J.C., Sharma, S.K., 2010a. Diversity in seed and flour properties in field pea (*Pisum sativum*) germplasm. Food Chem. (in press).

Singh, N., Kaur, S., Isono, N., Noda, T., 2010b. Genotypic diversity in physico-chemical, pasting and gel textural properties of chickpea (*Cicer arietinum* L.). Food Chem. 122, 65–73.

Singh, U., 1999. Cooking quality of pulses. J. Food Sci. Technol. 36, 1–4.

Su, H.S., Lu, W., Chang, K.C., 1997. Microstructure and physicochemical characteristics of starches in soybean and their bean paste products. Lebensm-Wiss. U-Technol. 31, 265–273.

Tester, R.F., Morrison, W.R., 1990a. Swelling and gelatinization of cereal starches. I. Effects of amylopectin, amylose, and lipids. Cereal Chem. 67, 551–557.

Tester, R.F., Morrison, W.R., 1990b. Swelling and gelatinization of cereal starches. II. Waxy rice starches. Cereal Chem. 67, 558–563.

Tester, R.F., Morrison, W.R., 1994. Properties of damaged starch granules. Composition and swelling of fractions of wheat-starch in water at various temperatures. J. Cereal Sci. 20, 175–181.

Tester, R.F., Karkalas, J., Qi, X., 2004. Starch structure and digestibility enzyme–substrate relationship. World. Poult. Sci. J. 60, 186–195.

Tsai, M.L., Li, C.F., Lii, C.Y., 1997. Effects of granular structures on the pasting behaviours of starches. Cereal Chem. 74, 750–757.

Wankhede, D.B., Ramtehe, R.S., 1982. Studies on isolation and physicochemical properties of starch from moth bean (*Phaseolus acentifolius*). Starch 34, 189–192.

Wu, H.C.H., Sarko, 1978. Packing analysis of carbohydrates and polysaccharides. 9. Double-helical molecular-structure of crystalline A-amylose. Carbohydr. Res. 61, 27–40.

Zhang, G.Y., Ao, Z.H., Hamaker, B.R., 2006. Slow digestion property of native cereal starches. Biomacromolecules 7, 3252–3258.

Zhou, M., Robards, K., Glennie-Holmes, M., Helliwell, S., 1998. Structure and pasting properties of oat starch. Cereal Chem. 75, 273–281.

Zhou, Y., Hoover, R., Liu, Q., 2004. Relationship between α-amylase degradation and the structure and physicochemical properties of legume starches. Carbohydr. Polym. 57, 299–317.

Functional and physicochemical properties of legume fibers

5

Uma Tiwari, Enda Cummins
UCD School of Agriculture, Food Science and Veterinary Medicine,
University College Dublin, Dublin, Ireland

5.1 Introduction

Legumes include pulses and other well-known fruits that bear legume fruits, including, but not limited to soybean, lupins, groundnuts and clover (Berrios et al., 2008). Legume (including pulse) fibers are reputed to have several beneficial health effects including delaying the release of carbohydrates (and therefore providing steady and sustained energy release), lowering of blood lipids, prevention of colon cancer, increasing the fecal transit time and improving digestion. Legume fibers enhance nutritional, biological and physicochemical properties by decreasing the transition time through the small intestine and also affecting innate immune responses of the gut mucosa both directly and indirectly (Tharanathan and Mahadevamma, 2003; Rochfort and Panozzo, 2007; Tosh and Yada, 2010). The potential health benefits of

Pulse Foods: Processing, Quality and Nutraceutical Applications. DOI: 10.1016/B978-0-1238-2018-1.00007-0

legume fibers have resulted in great interest among food scientists and other researchers to incorporate these fibers into human food (Dahl et al., 2003; Dalgetty and Baik, 2006; Pittaway et al., 2007). Dietary fiber usually refers to a complex mixture consisting of non-digestible components of food (Lunn and Buttriss, 2007; Tosh and Yada, 2010). Dietary fibers consist of different types of fiber (TDF) which have a different chemical composition and physicochemical properties that results in differing physiological effects. The value of consuming dietary fiber is largely due to the various physiological effects that can have important health benefits (Cummings, 1973).

The term "dietary fiber" was first coined by Hipsley (1953) to describe the non-digestible components of plants that make up the plant cell wall and include cellulose, hemicellulose and lignin. In later years, several researchers started to realize the health benefits associated with the consumption of dietary fibers, for example reduction in artherosclerosis (Trowell, 1972a), heart diseases (Trowell, 1972b) and gastrointestinal diseases (Burkitt et al., 1972). In the year 1976, the definition of dietary fiber was expanded to include other non-digestible polysaccharides, such as gums and mucilages (Trowell et al., 1976; DeVries, 2004; Lunn and Buttriss, 2007). In the UK, the term "dietary fiber" has been replaced by the term non-starch polysaccharides (NSPs): polysaccharides that occur in the diet, but excluding starch (Englyst and Cummings, 1990). To date, there are many "official" definitions for fiber that are used by various authoritative bodies around the world. The American Association of Cereal Chemists (AACC, 2001) defined dietary fiber as:

> *the edible parts of plants or analogous carbohydrates that are resistant to digestion and absorption in the human small intestine with complete or partial fermentation in the large intestine. Dietary fiber includes polysaccharides, oligosaccharides, lignin, and associated plant substances. Dietary fibers promote beneficial physiological effects including laxation, and/or blood cholesterol attenuation, and/or blood glucose attenuation.*

Considering the nutritional importance of increased total fiber content and the associated improved product quality, fortification of food with legume fiber has gained much commercial interest (Dalgetty and Baik, 2006). This chapter looks at legume fibers and their functional and physicochemical properties while evaluating the effects of processing technologies on the quantity and quality of legume fibers.

5.2   Legume dietary fibers

The dietary fibers of legumes can be classified according to their solubility, i.e. soluble dietary (SDF) or insoluble dietary fibers (IDF). The two broad categories of dietary fibers (SDF and IDF) are often determined analytically, depending on their solubility in water and buffer solutions. Soluble DFs (soluble-fiber polysaccharides) are widely used as food additives (thickeners, stabilizers, emulsifiers and gelling agents) (Harris and Ferguson, 1999). Many soluble DFs are quickly and extensively degraded and fermented in the large intestine, whereas insoluble DFs are often degraded slowly and partially degraded DFs are fermented (Klurfeld, 1990). Studies show a consistent association between insoluble DFs and protection against cancer (Klurfeld, 1990; Harris and Ferguson, 1999) when compared to soluble DFs.

5.2.1   Variation in dietary fiber content

The dietary fiber content of legume and legume products may vary according to the plant variety, environmental conditions, agronomic practices, environmental interaction and various food-processing operations (domestic and industrial) (Vidal-Valverde and Frias, 1991; Khatoon and Prakash, 2004; Martín-Cabrejas et al., 2006; Wang et al., 2008). Stoughton-Ens et al. (2010) demonstrated a strong genotypic and environment effect on the dietary fiber content of field peas. The level of SDF and IDF also varies depending upon variety of legume. The IDF content of most legumes is higher compared to SDF content. For example, a recent study found the IDF content of green peas and chickpeas was 28 and 20%, respectively, whereas their SDF content was 1.73 and 1.51%, respectively (Mallillin et al., 2008). Legumes (including pulses) are well known for human consumption in which the total DFs content ranges from 8 to 27.5%, with SDFs in the range 3.3–13.8% (Guillon and Champ, 2002). Table 5.1 shows the dietary fiber composition of various legumes.

The composition of the dietary fiber fractions within the legume may depend on its localization in the hull (outer fiber) or the cotyledons (inner fiber) (Pfoertner and Fischer, 2001; Guillon and Champ, 2002). Guillon and Champ (2002) reported the obvious difference in the cellulosic and non-cellulosic polysaccharides of inner and

Table 5.1 Dietary fiber composition of legumes

Legume	Botanical name	IDF (%)	SDF (%)	TDF (%)	IDF:SDF	Country	Reference
Green pea	Pisum sativum	11.3	8.7	20	1.30	USA	Dalgetty and Baik (2003)
Green pea	Pisum sativum	20.3	1.73	22.03	11.73	Brazil	de Almeida Costa et al. (2006)
Green pea	Pisum sativum	28	2.1	29.7	13.14	Philippines	Mallillin et al. (2008)
Chickpea	Cicer arietinum	10	8.4	18.4	1.19	USA	Dalgetty and Baik (2003)
Chickpea	Cicer arietinum	19.54	1.51	21.05	12.94	Spain	Martin-Cabrejas et al. (2006)
Chickpea	Cicer arietinum	24.9	1.3	26.2	19.15	Philippines	Mallillin et al. (2008)
Cowpea	Vigna catjang	25.92	1.03	28.5	25.2	India	Khatoon and Prakash (2004)
Cowpea	Vigna unguiculata	30.32	0.9	31.22	33.7	Cuba	Martin-Cabrejas et al. (2008)
Jack bean	Canavalia ensiformis	31.74	1.51	33.25	21.0	Cuba	Martin-Cabrejas et al. (2008)
Dolichos	Lablab purpureus	39.9	2.12	42.02	18.8	Cuba	Martin-Cabrejas et al. (2008)
Mucuna	Stizolobium niveum	39.76	2.9	42.66	13.7	Cuba	Martin-Cabrejas et al. (2008)
Soybean	Glycine max	52.06	2.71	54.77	19.2	Cuba	Marin-Cabrejas et al. (2008)
Green gram	Phaselous aureus	15.11	1.29	17.5	11.7	India	Khatoon and Prakash (2004)
Bengal gram	Cicer arietinum	27.84	1.42	28.8	19.6	India	Khatoon and Prakash (2004)
Horse gram	Dolichos biflorus	27.82	1.13	28.8	24.6	India	Khatoon and Prakash (2004)
Dry field beans	Dolichos lablab	23.83	0.61	24.7	39.1	India	Khatoon and Prakash (2004)
French beans	Phaseolus vulgaris	20.35	2.37	21.5	8.6	India	Khatoon and Prakash (2004)
Broad beans	Vicia faba	27.68	1.37	28.5	20.2	India	Khatoon and Prakash (2004)
Lentils	Lens esculenta	15.3	0.71	16.7	21.5	India	Khatoon and Prakash (2004)

IDF, insoluble dietary fiber; SDF, soluble dietary fiber; TDF, total dietary fiber.

outer layers. The cell walls of the cotyledons are non-lignified and contain a range of polysaccharides, including pectic substances ($\approx$55%), cellulose ($\approx$9%) and non-starchy non-cellulosic glucans (in the range 6–12%) (Brillouet and Carré, 1983; Brillouet and Riochet, 1983; van Laar et al., 1999; Guillon and Champ, 2002), while the seed coat contains large quantities of cellulose (ranging from 35 to 57%) and smaller amounts of hemicelluloses and pectins (Brillouet and Riochet, 1983; van Laar et al., 1999).

Physicochemical properties, such as water-retention capacity and oil-binding capacity, also differ with variation in the composition of fibers. For instance, the water-retention capacity of cotyledon fiber fractions is higher due to its soluble fiber which exhibits greater binding of water than hull fibers, whereas the oil-binding capacity of cotyledon fiber is in the same range as that for hull fibers (Pfoertner and Fishcher, 2001; Gullion and Champ, 2002). These physiochemical properties are discussed in more detail in Section 5.4. Cotyledon and hull fibers also exhibit variation in color and flavor which may influence organoleptic properties when substituted in food and food products. Pfoertner and Fishcher (2001) demonstrated that pea cotyledon fibers are white in color with a neutral beany flavor and its hull fiber is creamy white with a neutral flavor. In contrast, lupin cotyledon fibers are white in colour with a nearly neutral flavor whereas its hull fiber is creamy white with a slight neutral nutty flavor (Pfoertner and Fishcher, 2001).

5.2.2 Insoluble fibers

Insoluble fibers constitute the tough, fibrous parts of the plants and have passive water-holding properties that can reduce the risk of human disease and disorders including diverticular disease, hemorrhoids and constipation (Eastwood et al., 1980; Cummings, 1981). The insoluble fiber speeds up the passage of foods through the stomach and intestines and is readily converted to butyrate in the colon, therefore reducing the risk of colon cancer (Johnson et al., 2006). Many researchers evaluated that IDF is predominant in legumes ranging from 10 to 15% for lentil, chickpea and dry peas (Berrios et al., 2010). The three main components of insoluble fiber are cellulose, hemicellulose and lignin (Lunn and Buttriss, 2007). Cellulose is a polymer of β-(1-4)-d-Glc with hydrogen bonding, while hemicellulose consists of long chains of cellulose with pyranosyl bonding. Digestibility depends upon the amount and ratio of cellulose to hemicellulose, which affects the physical structure of the fiber and its transit time in the gut (Cummings, 1973). Lignin, a polyfunctional polymer, is intimately formed with the cellulose of plant cell walls and is very resistant to digestion, even with strong acid. Lignin is intricately tied to the dietary fiber polysaccharides in foods and increases resistance to digestion (AACC, 2001).

Insoluble dietary fibers mainly comprise of glucose, uronic acids, and arabinose as the main carbohydrate constituents, followed by xylose, galactose/rhamnose, and mannose, which are found in minor amounts (Martín-Cabrejas et al., 2004). The level of IDF constitutes a major portion of the legume DFs compared to SDFs (see Table 5.1). Su and Chang (1995) reported a higher level of IDF fraction (72–90%) in raw dry beans compared to soluble fibers. Wang et al. (2009) observed the lower soluble fraction of total dietary fiber (TDF) (10.6–13.9%) compared to insoluble dietary fibers in lentils. Similarly, Wang et al. (2008) demonstrated that IDF in field peas was the major portion of the TDF and ranged from 88.6% to 90.2%, while SDF ranged from 9.5% to 11.1% of the TDF. The concentration of IDF in legumes is also cultivar dependent. For example, Berrios et al. (2010) demonstrated that chickpeas had a high concentration of IDF and SDF while Mouser et al. (1983) reported higher values for SDF (2.04 and 4.17 g 100 g^{-1}) and lower IDF values (4.2 and 7.0 g 100 g^{-1}) for chickpeas and lentils, respectively. Bednar et al. (2001) noted that different legume samples (black bean, red kidney bean, lentil, navy bean, black-eyed pea, split pea and northern bean) comprise 92.2–100% of IDF and 0–7.8% of SDF. In addition, Stoughton-Ens et al. (2010) indicated that green cotyledon field peas contain higher levels of IDF and TDF when compared to yellow cotyledon field pea.

5.2.3 Soluble fibers

Soluble fiber dissolves in water to form a gel (i.e. viscous solution) and maximizes the viscosity of the food matrix and thus slows down digestion (Guillon and Champ, 2002). Soluble fibers consist of pectins, some hemicelluloses, gums, mucilages and storage polysaccharides. Soluble fiber can help lower postprandial blood glucose, insulin levels and serum cholesterol (Jenkins et al., 2000). Anderson et al. (1983) showed bean supplementation lowered low-density lipoprotein cholesterol more than high-density lipoprotein cholesterol concentrations, thereby helping in the prevention of coronary heart disease. Khatoon and Prakash (2004) studied the dietary fiber composition of eight whole legumes, namely Bengal gram (*Cicer arietinum*), broad beans (*Vicia faba*), cowpea (*Vigna catjang*), field beans (*Dolichos lablab*), green gram (*Phaseolus aureus Roxb*), horse gram (*Dolichos biflorus*), lentils (*Lens esculenta*) and French beans (*Phaseolus vulgaris*). They observed that the soluble fiber fraction of legumes ranged from 0.61% to 2.37% of total dietary fiber, with the highest being in French beans and the lowest in lentils. Similarly, Berrios et al. (2010) observed that the concentration of soluble fiber was significantly lower ($P < 0.05$) in lentils, chickpeas and dry peas, ranging from 0.27 to 0.75%. Mallillin et al. (2008) showed that the insoluble fiber content of legumes was significantly greater than that of soluble fiber (see Table 5.1). The ratio of insoluble and soluble fiber may play an important role in influencing potential health benefits (Jenkins et al., 2000; Aguilera et al., 2009). Incorporation of soluble fiber from dry beans may significantly lower cholesterol and reduce blood sugar concentrations (Anderson et al., 1994).

Incorporation of insoluble or soluble fiber may also influence the physiological benefits provided by high-fiber foods. Studies have shown that the incorporation of fiber components in food and food products contributes to the texture, gelling, thickening, emulsifying and stabilizing properties of certain foods (Dreher, 1987).

5.3 Factors affecting levels of dietary fibers

Processing improves the sensorial and organoleptic properties (including the mouth-feel characteristics and smoothness) of food and food products which, in turn, modifies the composition and availability of nutrients. Legumes are ordinarily processed prior to human consumption. Processing of legumes on a production scale (including milling and grinding, soaking, steaming, roasting, boiling, cooking, freezing, canning and dehydration) or other home-scale processing (like soaking, cooking, roasting, sprouting, milling and fermentation) influences the level and physicochemical properties of legume fibers. The emergence of new fiber sources along with new processing methods for improving their functionality may widen the use of fibers in the food industry (Guillon and Champ,

2000). Thus, enriching food products with legume fibers (both soluble and insoluble) can alter the sensory properties and texture of the final product while inferring added health benefits. Apart from processing, storage of legumes in dry conditions is also reported to modify dietary fiber content, possibly due to the dehydration or oxidative reactions (Selvendran et al., 1987).

5.3.1 Thermal processing

Legumes regularly undergo thermal treatment during domestic and industrial cooking. Heat generally changes the ratio of soluble to insoluble fiber. Similarly, drying of legumes, such as peas, can strongly influence the physicochemical properties of the fiber by collapsing the cell walls and thus changing the porosity of the fiber and its hydration properties (Guillon and Champ, 2000). Mahadevamma and Tharanathan (2004) investigated the effect of various processing techniques on the dietary fiber composition of Bengal gram (chickpea), black gram, green gram and red gram (pigeon pea) dhals (dehulled splits). They observed that the TDF contents of the samples subjected to various processing treatments were more or less comparable, although the SDF of some processed legumes increased (Table 5.2). Similarly, Mangala and Tharanathan (1999) reported an increase in soluble as well as total DF content

Table 5.2 Insoluble, soluble and total DF content in processed Bengal gram and green gram dhal flour (calculated as % dry matter)

Processing	Bengal gram			Green gram		
	IDF	SDF	TDF	IDF	SDF	TDF
Raw whole grain	19.05	2.90	21.95	13.00	3.21	16.21
Dhal flour	10.57	2.58	13.05	8.99	3.00	11.98
Traditional cooked	11.48	2.65	14.63	8.80	3.71	12.50
Deep-fat fried	8.94	3.00	11.93	11.03	0.97	11.95
Autoclaved	10.96	1.91	12.87	10.97	1.99	12.96
Popped or roasted[a]	5.93	3.98	9.91	13.92	3.98	17.91
Germinated[a]	23.89	3.96	27.85	15.94	5.98	21.92
Pressure cooked	9.82	3.92	13.75	10.56	2.00	12.56
Roller dried	5.95	3.69	9.94	5.93	1.99	7.92
Extrusion cooked	11.25	2.50	13.70	9.05	3.00	12.04

[a]Samples processed with husk.
Adapted from Mahadevamma and Tharanathan (2004).

in roller-dried cereals. The overall DF of germinated samples was significantly higher than that of the raw legumes due to the presence of husk in the whole gram, which was ground together for sample preparation. Husks are generally rich in several non-starch polysaccharides and lignin, which contribute to the overall DF content.

Pérez-Hidalgo et al. (1997) studied the effect of soaking, cooking and frying on the fiber content of chickpeas. They observed a total dietary fiber increase of 49.5% (from 16.8 to 25.1%) after cooking and decrease of 21.4% (from 16.8 to 13.2%) after frying. In addition, they observed an increase in the insoluble fiber fraction after cooking by 108% with no significant change in the level of insoluble dietary fiber after frying. The chemical bases for changes in the dietary fiber content of foods during cooking remain unclear. It has been proposed that the formation of resistant starch and the Maillard reaction products, together with condensed tannin–protein products, could contribute to the increased dietary fiber as dry matter is lost during cooking (not all cooking water is absorbed by samples) (Mongeau and Brassard, 1995).

Heat treatment during cooking may modify the structure of both the cell wall and storage of legume polysaccharides possibly by affecting the intactness of tissue histology and disrupting the protein–carbohydrate integration, thus reducing the solubility of dietary fiber (Siljestrom et al., 1986). The increase in IDF could also be due to protein–fiber complexes formed after possible chemical modification induced by the cooking of dry seeds (Bressani, 1993).

5.3.2 Milling and grinding

Legume milling efficiently separates or dehusks the outer layer from the cotyledons to produce a finely ground flour. During the milling process, the removal of the hull increased the soluble fiber fraction in the dehulled grain fraction with a decrease in total and insoluble fiber (Ramulu and Rao, 1997). This may be due to the fact that the hull contains the majority of legume fibers with a higher level of IDF. Dalgetty and Baik (2003) demonstrated the dehulling and milling effects on the fiber content of peas, lentils and chickpeas. They observed a decrease in TDF and IDF with an increase in SDF content for peas, lentils and chickpea flour which is mainly due to the hull removal during milling. Dehulling reduces both insoluble and total dietary fibers. The decrease observed in TDF content due to dehulling resulted mainly from a decrease in IDF content. Wang et al. (2009) studied the influence of cooking and dehulling on the nutritional composition of lentil varieties. They concluded that cooking significantly affects soluble, insoluble and total dietary fiber of cooked lentils compared to the dehulling process. Wang et al. (2009) noted the percentage of SDF to TDF in dehulled lentils was significantly higher (11.4 to 16.5%) compared to raw lentil (10.8 to 13.9%). In contrast, the ratio of IDF to TDF in the dehulled lentils was significantly lower (83.5 to 86.9%) than in the raw lentil (86.9 to 89.4%).

5.3.3 Soaking and fermentation

Soaking often softens the kernel due to water absorption, which in turn reduces the cooking time. Soaking of legumes is effective for the reduction of oligosaccharides. Rehman and Shah (2004) reported that soaking in tap water has no impact on the fiber content of black gram, chickpeas, lentils, red and white kidney beans, whereas soaking in $NaCO_3$ solution caused a sharp increase in hemicellulose (44.4–58.9%) and decrease in lignin contents (15.1–7.40%). A few authors report that soaking may degrade the insoluble fiber fraction because these fibers are closely linked to the food matrix (Gooneratne et al., 1994; Aguilera et al., 2009). In another study, Sefa-Dedeh and Stanley (1979) reported that an increased soaking time affected the microstructure of cowpeas and influenced the water absorption and dehulling properties as a result of a change in the water-holding capacity of cell walls of the seed.

Martín-Cabrejas et al. (2004) studied the effect of fermentation on TDF, IDF and SDF fractions of beans (*Phaseolus vulgaris* L.) and observed that both natural and lactic acid fermentation and autoclaving decreased TDF contents. They observed an increase in IDF from 69.8 to 80% whereas SDF decreased significantly from 30 to 19% of TDF. This could be due to a lower accessibility of microorganisms to the insoluble fiber components during fermentation. Martín-Cabrejas et al. (2004) also observed that lactic acid fermentation had a profound effect on dietary fibers compared to natural fermentation. Hence, fermentation of legumes alters the ratio of insoluble to soluble fibers. As discussed earlier, the ratio of insoluble to soluble fiber is an important variant related to structural, functional and also sensorial properties of food products containing legume fibers.

5.3.4 Boiling and roasting

Boiling and roasting are common processes used domestically to remove antinutritional factors and to prepare grain legumes as foods. Boiling and roasting may alter the dietary fiber content, depending on the level of heat treatment applied. Boiling will also leach out the indigestible complex sugars. Veena et al. (1995) demonstrated that processing treatments significantly decreased the SDF content and increased the IDF content of legumes. In addition, Azizah and Zainon (1997) studied the roasting effect on legumes and they observed a significant reduction in insoluble dietary fiber after heat treatment. Mahadevamma and Tharanathan (2004) observed a significant reduction in both insoluble and total dietary fiber with an increase in soluble fiber content of roasted Bengal gram and green gram (see Table 5.2).

5.3.5 Cooking

Household cooking may alter the composition of legume fibers and thus alter their physiological effects in the human body (Azizah and Zainon, 1997). Legume cooking and chemical modification cooking have a major influence on the fiber content of water-presoaked legumes. Domestic cooking and industrial cooking processes involve heat treatment which is the precursor for solubilization of fiber. During cooking of raw lentil varieties, Wang et al. (2009) observed the IDF content significantly increased from 10% to 26%, whereas SDF content decreased in cooked lentils compared to the raw lentils. Vidal-Valverde and Frias (1991) concluded that the cooking processes may vary and depend on the legume and can have the effect of decreasing the amount of dietary fiber. Legumes require relatively extended cooking times to soften the tissue, to be palatable and to degrade antinutritional factors, such as protease inhibitors, lectins, phytate and tannin. During cooking, several physicochemical changes occur in the dietary fiber constituent of the cell wall which result in the degradation and solubilization of pectic polysaccharides within the fiber matrix (Gooneratne et al., 1994). Such changes in pectic polysaccharide solubility may affect the nutritional response to the fiber like a hypocholesterolemic agent (Shutler et al., 1987) or improve glycemic response (Wolever, 1990) during gut transit (Gooneratne et al., 1994). Kutoš et al. (2003) observed complex changes in dietary fiber content due to thermal processing of beans, which depended on the bean cultivar, the processing method and duration. Figure 5.1 details the physiological effects of dietary fiber in the human gut.

Several authors reported the effects on dietary fiber content as a result of different cooking methods for a range of legumes (Vidal-Valverde and Frias, 1991; Gooneratne et al., 1994; Periago et al., 1996). During cooking, a reduction in content and softening of the soluble fibers occurs (Vidal-Valverde and Frias, 1991). However, Kutoš et al. (2003) found that SDF for beans increased during cooking. Similarly, Mahadevamma and Tharanathan (2004) observed a significant increase in soluble fiber content of pressure-cooked Bengal gram and green gram (see Table 5.2). Wang et al. (2009) observed a significant decrease in IDF during cooking of eight lentil varieties (133.9–161.1 g kg^{-1} dry matter) compared to the uncooked raw samples (114.1–128.3 g kg^{-1} dry matter). Both pressure cooking and microwave cooking may alter the physicochemical properties of dietary fiber, especially the soluble fiber which is

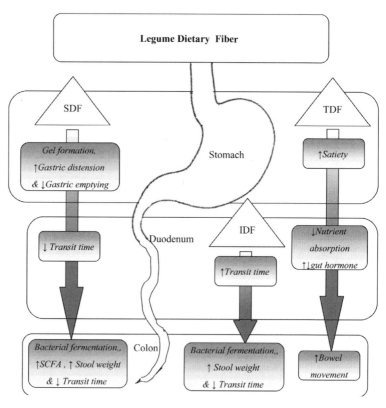

Figure 5.1 Physiological effects of dietary fiber in the human gut. ↑ Indicates an increase, ↓ indicates a decrease.

more susceptible to degradation (Plaami, 1997). In a study, Ramulu and Rao (1997) demonstrated the effect of pressure cooking on a selection of four legumes; this showed a significant increase in their dietary fiber contents. However, Khatoon and Prakash (2004) compared the effect on dietary fiber of microwave cooking and pressure cooking and observed that during the latter the reduction of dietary fiber was higher.

5.3.6 Drying and dehydration

Industrial processing, such as drying and dehydration, of food legumes can be used to alter the dietary and physicochemical properties of the fiber. Thermal dehydration is a process which includes soaking and cooking followed by drying. The process is reported to influence both SDF and IDF fractions of cooked chickpea and beans (Aguilera et al., 2009). Aguilera et al. (2009) reported a significant increase in the TDF from 12% to 27% with a significant increase in SDF content of chickpeas by 104% following the thermal dehydration process. This increase could be due to the branched structure of pectins which are most susceptible to being lost during the dehydration process (Aguilera et al., 2009). Martín-Cabrejas et al. (2008) observed different behavior of SDF during thermal dehydration. They observed a 32% increase for chickpeas and a 27% decrease in the lentil soluble fiber fraction. Table 5.3 shows the changes in dietary fiber fraction during industrial dehydration of legumes. Changes in the composition of dietary fiber might be used to alter the dietary and functional characteristics of the fiber, especially in chickpeas, whereas the ratio IDF/SDF was reported to reduce significantly.

5.3.7 Canning

Li et al. (2002) studied the dietary fiber (SDF and IDF) content in varying proportions in canned and drained chickpeas, cowpeas, pinto beans and red kidney beans. The latter showed the greatest amount of IDF, ranging from 4.11 to 5.79%, whereas SDF ranged from 0.09 to 1.36%. Similary, Kutoš et al. (2003) studied the effect of canning on soluble, insoluble and total dietary fiber contents on common beans (*Phaseolus vulgaris* L). They concluded that the canning process reduced the SDF from $3.5\,g\,100\,g^{-1}$ in raw beans to $2.3\,g\,100\,g^{-1}$ in canned beans, probably due to the processing and treatments during the canning process.

Table 5.3 Changes in dietary fiber fraction (g kg^{-1} dry matter) during the industrial dehydration of legumes

Legume	IDF	SDF	TDF	IDF:SDF	
Chickpea					Martín-Cabrejas et al. (2006)
Raw	195.4	15.1	210.5	13	
Soaked	215.1	15.4	230.5	14	
Soaked + cooked	233.1	14.6	247.7	16	
Soaked + cooked + dehydrated	265.8	19.9	285.7	13	
Lentil					Martín-Cabrejas et al. (2006)
Raw	216.3	26.8	243.1	8	
Soaked	217.3	17.6	234.9	12	
Soaked + cooked	282.5	21.8	304.3	13	
Soaked + cooked + dehydrated	274.0	19.6	293.6	14	
Chickpea					Aguilera et al. (2009)
Raw	204.8	9.6	214.4	21	
Soaked	197.6	12.8	210.4	15	
Soaked + cooked	235.1	15.0	250.0	16	
Soaked + cooked + dehydrated	252.9	19.6	272.5	13	
White bean					Aguilera et al. (2009)
Raw	211.4	58.2	269.6	4	
Soaked	212.0	65.8	277.8	3	
Soaked + cooked	218.9	77.6	296.5	3	
Soaked + cooked + dehydrated	230.8	79.1	309.9	3	
Pink-mottled cream bean					Aguilera et al. (2009)
Raw	163.7	54.3	218.1	3	
Soaked	147.6	64.4	212.0	2	
Soaked + cooked	171.0	73.7	244.9	2	
Soaked + cooked + dehydrated	189.5	75.2	264.7	3	

5.3.8 Extrusion cooking

Extrusion technology has been used in the development of expanded, novel, value-added legume-based ready to eat snacks. Extrusion improves the nutritional content of formulated legume flour which, in turn, improves the dietary fiber (Berrios et al., 2002, 2008, 2010; Patil et al., 2007). Extruded products are good sources of dietary fibers, and extrusion cooking is a suitable process for the production of fiber-enriched products. Extrusion cooking has been considered as an alternative way of modifying the functionality of dietary fiber (Camire et al., 1990). Berrios et al. (2002) observed a decrease in IDF and increase in SDF during extrusion cooking of bean flour. This shows that the extrusion processing causes a redistribution of the IDF to SDF fractions. Similar observations were reported by other researchers (Lintas et al., 1995; Gualberto et al., 1997). This fiber fraction redistribution possibly resulted from hemicellulose depolymerization (Berrios et al., 2002). Berrios et al. (2010) studied the changes in insoluble and soluble fiber fractions during the extrusion of chickpea, lentil and dry pea flour. They observed a significant decrease in the insoluble fiber for all three legume flours with the exception of chickpea flour. Similarly, Mahadevamma and Tharanathan (2004) also showed that the TDF content decreased for whole grain by 37.4% for chickpea and 26% for green gram following extrusion cooking. Therefore, the influence of extrusion cooking on dietary fiber content depends on the type of legume and other processing methodologies involved (Berrios et al., 2010).

Martín-Cabrejas et al. (1999) studied the effect of extrusion cooking on dietary fiber content and properties of hard-to-cook beans (*Phaseolus vulgaris* L). Extrusion cooking of fresh beans caused a decrease in IDF from 15.5 to 10.8% with a significant increase in SDF from 9.2% for extruded beans compared to 3.7% in raw common beans. Extrusion processing parameters, such as temperature and moisture content, greatly influence the ratio of soluble and insoluble fiber fractions. Martín-Cabrejas et al. (1999) observed the greatest increase of SDF with 30% moisture content. In general, during extrusion there is no significant change in total dietary fiber; however, there are significant changes in the soluble to insoluble ratio, probably as a result of a solubilization and redistribution of insoluble fibers to soluble ones. The redistribution effect of dietary fiber is also reported for the extrusion of whole grain wheat flour (Siljestrom et al., 1986).

5.4 Physicochemical properties of legume fibers

Legume fibers possess unique physicochemical or functional properties responsible for eliciting physiological responses. These properties include water- and oil-holding capacities (Lopéz et al., 1997; Leterme et al., 1998) and cation exchange capacity (Torre et al., 1992). The physicochemical properties of legume fiber are important for functional behaviors in food use and for diet-related health effects. For example, Chau and Cheung (1999) observed a linear correlation between the high-density lipoprotein (HDL):total cholesterol ratio and the bulk density, water-holding capacity, and cation exchange capacity of the legume insoluble fibers prepared from *Phaseolus angularis*, *Phaseolus calcaratus* and *Dolichos lablab* seeds. Functional properties such as swelling, water-holding capacity, fat- or oil-retention capacity, bulk density, rheological properties, porosity, particle size and ion exchange capacity affect the nutritional and sensory characteristics of the food, and play an important role in the physical behavior of food or its ingredients during preparation, processing and storage. In general, legume fiber is added to foods not only to increase the fiber content for health benefits but also to improve functional and organoleptic properties such as water and oil retention, viscosity, texture and mouth feel. Legume fibers have unique functional properties which can be used in several food products. For example, legume fibers can be used to enhance or modify the texture of food products through increased fat or water retention (Tosh and Yada, 2010). Tables 5.4 and 5.5 show some of the physicochemical properties of legume

Table 5.4 Physicochemical properties of legume fiber

Legume fiber	Bulk density ($g\ ml^{-1}$)	Swelling capacity ($ml\ g^{-1}$)	Water-holding capacity ($ml\ g^{-1}$)	Oil-binding capacity ($ml\ g^{-1}$)	Reference
Pea hull	0.75	1.88	5.5	1.51	Dalgetty and Baik (2003)
Lentil hull	0.81	2.38	3.6	1.63	Dalgetty and Baik (2003)
Chickpea hull	0.73	3.61	6.2	1.76	Dalgetty and Baik (2003)
Pea IDF	0.21	5.56	13.4	6.93	Dalgetty and Baik (2003)
Lentil IDF	0.36	8.04	11.1	4.01	Dalgetty and Baik (2003)
Chickpea IDF	0.34	4.28	10.1	4.25	Dalgetty and Baik (2003)
Pea SDF	0.80	–	–	1.15	Dalgetty and Baik (2003)
Lentil SDF	0.80	–	–	0.89	Dalgetty and Baik (2003)
Chickpea SDF	0.84	–	–	1.14	Dalgetty and Baik (2003)
Mung bean hulls (>35)[a]	0.45	9.2	4.44	1.83	Huang et al. (2009)
Mung bean hulls (35–50)[a]	0.52	6.1	3.52	1.5	Huang et al. (2009)
Mung bean hulls (<50)[a]	0.64	5.51	3.13	1.49	Huang et al. (2009)

[a]Particle size (mesh).

Table 5.5 Physicochemical properties[a] of three legume insoluble fibers and cellulose

Fiber source	Bulk density (g l^{-1})	Water-holding capacity (g kg^{-1})	Cation exchange capacity (meq kg^{-1})
Cellulose	710 ± 12.7^w	5590 ± 144^w	40.0 ± 0.83^w
Phaseolus angularis	380 ± 7.07^x	7830 ± 63.3^x	380 ± 7.07^x
Phaseolus calcaratus	360 ± 11.3^x	7570 ± 42.4^x	320 ± 6.94^y
Dolichos lablab	420 ± 5.66^y	7690 ± 56.6^x	450 ± 10.3^z

[a]Means of triplicate (± standard derivation). Values in the same column with different superscripts (w, x, y or z) are significantly different (Tukey, $P < 0.05$). Adapted from Chau and Cheung (1999) (with permission).

fibers. Growing consumer awareness and the potential health benefits of legume fibers have resulted in the incorporation of legume fibers into bakery products including breakfast cereals, pasta and noodles, beverages, meat products and dairy products (Sandström et al., 1994; Gelroth and Ranhotra, 2001; Dalgetty and Baik, 2006). The health benefits of legume fibers in addition to their functional properties (water-holding capacity, synergism with sweeteners and fat replacement properties) have created renewed interest in their utilization, particularly in the food and nutraceutical industry (Pszczola, 1998; McKee and Latner, 2000). In addition to functional properties, the physicochemical properties of legume fibers can be modified by food-processing operations, such as grinding or extrusion, to improve functionality. For example, extrusion cooking has been considered as an alternative way of modifying the functionality of dietary fiber (Qian and Ding, 1996; Camire et al., 1997). Martín-Cabrejas et al. (2003) observed an increase in soluble dietary fiber in germinated peas thus influencing insoluble/soluble fiber ratios, which are important from both dietary and functional perspectives.

5.4.1 Bulk density

Bulk density is defined as weight of fiber per unit volume, often expressed as g ml^{-1}, and is a good index of structural changes (Sreerama et al., 2009). The common method of estimating bulk density of fibers is by placing a known quantity (2 g) of fiber into a graduated syringe and applying sufficient pressure to pack the content in the syringe while recording the final volume (Parott and

Thrall, 1978). Chau and Cheung (1999) investigated the effects of the insoluble dietary fibers prepared from *Phaseolus angularis*, *Phaseolus calcaratus* and *Dolichos lablab* seeds relative to cellulose on control of cholesterol absorption in hamsters. They observed that the bulk density of *Dolichos lablab* insoluble fiber was significantly ($P < 0.05$) higher, whereas the bulk density of all three insoluble fibers was lower than that of the cellulose (see Table 5.5). Further, a linear correlation was identified between the HDL:total cholesterol ratio and the bulk density ($r = -0.99$, $P < 0.05$) from the three legume insoluble fibers (Chau and Cheng, 1999). Dalgetty and Baik (2003) concluded that hull and soluble fiber contain higher densities of soluble fiber than insoluble fibers. The bulk density of pea hull, insoluble and soluble fibers was shown to be 0.75, 0.21 and 0.80 g ml^{-1}, respectively. They also noted that the lentil hull and insoluble fibers had the highest bulk density with 0.81 and 0.21 g ml^{-1}, respectively, while chickpeas soluble fiber was the densest of the soluble fibers with 0.83 g ml^{-1}. Huang et al. (2009) demonstrated that smaller fiber particles are shown to have a higher bulk density and may lower the ability of the fiber to absorb water and oil. They observed a higher bulk density of 0.64 g ml^{-1} for mung bean hulls with particle size of <50 mesh, whereas a lower bulk density of 0.45 g ml^{-1} was observed with a particle size >0.35 g ml^{-1}.

5.4.2 Swelling capacity

The swelling capacity of a fiber can be defined as the ratio of the volume occupied when the sample is immersed in excess water (after equilibration) to the actual weight (Equation 5.1). It measures the ratio of volume occupied when the sample is immersed in excess water after equilibration to the actual weight (Raghavendra et al., 2006). A sample size of 100–200 mg is generally hydrated with water (10 ml) for 18 h with no external stress except gravity (Robertson et al., 2000; Tosh and Yada, 2010). Robertson et al. (2000) reported an improved recording of the swelling capacity with a larger sample size (i.e. 200 mg) especially for the fibers with low swelling ability, such as pea hulls. Dalgetty and Baik (2003) hydrated a sample size of 500 mg with 50 ml of water and equilibrated for 12 h at 24 °C to determine the swelling capacity of pea, lentil and chickpea fibers. A higher swelling capacity of a fiber increases the viscosity and decreases the gastrointestinal transit time. Swelling is a diffusion phenomenon driven by the affinity of the molecules of the swelling material for the molecules of the containing fluid. The swelling power of insoluble fiber is higher compared to soluble fibers due to the chemical composition. Insoluble fibers also contain higher amounts of hemicellulosic sugars (arabinose and xylose) (Dalgetty and Baik, 2003).

$$\text{Swelling capacity (ml g}^{-1}) = \frac{\text{Volume occupied by sample}}{\text{Original sample weight}} \qquad (5.1)$$

Dalgetty and Baik (2003) observed that the source of fiber also influenced the swelling capacity. For example, the swelling power of insoluble fiber was 4.3–5.6% higher compared to hull for lentil, chickpea and pea. In their study, the swelling capacity of chickpea hulls (3.61 ml g^{-1}) was higher than that for pea (1.88 ml g^{-1}) or lentil (2.38 ml g^{-1}) hulls, whereas for the insoluble cotyledon fractions of these legumes, lentils exhibited a higher swelling capacity (8.04 ml g^{-1}) than peas (5.56 ml g^{-1}) or chickpeas (4.28 ml g^{-1}) (Dalgetty and Baik, 2003). Furthermore, the swelling capacity of fibers is strongly influenced by the particle size. Huang et al. (2009) noted a decrease in swelling capacity of mung bean hulls with a decrease in particle size. The decrease in swelling capacity was estimated to go from 9.2 ml g^{-1} to 5.51 ml g^{-1} with a decrease in particle size from >35 mesh to <50 mesh for mung bean hull. The above is in agreement with the study of Auffret et al. (1994), who concluded that grinding also reduces the swelling capacity of fibers, probably by alteration and collapsing of the fiber matrix.

5.4.3 Water-holding capacity/water-binding capacity

The water-holding and binding capacities have a particularly important relevance to the physiological actions and technological properties of dietary fiber. Wolever (1990) stated that the ability of fibers to swell and to bind water is highly indicative of their physiological role in gut function and control of postprandial glucose levels. Further, the fecal-bulking capacity of dietary fiber is related to its water-holding capacity (Davidson and McDonald, 1998). Insoluble fiber consisting of lignin, cellulose and hemicellulose usually has high water-holding capacity and contributes to increased fecal bulk and frequency of defecation (Eastwood et al., 1980;

Cummings, 1981; Campos-Vega et al., 2009). The water-holding and binding capacities are defined as the weight of water retained by 1 g dry material IDF or water held by the legume fiber by soaking the legume sample (Chau and Cheung, 1999). Rey and Labuza (1981) defined water-binding capacity (WBC) as the water contents measured after an external force (e.g. centrifugation) has been applied, whereas water-holding capacity (WHC) is measured without stress. The terms water-binding and water-holding capacity may also be used interchangeably, as both reflect the ability of a fiber to swell (Plaami, 1997). Water can be bound to several fibers in several ways (Fig. 5.2). Chaplin (2003) reported that the fiber may interact with water by means of polar and hydrophobic interactions coupled with hydrogen bonding, which vary with the flexibility of the fiber surface.

The WHC of legume fibers is also affected by the ratio of lignin to polysaccharides (Górecka et al., 2000). Górecka et al. (2000)

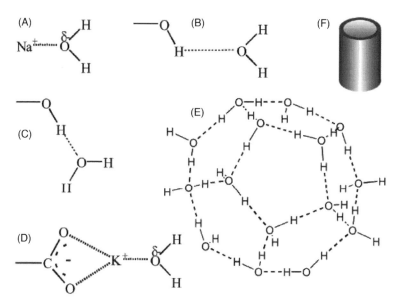

Figure 5.2 Fiber can directly bind water in a number of ways: (A) polar effects involving both anions and cations orient and attract water; (B) weak hydrogen bonding involves bent or extended hydrogen bonds; (C) strong hydrogen bonding involves straight, directed and short hydrogen bonds; (D) ionic interactions involve strongly held and oriented water, shown through interactions between a carboxylate group and potassium ion; (E) hydrophobic (surface) effects involve water clathrate formation at surfaces where water molecules form a flattish network of pentagons and hexagons; (F) enclosure of water involving capillary action (not to scale). (From Chaplin, 2003, with permission.)

observed that the WHC of lupins' (*L. luteus* and *L. albus*) hull is dependent both on variety and on pH. They observed that both lupin hull varieties demonstrated the highest WHC at pH 8.7, while the lowest value was observed in acid conditions (pH = 1.8). Similarly, Dalgetty and Baik (2003) observed higher WHC for pea insoluble dietary fibers of 13.4 ml g^{-1} compared to lentil and chickpea insoluble dietary fiber with 11.1 and 10.1 ml g^{-1}, respectively. Varietal variation might be due to different water-holding properties of legume fibers. For instance, the lignin fraction in legume fibers has hydrophobic properties and binds a significantly lower amount of water in comparison to hydrophilic polysaccharides (Bell and Shires, 1982), whereas Labuza (1986) indicated pectins as a major fiber fraction responsible for water binding.

Ghavidel and Prakash (2006) observed a higher water-absorption capacity of green gram (*Phaseolus aureus*), cowpea (*Vigna catjang*), lentil (*Lens culinaris*) and Bengal gram (*Cicer arietinum*) compared to dehulled samples and concluded that these variations may be due to the ability of fiber to bind and hold water (Bollinger, 1999). Betancur-Ancona et al. (2004) studied the WHC of jack bean (*Canavalia ensiformis*) and lima bean (*Phaseolus lunatus*) fibrous residues, noted for their low WHC of 39.5 and 26.5%, respectively, for jack and lima beans. They concluded that the low WHC was due to the low content of soluble fibers. However, the jack bean had a higher WHC compared to lima bean due to a higher level of soluble fiber in the jack bean (3.38%) compared to the lima bean (0.77%). Huang et al. (2009) reported a reduction in the WHC from 4.44 ml g^{-1} to 3.13 ml g^{-1} for mung bean hull with decreasing particle size from >35 mesh to <50 mesh. However, Auffret et al. (1994) reported that, after grinding, pea hulls had increased WBC and WHC due to an increase in surface area and pore volume.

5.4.4 Oil-binding capacity

The oil-binding capacity is another functional property which helps in the stabilization of high-fat products and emulsions. The fat- or oil-retention property is the hydrophobic bonding ability of the fiber molecules. The retention of oil may depend on different processing methods (Gelroth and Ranhotra, 2001). Oil-holding capacity can be determined by stirring a known quantity (e.g. 1.0 g) of sample fiber into 10 ml of vegetable oil (e.g. corn or sunflower) for a certain time (e.g. 1 min) and then centrifuging at 2200 g for 30 min (Chau et al., 1997). Oil-holding capacity can be expressed as grams of oil held per gram of fiber. Some researchers follow a similar protocol for measuring the fat/oil retention of fibers that are used for water-retention capacity (including sample centrifugation) with oil instead of water (Chau et al., 1997; Abdul-Hamid and Luan, 2000). Therefore, the terms oil-holding capacity and oil-binding capacity are used interchangeably, although the methods vary as in the case of water-holding and water-binding capacity (Tosh and Yada, 2010).

The oil-holding capacity of jack beans was 23% and 18% for lima bean fibrous residues as reported by Betancur-Ancona et al. (2004); this variation may be as a result of low oil absorption due to the presence of soluble fiber. Similarly, López et al. (1997) reported that more refined particles may decrease the oil-holding capacity due to the surface properties and hydrophobic nature of the particles causing physiological intestinal traffic as the fecal mass descends. Huang et al. (2009) did not observe a significant effect of particle size on oil-holding capacity of mung bean hull legume fibers. Pfoertner and Fischer (2001) observed the oil retention of cotyledons and the hull of peas, soya and lupin were in the same range. Some studies also reported that the presence of lignin and hemicelluloses in fiber fractions enhances the oil-holding capacity. Ang (1991) showed that longer length fibers retain more oil than shorter length cellulose fibers. Legume fibers (*C. ensiformis* and *P. lunatus*) may have low oil-binding capacity, probably due to their deficient lignin content, which traps the oil on the fiber (Betancur-Ancona et al., 2004). The oil-holding capacity differed with varying hydrophobic bonding abilities of the fiber molecule. For example, the mung bean hull fiber oil-holding capacity is lower than that of cellulose (Huang et al., 2009). Anderson and Berry (2001) observed that the inner pea fiber has the potential to retain maximum amounts of fat during high temperature heating, possibly due to its gelling and swelling properties.

5.4.5 Cation exchange capacity

The cation exchange capacity of legume fibers can be determined by converting the cationic functional groups present in the legume into their acidic form by stirring overnight at 4 °C in hydrochloric acid followed by extensive washing. The acidic and washed sample is titrated with potassium hydroxide (Ralet et al., 1993). The cation exchange capacity is expressed as the number of milliequivalents per kilogram of dry legume fiber sample. The cation exchange capacity of fibers depends on the presence of several functional groups which are responsible for exchange ability. This property depends primarily on the presence of phenol groups (particularly of the lignin fraction), and carboxyl groups from weak uronic acids of the pectin and hemicellulose fraction, glucuronoxylans (McConell et al., 1974; Gordon, 1989; Górecka et al., 2000). These functional groups have the capacity to bind heavy metal ions and therefore act as a cation exchanger. Chau and Cheung (1999) observed a strong positive correlation between the uronic acid contents and the cation exchange capacities ($r = 0.95$) of the fibers (see Table 5.5). Some essential micronutrients, such as zinc, may be lost as a result of this binding, and this property is sometimes considered a disadvantage of fiber-rich foods (James et al., 1978; Robertson et al., 1979; Tharanathan, 2002). The cation exchange capacity of fibers can lead to the increased fecal excretion of minerals and electrolytes (Toma and Curtis, 1986; McDougall et al., 1996). However, the higher cation exchange capacity of fibers may help in binding of heavy metals. Hence, dietary-rich fiber, mineral-depleted diets may cause nutritional problems. Increased mineral content of dietary fiber-rich foods such as fruits and vegetables may offset the above risk (McDougall et al., 1996).

Table 5.5 shows that the cation exchange capacities of all three legume insoluble fibers (*Phaseolus angularis*, *Phaseolus calcaratus*, *Dolichos lablab*) were significantly different from each other and are significantly higher than that of the cellulose (Chau and Cheung, 1999). Fibers with a high cation exchange capacity might destabilize, entrap and disintegrate the micelles (emulsion of lipid) by forming fiber–micelle complexes which act as barriers to the diffusion or absorption of micelles (Furda, 1990). Thus, the lipids, including cholesterol, could not be effectively absorbed and utilized in the human body, leading to the cholesterol-lowering effect of fibers. Górecka et al. (2000) observed a higher cation exchange capacity for dietary fiber of lupins (*L. luteus* and *L. albus*), which was within the range of 0.150 to 0.750 mEq g^{-1} dry matter. They also observed that the cation exchange capacity of dietary fiber is higher in alkaline conditions (pH = 8.7), with a lower value in acidic conditions (pH = 1.8).

5.5 Physiological activity of legume fibers

Physiological activity and the technological application of fiber depend on its composition and functional properties. The physiological functions of most dietary fibers, including legume fibers, are often attributed to their physicochemical properties, water-holding capacity, swelling, rheological, fat-binding properties and susceptibility to bacterial degradation or fermentation (Dikeman and Fahey, 2006). In fact, the health effects of soluble fiber (lowering cholesterol and the rate of glucose absorption and postprandial plasma glucose concentrations) have been associated with its viscosity (Dikeman and Fahey, 2006). Physicochemical properties of dietary fiber are essential for functional characteristics of foods and result in a number of physiological effects (de Almeida Costa et al., 2006). It has been reported that diets naturally high in dietary fiber can bring about five main physiological consequences: (1) improvements in gastrointestinal health; (2) improvements in glucose tolerance and the insulin response; (3) reduction of hyperlipidemia, hypertension and other CHD risk factors; (4) reduction in the risk of developing some cancers; and (5) increased satiety and hence some degree of weight management (Turnbull et al., 2005). Table 5.6 shows some health and physiological benefits of legume fibers from various sources. The gastrointestinal functions are associated with water-holding capacity, viscosity, bulk, fermentability and the ability to bind bile acids (Schneeman, 2001; Tosh and Yada, 2010). The mechanisms by which fibers increase viscosity of the digesta or physically trap components depend largely on the functional properties of the fiber rather than its chemical composition, whereas susceptibility to fermentation in the colon is dependent on chemical structure (Morris, 2001; Tosh and Yada, 2010).

The water-binding capacity (WBC) and viscosity are two important physicochemical properties of a dietary fiber that are linked to its physiological effects in the human upper gastrointestinal tract. Turnbull et al. (2005) employed *in vitro* measurement of the physicochemical properties of dietary fibers of Australian sweet lupin (*Lupinus angustifolius*), soy kernel fiber, pea hull, cellulose and wheat fiber under conditions simulating the human gastrointestinal tract (Table 5.7). Turnbull et al. (2005) observed the unusually high WBC of lupin kernel fiber compared to other fibers investigated.

Table 5.6 Physiological and health benefits of legume fibers

Fiber	Amount	Period	Test	Product	Health/physiological effects	Reference
Pea and soybean fiber	10 g of TDF	1 day	Human, 6 males	Meal	Decrease in cholesterol	Dubois et al. (1993)
Pea fiber	20 g	2 weeks	5 men, 6 women	Nutrio P-Fibre 150, DDS Nutrio	Total cholesterol lowered by (4.36 to 4.19) 0.17 mmol l^{-1} & VLDL triglyceride concentration lowered (0.41 to 0.31) 0.10 mmol l^{-1}	Sandström et al. (1994)
Green lentils NSP	11.8 g	3 weeks	19 men	Loaves, cakes, soups	Fecal weight ↑ from 131 to 189 g d^{-1}; serum lipids are unchanged	Stephen et al. (1995)
Pea fiber	4 g/d	6 weeks	114 M/F	Food fortified with pea fiber	Bowel movement ↑ 18.7 to 20.1; improved bowel function and satiety	Dahl et al. (2003)
Lupin kernel fiber	17–30 g d^{-1} fiber	1 month	38 men		↑defecation by 0.13 events d^{-1} & ↑ fecal output by 21%, while ↓ transit time by 17% and fecal pH by 0.26 units. Increased healthy bowel function and reduced colon cancer risk	Johnson et al. (2006)
Lupin kernel fiber	≈20 g d^{-1} fiber	28 days	18 males	Canned chickpeas;	Beneficial to colon health	Smith et al. (2006)
Chickpea fiber	27 g	5 weeks	27 adults	30% chickpea flour in bread or shortbread biscuit	Reductions in serum total cholesterol of 0.25 mmol l^{-1} and 0.2 mmol l^{-1} for LDL-C	Pittaway et al. (2007)

Table 5.7 Effect of simulated gastrointestinal conditions on the water-binding capacity of the dietary fiber ingredients

	Water-binding capacity (g water g^{-1} dry solids)[a]			
		Simulated gastrointestinal conditions		
	Pre-ingestion (control)	Mouth	Stomach	Duodenum
Lupin kernel fiber	10.97 ± 0.54	8.47 ± 0.30[c]	10.50 ± 0.31	11.07 ± 0.06
Soy kernel fiber	7.21 ± 0.17[b]	6.68 ± 0.05[b,c]	6.24 ± 0.23[b,c]	8.88 ± 0.21[b,c]
Pea hull	5.18 ± 0.07[b]	4.21 ± 0.10[b,c]	4.35 ± 0.12[b,c]	5.29 ± 0.18[b]
Cellulose	4.79 ± 0.09[b]	4.42 ± 0.06[b,c]	4.69 ± 0.06[b]	4.83 ± 0.11[b]
Wheat fiber	4.42 ± 0.03[b]	4.16 ± 0.05[b]	4.23 ± 0.15[b]	4.16 ± 0.04[b,c]

[a] Values are arithmetic means of triplicate samples ± standard deviation.
[b] Values in the same column are significantly different from lupin kernel fiber under the same conditions (Tukey's test, $P < 0.05$, logarithmic transformation of the data).
[c] Values in the same row are significantly different from the same fiber under pre-ingestion (control) conditions (Tukey's test, $P < 0.05$, logarithmic transformation of the data).
Adapted from Turnbull et al. (2005) with permission.

This suggests that lupin kernel fiber has the potential to bulk and dilute the contents of the upper gastrointestinal tract, delay gastric emptying and slow orocecal transit, thus prolonging the intestinal phase of nutrient processing and absorption. In turn, these predicted gastrointestinal effects may lead to benefits for the dietary control of non-insulin-dependent diabetes mellitus (Brand Miller and Foster-Powell, 1999) and obesity (Burton-Freeman, 2000). Cherbut et al. (1991) reported that pea hull fiber induced long orocecal transit times *in vivo*. Table 5.7 highlights that physiochemical properties of fibers that may be modified by the conditions in the human upper gastrointestinal tract. Table 5.8 details the relationship between functional properties and physiological activity of legumes.

5.6 Conclusions

Legume fibers are used as functional ingredients for their beneficial health effects. Many researchers and food processors are working to meet consumer demand by incorporating legume fibers in different food products (e.g. bread) as new functional ingredients. Several literature sources indicate that, in legumes, the levels of insoluble fibers are higher compared to the soluble fibers. This may speed the passage of foods through the stomach and intestines and reduce the risk of

Table 5.8 Relationship between functional properties and physiological activity of legumes

Physiochemical properties	Physiological benefits	References
Water-holding capabilities	Facilitate an easy passage through the colon, thus decreasing the stool transit time Fecal bulking	Stephen and Cummings (1979)
Viscosity	The high viscosity of the DF helps in absorption (and removal) of some of the toxic metabolites generated in the colon, and as a result nullifying the onset of any painful symptoms of diverticular disease, and it also helps in preventing constipation The increased viscosity of the digests modulates serum glucose and lipid levels The decrease in transit time reduces the contact time of carcinogens with the colonic mucosa and thereby offers prevention of colonic cancer	Spiller (1992); Aldoori et al. (1998)
Cation exchange capacity	Some of the acidic polysaccharides (pectins) of DF have free uronic acid carboxyls, which have the capacity to bind heavy metal ions and therefore act as cation exchangers	Bhatt et al. (1987)
	Some of the essential micronutrients such as zinc may be lost as a result of this binding, and this property sometimes is not considered as a beneficial advantage of some fiber-rich foods	James et al. (1978); Robertson et al. (1979)

colorectal cancer. An increase in legume fiber in the human diet helps in altering the gastrointestinal functions, lowering blood lipid and glucose levels. In addition, legume fibers exhibit unique physicochemical properties which are influential in determining their beneficial physiological health effects. With strategic use, as a stand-alone food or food ingredient, and with carefully controlled processing conditions, legume fiber can play a key role in human health and nutrition.

References

Abdul-Hamid, A., Luan, Y.S., 2000. Functional properties of dietary fiber prepared from defatted rice bran. Food Chem. 68, 15–19.

Aguilera, Y., Martín-Cabrejas, M.A., Benítez, V., Mollá, E., López-Andréu, F.J., Esteban, R.M., 2009. Changes in carbohydrate fraction during dehydration process of common legumes. J. Food Compos. Anal. 22, 678–683.

Aldoori, W.H., Giovannucci, E.L., Rockett, H.R., Sampson, L., Rimm, E.B., Willett, W.C., 1998. A prospective study of dietary fiber types and symptomatic diverticular disease in men. J. Nutr. 128, 714–719.

American Association of Cereal Chemists (AACC), 2001. The definition of dietary fiber. (Report of the Dietary Fiber Definition Committee to the Board of Directors of the AACC). Cereal Foods World, 46, 112–126.

Anderson, E.T., Berry, B.W., 2001. Effects of inner pea fiber on fat retention and cooking yield in high fat ground beef. Food Res. Int. 34, 689–694.

Anderson, J.W., Chen, W.-J.L., 1983. Legumes and their soluble fiber: effects on cholesterol-rich lipoproteins. In: Unconventional sources of dietary fiber. American Chemical Society, pp. 49–59.

Anderson, J.W., Smith, B.M., Gustafdson, N.J., 1994. Health benefits and practical aspects of high-fiber diets. Am. J. Clin. Nutr. 59, 1242S–1247S.

Ang, J.F., 1991. Water retention capacity and viscosity effect of powdered cellulose. J. Food Sci. 56, 1682–1684.

Auffret, A., Ralet, M.C., Guillon, F., Barry, J.L., Thibault, J.F., 1994. Effect of grinding and experimental conditions on the measurement of hydration properties of dietary-fibers. LWT – Food Sci. Technol. 27, 166–172.

Azizah, A.H., Zainon, H., 1997. Effect of processing on dietary fiber contents of selected legumes and cereals. Malay. J. Nutr. 3, 131–136.

Bednar, G.E., Patil, A.R., Murray, S.M., Grieshop, C.M., Merchen, N.R., Fahey, G.C., Jr., 2001. Starch and fiber fractions in selected food and feed ingredients affect their small intestinal digestibility and fermentability and their large bowel fermentability *in vitro* in a canine model. J. Nutr. 131, 276–286.

Bell, J.M., Shires, A., 1982. Composition and digestibility by pigs of hull fractions from rapeseed cultivars with yellow or brown seed coats. Can. J. Anim. Sci. 62, 557–565.

Berrios, J. De J., Camara, M., Torija, M.E., Alonso, M., 2002. Effect of extrusion cooking and sodium bicarbonate addition on the carbohydrate composition of black bean flours. J. Food Process. Preserv. 26, 113–128.

Berrios, J. De J., Morales, P., Cámara, M., Sánchez-Mata, M.C., 2010. Carbohydrate composition of raw and extruded pulse flours. Food Res. Int. 43, 531–536.

Berrios, J.DeJ., Tang, J. and Swanson, B.G., 2008. Extruded legumes. US patent application publication no. 2008/0145483 A1.

Betancur-Ancona, D., Peraza-Mercado, G., Moguel-Ordoñez, Y., Fuertes-Blanco, S., 2004. Physicochemical characterization of lima bean (*Phaseolus lunatus*) and jack bean (*Canavalia ensiformis*) fibrous residues. Food Chem. 84, 287–295.

Bhatt, R.U., Salimath, P.V., Tharanathan, R.N., 1987. A mucilaginous acidic polysaccharide from black gram (*Phaseolus mungo*): structure–function characteristics. Carbohydr. Res. 161, 161–166.

Bollinger, H., 1999. Functional food – use of dietary fiber as a multifunctional component. Food Market. Technol. 13, 10–12.

Brand Miller, J., Foster-Powell, K., 1999. Diets with a low glycemic index: from theory to practice. Nutr. Today 34, 64–72.

Bressani, R., 1993. Grain quality of common beans. Food Rev. Int. 9, 237–297.

Brillouet, J.-M., Carré, B., 1983. Composition of cell walls from cotyledons of *Pisum sativum*, *Vicia faba* and *Glycine max*. Phytochemistry 22, 841–847.

Brillouet, J.-M., Riochet, D., 1983. Cell wall polysaccharides and lignin in cotyledons and hulls of seeds from various lupin (*Lupinus* L.) species. J. Sci. Food Agric. 34, 861–868.

Burkitt, D.P., Walker, A.R.P., Painter, N.S., 1972. Effect of dietary fiber on stools and transit times, and its role in the causation of disease. Lancet 300, 1408–1412.

Burton-Freeman, B., 2000. Dietary fiber and energy regulation. J. Nutr. 130, S272–S274.

Camire, M.E., Camire, A.L., Krumhar, K., 1990. Chemical and nutritional changes in foods during extrusion. Crit. Rev. Food Sci. Nutr. 29, 35–57.

Camire, M.E., Viollette, D., Dougherty, M.P., McLaughlin, M.A., 1997. Potato peel dietary fiber composition: effects of peeling and extrusion cooking processes. J. Agric. Food Chem. 45, 1404–1408.

Campos-Vega, R., Reynoso-Camacho, R., Pedraza-Aboytes, G., 2009. Chemical composition and *in vitro* polysaccharide fermentation of different beans (*Phaseolus vulgaris* L.). J. Food Sci. 74(7), T59–T62.

Chaplin, M.F., 2003. Fiber and water binding. Proc. Nutr. Soc. 62, 223–227.

Chau, C.-F., Cheung, P.C.-K., 1999. Effects of the physico-chemical properties of three legume fibers on cholesterol absorption in hamsters. Nutr. Res. 19, 257–265.

Chau, C.F., Cheung, K., Wong, Y.S., 1997. Functional properties of protein concentrates from three Chinese indigenous legume seeds. J. Agric. Food Chem. 45, 2500–2503.

Cherbut, C., Salvador, V., Barry, J.-L., Doulay, F., Delort-Laval, J., 1991. Dietary fiber effects on intestinal transit in man: involvement of their physico-chemical properties. Food Hydrocoll. 5, 15–22.

Cummings, J.H., 1973. Dietary fiber. Gut 14, 69–81.

Cummings, J.H., 1981. Short chain fatty acids in human colon. Gut 22, 763–769.

Dahl, W.J., Whiting, S.J., Healey, A., Zello, G.A., Hildebrandt, L., 2003. Increased stool frequency occurs when finely processed pea hull fiber is added to usual foods consumed by elderly residents in long-term care. J. Am. Diet. Assoc. 103, 1199–1202.

Dalgetty, D.D., Baik, B.-K., 2003. Isolation and characterization of cotyledon fibers from peas, lentils, and chickpeas. Cereal Chem. 80, 310–315.

Dalgetty, D.D., Baik, B.-K., 2006. Fortification of bread with hulls and cotyledon fibers isolated from peas, lentils, and chickpeas. Cereal Chem. 83, 269–274.

Davidson, M.H., McDonald, A., 1978. Fiber: forms and functions. Nutr. Res. 18, 617–624.

de Almeida Costa, G.E., Queiroz-Monici, K.d.S., Reis, S.M.P.M., de Oliveira, A.C., 2006. Chemical composition, dietary fiber and resistant starch contents of raw and cooked pea, common bean, chickpea and lentil legumes. Food Chem. 94, 327–330.

DeVries, J.W., 2004. On defining dietary fiber. Proc. Nutr. Soc. 62, 37–43.

Dikeman, C.L., Fahey, Jr., G.C., 2006. Viscosity as related to dietary fiber: a review. Crit. Rev. Food Sci. Nutr. 46, 649–663.

Dreher, M.L., 1987. Handbook of dietary fiber: an applied approach. Marcel Dekker, New York.

Dubois, C., Cara, L., Armand, M., 1993. Effects of pea and soybean fiber on postprandial lipemia and lipoproteins in healthy adults. Eur. J. Clin. Nutr. 47, 508–520.

Eastwood, M.A., Brydon, W.G., Tadesse, K., 1980. Effect of fiber on colonic function. In: Spiller, G.A., Kay, R.M. (Eds.), Medical aspects of dietary fiber. Plenum Press, New York, pp. 1–26.

Englyst, H.N., Cummings, J.H., 1990. Non-starch polysaccharides (dietary fiber) and resistant starch. In: Furda, I., Brine, C.J. (Eds.),

New developments in dietary fiber. Plenum Press, New York, pp. 205–225.

Furda, I., 1990. Interaction of dietary fiber with lipids mechanistic theories and their limitations. In: Furda, I., Brine, C.J. (Eds.), New developments in dietary fiber. Plenum, New York, pp. 67–82.

Gelroth, J., Ranhotra, G.R., 2001. Food uses of fiber. In: Cho, S.S., Dreher, M.L. (Eds.), Handbook of dietary fiber. Marcel Dekker, New York.

Ghavidel, R.A., Prakash, J., 2006. Effect of germination and dehulling on functional properties of legume flours. J. Sci. Food. Agric. 86, 1189–1195.

Gooneratne, J., Majsak-Newman, G., Robertson, J.A., Selvendran, R.R., 1994. Investigation of factors that affect the solubility of dietary fiber, as nonstarch polysaccharides, in seed tissues of mung bean (*Vigna radiata*) and black gram (*Vigna mungo*). J. Agric. Food Chem. 42, 605–611.

Gordon, D.T., 1989. Functional properties vs physiological action of total dietary fiber. Cereal Foods World 34, 517–525.

Górecka, D., Lampart-Szczapa, E., Janitz, W., Sokolowska, B., 2000. Composition of fractional and functional properties of dietary fiber of lupines (*L. luteus* and *L. albus*). Nahrung 44, 229–232.

Gualberto, D.G., Bergman, C.J., Kazemzadeh, M., Weber, C.W., 1997. Effect of extrusion processing on the soluble and insoluble fiber, and phytic acid contents of cereal brans. Plant foods Hum. Nutr. 51, 187–198.

Guillon, F., Champ, M., 2000. Structural and physical properties of dietary fibers and consequences of processing on human physiology. Food Res. Int. 33, 233–245.

Guillon, F., Champ, M.M.-J., 2002. Carbohydrate fractions of legumes: uses in human nutrition and potential for health. Br. J. Nutr. 88(3), S293–S306.

Harris, P.J., Ferguson, L.R., 1999. Dietary fibers may protect or enhance carcinogenesis. Mutat. Res. 443, 95–110.

Hipsley, E.H., 1953. Dietary "fiber" and pregnancy toxaemia. Br. Med. J. ii, 420–422.

Huang, S.-C., Lia, T.-S., Cheng, T.-C., Chan, H.Y., Hwang, S.-M., Hwang, D.-F., 2009. *In vitro* interactions on glucose by different fiber materials prepared from mung bean hulls, rice bran and lemon pomace. J. Food Drug Anal. 17, 307–314.

James, W.P.T., Branch, W.J., Southgate, D.A.T., 1978. Calcium binding by dietary fiber. Lancet i, 638–639.

Jenkins, D.J.A., Kendall, C.W.C., Vuksan, V., 2000. Viscous fibers, health claims, and strategies to reduce cardiovascular disease risk. Am. J. Clin. Nutr. 71, 401–402.

Johnson, S.K., Chua, V., Hall, R.S., Baxter, A.L., 2006. Lupin kernel fiber foods improve bowel function and beneficially modify some putative faecal risk factors for colon cancer in men. Br. J. Nutr. 95, 372–378.

Khatoon, N., Prakash, J., 2004. Nutritional quality of microwave-cooked and pressure-cooked legumes. Int. J. Food Sci. Nutr. 55, 441–448.

Klurfeld, D.M., 1990. Insoluble dietary fiber and experimental colon cancer: are we asking the proper questions? In: Kritchevsky, D., Bonfield, C., Anderson, J.W. (Eds.), Dietary fiber: chemistry, physiology and health effects. Plenum, New York, pp. 403–417.

Kutoš, T., Golob, T., Kač, M., Plestenjak, A., 2003. Dietary fiber of dry and processed beams. Food Chem. 80, 231–235.

Labuza, T.P., 1986. Comparison of binding of fruit vegetable and cereal fibers. Cereal Food World 31, 599.

Leterme, P., Froidmont, E., Rossi, F., Théwis, A., 1998. The high water-holding capacity of pea inner fibers affects the ileal flow of endogenous amino acids in pigs. J. Agric. Food Chem. 46, 1927–1934.

Li, B.W., Andrews, K.W., Pehrsson, P.R., 2002. Individual sugars, soluble, and insoluble dietary fiber contents of 70 high consumption foods. J. Food Compos. Anal. 15, 715–723.

Lintas, C., Cappelloni, M., Montalbano, S., Gambelli, L., 1995. Dietary fiber in legumes: effect of processing. Eur. J. Clin. Nutr. 49, 298–302.

López, G., Ros, G., Rincón, F., Periago, M., Martínez, C., Ortuño, J., 1997. Propiedades funcionales de la fibra dietética. Mecanismos de acción en el tracto gastrointestinal. Arch. Latinoam. Nutr. 47, 203–207.

Lunn, J., Buttriss, J.L., 2007. Carbohydrates and dietary fiber. Nutr. Bull. 32, 21–64.

Mahadevamma, S., Tharanathan, R.N., 2004. Processing of legumes: resistant starch and dietary fiber contents. J. Food Qual. 27, 289–303.

Mallillin, A.C., Trinidad, T.P., Raterta, R., Dagbay, K., Loyola, A.S., 2008. Dietary fiber and fermentability characteristics of root crops and legumes. Br. J. Nutr. 100, 485–488.

Mangala, S.L., Tharanathan, R.N., 1999. Structural studies of resistant starch derived from processed (autoclaved) rice. Eur. Food Res. Technol. 209, 38–42.

Martín-Cabrejas, M.A., Jaime, L., Karanja, C., 1999. Modifications to physicochemical and nutritional properties of hard-to-cook beans

(*Phaseolus vulgaris* L.) by extrusion cooking. J. Agric. Food Chem. 47, 1174–1182.

Martín-Cabrejas, M.A., Ariza, N., Esteban, R., Mollá, E., Waldron, K., López-Andréu, F.J., 2003. Effect of germination on the carbohydrate composition of the dietary fiber of peas (*Pisum sativum* L.). J. Agric Food.Chem. 51, 1254–1259.

Martín-Cabrejas, M.A., Sanfiz, B., Vidal, A., Mollá, E., Esteban, R.M., López-Andréu, F.J., 2004. Effect of fermentation and autoclaving on dietary fiber fractions and antinutritional factors of beans (*Phaseolus vulgaris* L.). J. Agric. Food Chem. 52, 261–266.

Martín-Cabrejas, M.A., Aguilera, Y., Beníítez, V., Mollá, E., López-Andréu, F.J., Esteban, R.M., 2006. Effect of industrial dehydration on the soluble carbohydrates and dietary fiber fractions in legumes. J. Agric. Food Chem. 54, 7652–7657.

Martín-Cabrejas, M.A., Díaz, M.F., Aguilera, Y., Benítez, V., Mollá, E., Esteban, R.M., 2008. Influence of germination on the soluble carbohydrates and dietary fiber fractions in non-conventional legumes. Food Chem. 107, 1045–1052.

McConnell, A.A., Eastwood, M.A., Mitchell, W.D., 1974. Physical characteristics of vegetable foodstuffs that could influence bowel function. J. Sci. Food Agric. 25, 1457–1464.

McDougall, G.J., Morrison, I.M., Stewart, D., Hillman, J.R., 1996. Plant cell walls as dietary fiber: range, structure, processing and function. J. Sci. Food Agric. 70, 133–150.

McKee, L.H., Latner, T.A., 2000. Underutilized sources of dietary fiber: a review. Plant Foods Hum. Nutr. 55, 285–304.

Mongeau, R., Brassard, R., 1995. Importance of cooking temperature and pancreatic amylase in determination of dietary fiber in dried legumes. J. AOAC Int. 78, 1444–1449.

Morris, E.R., 2001. Assembly and rheology of non-starch polysaccharides. In: McCleary, B.V., Prosky, L. (Eds.), Advanced dietary fiber technology. Blackwell Science, Oxford, pp. 30–41.

Mouser, F., Suckow, P., Kulikowski, W., 1983. Ahalytshe Bestimmung vonTrebern fur ballaststoffen in Brot. Obst und Gemuse. Getreide. Mehl. Brot. 37, 380–382.

Parrott, M.E., Thrall, B.E., 1978. Functional properties of various fibers: physical properties. J. Food Sci. 43, 759–764.

Patil, R.T., Berrios, J.D.J., Tang, J., Swanson, B.G., 2007. Evaluation of the methods for expansion properties of legume extrudates. Appl. Eng. Agric. 23, 777–783.

Pérez-Hidalgo, M.A., Guerra-Hernandez, E., Garcia-Villanova, B., 1997. Dietary fiber in three raw legumes and processing effect on chick peas by an enzymatic–gravimetric method. J. Food Compos. Anal. 10, 66–72.

Periago, M.J., Ros, G., Casas, J.L., 1997. Non-starch polysaccharides and *in vitro* starch digestibility of raw and cooked chick peas. J. Food Sci. 62, 93–96.

Pfoertner, H.N., Fischer, J., 2001. Dietary fibers of lupins and other grain legumes. In: McCleary, B.V., Prosky, L. (Eds.), Advanced dietary fiber technology. Blackwell Science, Oxford, pp. 361–366.

Pittaway, J.K., Ahuja, K.D.K., Robertson, I.K., Ball, M.J., 2007. Effects of a controlled diet supplemented with chickpeas on serum lipids, glucose tolerance, satiety and bowel function. J. Am. Coll. Nutr. 26, 334–340.

Plaami, S.P., 1997. Content of dietary fiber in foods and its physiological effects. Food Rev. Int. 13, 29–76.

Pszczola, D.E., 1998. Fiber has a strong supporting role in nutraceutical movement. Food Technol. 52, 90–96.

Qian, J., Ding, X., 1996. Effect of twin-screw extrusion on the functional properties of soya fiber. J. Sci. Food Agric. 71, 64–68.

Raghavendra, S.N., Swamya, S.R.R., Rastogi, N.K., Raghavarao, K.S.M.S., Kumar, S., Tharanathan, R.N., 2006. Grinding characteristics and hydration properties of coconut residue: a source of dietary fiber. J. Food Eng. 72, 281–286.

Ralet, M.-C., Valle, G.D., Thibault, J.-F., 1993. Raw and extruded fiber from pea hulls. Part I: Composition and physico-chemical properties. Carbohydr. Polym. 20, 17–23.

Ramulu, P., Rao, R.U., 1997. Effect of processing on dietary fiber content of cereals and pulses. Plant Foods Hum. Nutr. 50, 249–257.

Rehman, Z.-U., Shah, W.H., 2004. Domestic processing effects on some insoluble dietary fiber components of various food legumes. Food Chem. 87, 613–617.

Rey, D.K., Labuza, T.P., 1981. Characterization of the effect of solute in water-binding and gel strength properties of carrageenan. J. Food Sci. 46, 786.

Robertson, J., Brydon, W.G., Tadesse, K., Wenham, P., Walls, A., Eastwood, M.A., 1979. The effect of raw carrot on serum lipids and colon function. Am. J. Clin. Nutr. 32, 1889–1892.

Robertson, J.A., de Monredon, F.D., Dysseler, P., Guillon, F., Amadò, R., Thibault, J.-F., 2000. Hydration properties of dietary fiber and resistant starch: a European collaborative study. Lebensmit. -Wissensch. Technol. 33, 72–79.

Rochfort, S., Panozzo, J., 2007. Phytochemicals for health, the role of pulses. J. Agric. Food Chem. 55, 7981–7994.

Sandström, B., Hansen, L.T., Sorenson, A., 1994. Pea fiber lowers fasting and postprandial blood triglyceride levels in humans. J. Nutr. 124, 2386–2396.

Schneeman, B.O., 2001. Dietary fiber and gastrointestinal function. In: McCleary, B.V., Prosky, L. (Eds.), Advanced dietary fiber technology. Blackwell Science, Oxford, pp. 168–176.

Sefa-Dedeh, S., Stanley, D.W., 1979. The relationship of microstructure of cowpeas to water absorption and dehulling properties. Cereal Chem. 56, 379–386.

Selvendran, R.R., Stevens, B.J.H., DuPont, M.S., 1987. Dietary fiber: analysis and properties. Adv. Food Res. 31, 117–209.

Shutler, S.M., Walker, A.F., Low, A.G., 1987. The cholesterol lowering effects of legumes. I: Effects of the major nutrients. Hum. Nutr. Food Sci. Nutr. 41F, 71–86.

Siljestrom, M., Westerlund, E., Bjorck, I., Holm, J., Asp, N.G., Theander, O., 1986. The effects of various thermal processes on dietary fiber and starch content of whole grain wheat and white flour. J. Cereal Sci. 4, 315–323.

Smith, S.C., Choy, R., Johnson, S.K., Hall, R.S., Wildeboer-Veloo, A.C.M., Welling, G.W., 2006. Lupin kernel fiber consumption modifies fecal microbiota in healthy men as determined by rRNA gene fluorescent in situ hybridization. Eur. J. Nutr. 45, 335–341.

Spiller, G.A. (Ed.), 1992. CRC handbook of dietary fiber in human nutrition. CRC Press, Boca Raton, FL.

Sreerama, Y.N., Sashikala, V.B., Pratape, V.M., 2009. Expansion properties and ultrastructure of legumes: effect of chemical and enzyme pre-treatments. LWT – Food Sci. Technol. 42, 44–49.

Stephen, A.M., Cummings, J.H., 1979. Water holding by dietary fiber in vitro and its relationship to faecal output in man. Gut 20, 722–729.

Stephen, A.M., Dahl, W.J., Sieber, G.M., van Blaricom, J.A., Morgan, D.R., 1995. Effect of green lentils on colonic function, nitrogen balance, and serum lipids in healthy human subjects. Am. J. Clin. Nutr. 62, 1261–1267.

Stoughton-Ens, M.D., Hatcher, D.W., Wanga, N., Warkentin, T.D., 2010. Influence of genotype and environment on the dietary fiber content of field pea (Pisum sativum L.) grown in Canada. Food Res. Int. 43, 547–552.

Su, H.L., Chang, K.C., 1995. Physicochemical and sensory properties of dehydrated bean paste products as related to bean varieties. J. Food Sci. 60, 764–794.

Tharanathan, R.N., 2002. Food-derived carbohydrates – structural complexity and functional diversity. Crit. Rev. Biotechnol. 22, 65–84.

Tharanathan, R.N., Mahadevamma, S., 2003. Grain legumes – a boon to human nutrition. Trends Food Sci. Technol. 14, 507–518.

Toma, R.B., Curtis, D.J., 1986. Dietary fiber: effect on mineral bioavailability. Food Technol. 40, 111–116.

Torre, M., Rodriguez, A.R., Saura-Calixto, F., 1992. Study of the interactions of calcium ions with lignin, cellulose, and pectin. J. Agric. Food Chem. 40, 1762–1766.

Tosh, S.M., Yada, S, 2010. Dietary fibers in pulse seeds and fractions: characterization, functional attributes, and applications. Food Res. Int. 43, 450–460.

Trowell, H.C., 1972a. Crude fiber, dietary fiber and atherosclerosis. Atherosclerosis 16, 138–140.

Trowell, H.C., 1972b. Ischemic heart disease and dietary fiber. Am. J. Clin. Nutr. 25, 926–932.

Trowell, H., Southgate, D.A.T., Wolever, T.M.S., Leeds, A.R., Gassull, M.A., Jenkins, D.J.A., 1976. Dietary fiber redefined. Lancet i, 967.

Turnbull, C.M., Baxter, A.L., Johnson, S.K., 2005. Waterbinding capacity and viscosity of Australian sweet lupin kernel fiber under *in vitro* conditions simulating the human upper gastrointestinal tract. Int. J. Food Sci. Nutr. 56, 87–94.

Van Laar, H., Tamminga, S., Williams, B.A., Verstegen, M.W.A., Engels, M., 1999. Fermentation characteristics of cell-wall sugars from soya bean meal, and from separated endosperm and hulls of soya beans. Anim. Feed Sci. Technol. 79, 179–193.

Veena, A., Urooj, A., Puttaraj, S., 1995. Effect of processing on the composition of dietary fiber and starch in some legumes. Nahrung-Food 39, 132–138.

Vidal-Valverde, C., Frias, J., 1991. Legume processing effects on dietary fiber components. J. Food Sci. 56, 1350–1352.

Wang, N., Hatcher, D.W., Gawalko, E.J., 2008. Effect of variety and processing on nutrients and certain anti-nutrients in field peas (*Pisum sativum*). Food Chem. 111, 132–138.

Wang, N., Hatcher, D.W., Toews, R., Gawalko, E.J., 2009. Influence of cooking and dehulling on nutritional composition of several varieties of lentils (*Lens culinaris*). LWT – Food Sci. Technol. 42, 842–848.

Wolever, T.M.S., 1990. Relationship between dietary fiber content and composition in foods and the glycemic index. Am. J. Clin. Nutr. 51, 72–75.

Functional and physicochemical properties of non-starch polysaccharides

6

Charles Brennan[1], Uma Tiwari[2]
[1]Department of Food and Tourism, Manchester Metropolitan University, Manchester, UK
[2]UCD School of Agriculture, Food Science and Veterinary Medicine, University College Dublin, Dublin, Ireland

6.1   Introduction

Non-starch polysaccharides (NSPs) or complex carbohydrates are the major part of dietary fiber (DF) and can be measured more precisely than total dietary fiber; they include cellulose, pectins, glucans, gums, mucilages, inulin and chitin (and exclude lignin) (Cummings and Englyst, 1995). NSPs are the principal components of the plant cell wall and constitute a major source of fiber in the diet (Selvendran and Robertson, 1990). Hence, the estimation of NSPs provides

Pulse Foods: Processing, Quality and Nutraceutical Applications. DOI: 10.1016/B978-0-1238-2018-1.00007-0

a good estimate of fiber from plant foods (Englyst et al., 1994). According to Cummings (1997), NSPs are defined as polysaccharides (DP ≥ 10), which are non-α-glucans that reach the human colon. The key aspect of NSPs is that they are plant materials that are not digested by the enzymes of the human digestive tract but remain fermentable in the large intestine. The non-digestibility of NSP in the small intestine of human subjects has been demonstrated in studies in ileostomy patients (Sandberg et al., 1981; Englyst and Cummings, 1985). However, the criterion of non-digestibility in the small intestine is nowadays the fundamental point of most DF definitions (Champ et al., 2003).

Non-starch polysaccharides are traditionally classified into soluble and insoluble fractions (Sasaki et al., 2004), although the exact differences in solubility are not always clear. It must be emphasized that this solubility may be determined under conditions which do not occur in the human small intestine (Topping, 1991). Nevertheless, the terms "soluble fiber" and "insoluble fiber" have entered into common usage and also serve to segregate NSPs on one of their best documented physiological effects – the lowering of plasma cholesterol, an established risk factor for cardiovascular diseases (CVD) (Brennan, 2005; Topping, 2007; Fernando et al., 2008). Physiologically, the NSPs can influence the bacterial flora of the colon (Plaami, 1997). This action is related to the fact that NSPs are potential substrates for colonic fermentation. Similar to bifidogenic compounds, these are food substances not digested by gastrointestinal enzymes that beneficially affect the host by selective stimulation of growth and/or activity of a limited number of colonic bacteria capable of producing short-chain fatty acids (SCFAs), mainly acetate, propionate and butyrate. Small amounts of branched SCFAs may also be formed from indigestible protein (Roberfroid, 2001; Campos-Vega et al., 2009).

6.2 NSP content of pulses

In general, pulses contain NSPs in the range of 5.5–19.6% with a higher level of insoluble fraction (Table 6.1). The NSP include celluloses, hemicelluloses, lignin, pectin, gums and mucilages. The insoluble fraction of NSP is normally found on the outer protective layer of plants, whereas soluble NSP is normally found in the inner parts of plants. The major neutral sugars identified in the soluble

Table 6.1 NSP content (g 100 g^{-1} DM) of pulses

Pulse	Insoluble NSP	Soluble NSP	Total NSP	Reference
White bean	12.73	4.54	17.27	Anderson and Bridges (1988)
	9.92	7.91	17.83	Bravo (1999)
Common bean	9.64	9.34	18.98	Marconi et al. (2000)
Pinto bean	11.45	8.15	19.6	Anderson and Bridges (1988)
	9.0	7.07	16.07	Bravo (1999)
Chickpea	7.6	2.0	9.6	Periago et al. (1997)
	5.37	3.41	8.78	Bravo (1999)
	8.76	2.82	11.58	Marconi et al. (2000)
Lentils, dried	9.29	1.32	10.61	Anderson and Bridges (1988)
Green lentil	8.1	2.0	10.1	Stephen et al. (1995)
	6.74	1.92	8.66	Bravo (1999)
Red lentil	4.0	1.5	5.5	Stephen et al. (1995)

and insoluble NSP fractions of horse gram were arabinose, xylose and glucose (Bravo et al., 1999). Variations occur in NSP content of pulses, which are mainly attributed to species, variety and other agronomic factors. Bravo (1999) observed higher levels of insoluble NSP content in beans, lentils and chickpeas compared to soluble NSP. They observed about 80% of total NSP as insoluble fraction in lentils. Gooneratne et al. (1994a) studied the distribution of NSP in mung bean tissue. They observed 0.4% in embryo, 2.5% in hull and 10.4% in cotyledon with a total fiber content of 13.3%, as NSP. Arabinose, galactose and uronic acid (galacturonic) account for 25% of the soluble NSP in mung bean, whereas pectic polysaccharides rich in arabinose were the major non-cellulosic polysaccharides in the insoluble fiber fraction, accounting for 42% of the NSP in mung bean (Gooneratne et al., 1994a).

6.3 Cellulose and hemicelluloses

Cellulose consists of unbranched polymers of β-linked glucose residues arranged in linear chains which exist as a crystalline array of many parallel, oriented chains – microfibrils – resulting in "amorphous" or paracrystalline regions of the microfibril where the β-1,4-glucan chains are less ordered (Rose and Bennett, 1999). Cellulose content varies depending on the species.

Hemicelluloses have β-(1-4)-linked backbones of xylose, mannose or glucose residues that can form extensive hydrogen bonds with cellulose. Hemicelluloses are cellulose-binding polysaccharides, which together with cellulose form a network. However, branches and other structural modifications in their structure prevent them from forming microfibrils by themselves. Xyloglucan and arabinoxylans are the most abundant hemicelluloses (Cosgrove, 2005). The hemicelluloses are more abundant in secondary walls than in the primary walls of both dicot and monocot species. Monocot species tend to have significantly more hemicellulose and less pectin than dicots, and also have mixed linkage glucans that make up a major proportion of monocot hemicellulose polysaccharides (Caffall and Mohnen, 2009). Srisuma et al. (1991) observed the cellulose content was the major component of the navy bean seed coat, ranging up to 60%, followed by the hemicellulose, ranging up to 20%, and a small amount of lignin, about 2%.

Similarily, Górecka et al. (2000) reported that cellulose was predominant in the hull fraction of lupin whereas in the flour hemicellulose was the predominant fraction. Sosulski and Wu (1988) determined that field pea hulls contained 82.3% total dietary fiber and indicated that cellulose (62.3%) was the predominant form with only 8.2% hemicellulose content, whereas Pérez-Hidalgo et al. (1997) reported that the main component of pulses was hemicellulose with $\approx$12% for chickpeas and kidney beans and $\approx$16% for lentils but their cellulose content was low, ranging from 6 to 8%. This clearly illustrates the possible variations that exist in relatively close species.

6.4   Pectin, gums and mucilages

The pectins, which are most abundant in the plant primary cell walls and the middle lamellae, represent a class of molecules defined by the presence of galacturonic acid (Caffall and Mohnen, 2009). Pectins include homogalacturonan and rhamnogalacturonan I. Homogalacturonan, which makes up the bulk of pectins, is composed of linear chains of galacturonic acid residues that are methyl-esterified and acetylated to different extents depending on the plant species (Willats et al., 2001). Bailoni et al. (2005) evaluated pectin content of the dietary fiber of legumes, reporting a range from 8 to 13%. Bravo (1999) demonstrated various pulse processing techniques and reported the presence of highest pectin content in soluble NSP fractions, except for lentils which appeared to have similar contents of soluble pectins and pectic substances associated with insoluble cell wall polysaccharides. Many processing factors such as boiling and thermal dehydration may degrade the pectin content of legumes (Cheung and Chau, 1998; Brennan and Tudorica, 2008; Aguilera et al., 2009). This may be due to dissolution of the middle lamellae and some breakdown of pectins through β-elimination during heat treatment (McDougall et al., 1996). Gums are used as thickening agent and increase the viscosity of food which has properties that are intermediate: it hydrates rapidly in cold water, but interacts with gel-forming polysaccharides (Harris and Smith, 2006). Mucilages are usually sticky substances and the soluble mucilages amount to over 40% of NSP composition, depending on the variety of legume seed. Mucilages are produced by the outer layer of the seed, i.e epidermis of seed (Harris and Ferguson, 1999). This mucilagenous mixture of compounds has been shown to be useful as a potential food ingredient with guar and locust bean gum (Tudoricã et al., 2002; Brennan et al., 2006).

6.5 Physiological effects of NSP

As indicated previously, the NSP component of pulses has been reported for its valuable contribution to a healthy diet in both humans and animals (Goodlad and Mathers, 1991; Stephen et al., 1995; Brennan, 2005). NSPs are the principal components of dietary fiber and the lack of small intestinal digestibility explains the majority of their principal physiological properties. The physicochemical properties of NSPs result in a number of physiological effects that have been related to certain health benefits (Brennan, 2005). NSPs have been claimed to modulate blood glucose and insulin responses to foods (Jenkins et al., 1981), to lower blood cholesterol (Lairon, 1996) and to have beneficial effects on the prevention and treatment of certain diseases like gallstones, diverticular disease, obesity, constipation or colon cancer (Cummings et al., 1992; Cummings and Englyst, 1995). Studies in healthy humans have shown that ingested NSP is essentially unchanged during passage through the stomach and small intestine (Englyst and Cummings, 1985, 1987). The potential function of NSP in appetite control is to modulate the rate of stomach emptying, with a prolonged residence time of the food in the stomach which makes you feel full. The role of NSP in stomach fullness is realized either by increasing viscosity or by forming a gel (Lyly et al., 2004; Lundin et al., 2008).

The function of NSP in human health is still not fully understood. In a human nutrition study, Stephen et al. (1995) reported that consuming approximately 12 g of soluble NSP from green lentils effectively increased fecal weight from $131\,g\,day^{-1}$ to $189\,g\,day^{-1}$. However, some authors argue that not enough evidence has been produced to demonstrate the role of NSP to reduce cholesterol and glucose level (Goodlad and Mathers, 1991). Cobiac et al. (1990) observed that intake of 12 g NSP from canned baked beans did not alter the plasma cholesterol or the glucose concentration in hypercholesterolemic men. Similarly, Key and Mathers (1993) noted that NSP digestibilities were 0.56 and 0.86 for wholemeal bread and beans, respectively, with no evidence that the dietary presence of beans affected digestibility of bread NSP. In an *in vitro* study Campos-Vega et al. (2009) suggested that the common bean is an excellent source of polysaccharides that can be fermented in the colon and produce short-chain fatty acids which exert health benefits.

6.6 Effect of processing on NSP

As mentioned earlier, processing of pulses affects dietary fiber content and functional properties of dietary fiber. Similar to other fiber fractions, NSP content and properties are also influenced by processing. Marconi et al. (2000) studied the effects of both conventional and microwave cooking on the NSP content of chickpeas and common beans. They observed that both of the cooking procedures produced a redistribution of the insoluble to soluble NSP fraction, without affecting the total NSP content. They also observed that the decrease in the insoluble NSP content and increase in the soluble NSP fractions were more pronounced after microwave cooking compared to conventional cooking based on the ratio between the insoluble and soluble fiber fractions. The ratio of insoluble and soluble fiber fractions reduced from $3.11\,g$ $100\,g^{-1}$ dry matter (raw chickpeas) to $2.80\,g\,100\,g^{-1}$ dry matter during traditional cooking and $2.49\,g\,100\,g^{-1}$ dry matter after microwave cooking. They observed a similar effect in bean samples, where the insoluble:soluble ratio was reduced from 1.03 g 100 g^{-1} dry matter in the raw beans, to 0.87 g 100 g^{-1} dry matter after traditional cooking and 0.76 after microwave cooking. This suggests that the depolymerization of cell wall polysaccharides could be more extensive during microwave cooking than during conventional cooking (Marconi et al., 2000). Redistribution and subsequent shift in insoluble:soluble ratio could be caused by a partial solubilization and depolymerization of hemicellulose and insoluble pectic substances (Vidal-Valverde and Frias, 1991; Lintas et al., 1995). Increased cooking causes redistribution of soluble and insoluble NSP of the legume flours which may influence the total NSP content (Gooneratne et al., 1994b; Cheung and Chau, 1998). Periago et al. (1997) reported that both domestic and industrial cooking altered the NSP of chickpea, possibly due to depolymerization of insoluble material and the loss of soluble component as a consequence of processing. The soluble NSP increased in industrially processed pulses at the expense of the insoluble fraction and the reduction of total NSP was probably due to the presence of other food ingredients (Bravo, 1999). Similarly, Marconi et al. (2000) reported that post-cooking of chickpea and common beans led to an increase in total NSP content, which may be due to a greater loss of non-fiber components. Rehman and Shah (2004) studied the effect of domestic processing on the cellulose, hemicellulose and lignin contents of black grams, chickpeas, lentils, and red and white kidney beans (Table 6.2). They observed that pressure cooking had a pronounced effect on cellulose, hemicellulose and lignin contents compared to the microwave and ordinary cooking. They also observed that the reduction in hemicellulose was higher compared to cellulose. The reduction in cellulose and hemicellulose is mainly due to breakdown of these products into simple carbohydrates (Rehman and Shah, 2004).

Periago et al. (1996) studied the effect of thermal treatment such as freezing, cooking and canning of peas. They observed that frozen cooked peas showed a higher content of total NSP than raw frozen peas whereas canned peas had slightly lower total NSP contents compared with frozen raw peas. The high temperature used in canning may lead to disruption of linkages in cell wall polysaccharides, which could lead to an increase in soluble NSP or to loss of NSP (Periago et al., 1996). In another study, Periago et al. (1996) reported

Table 6.2 Effect of different cooking methods on cellulose, hemicellulose and lignin contents (g $100\,g^{-1}$ on dry basis) of various food pulses

Pulse	Cellulose				Hemicellulose				Lignin			
	Raw	Ordinary cooking	Pressure cooking	Microwave cooking	Raw	Ordinary cooking	Pressure cooking	Microwave cooking	Raw	Ordinary cooking	Pressure cooking	Microwave cooking
Black gram	9.70±0.95[a]	7.65±0.41[b]	7.75±0.38[b]	7.55±0.42[b]	12.0±1.11[a]	9.24±0.62[b]	6.89±0.70[c]	9.10±0.88[b]	1.70±0.11[a]	1.90±0.08[b]	2.07±0.12[c]	1.80±0.13[b]
Chickpea	8.45±1.21[a]	6.70±0.48[b]	6.00±0.45[c]	6.90±0.44[b]	16.00±1.08[a]	12.2±0.73[b]	9.23±0.48[c]	12.1±0.75[b]	1.08±0.16[a]	1.18±0.12[a]	1.25±0.11[c]	1.15±0.16[a]
Lentil	8.10±1.08[a]	6.80±0.72[b]	6.60±0.49[c]	6.75±0.33[b]	20.3±1.32[a]	13.5±0.80[b]	12.6±0.81[c]	14.0±0.90[b]	1.42±0.15[a]	1.58±0.21[b]	1.19±0.13[c]	1.9±0.13[c]
Red kidney bean	6.22±0.48[a]	4.98±0.22[b]	4.44±0.21[c]	4.88±0.24[b]	20.2±1.17[a]	13.78±0.42[b]	12.1±0.55[c]	15.1±0.50[b]	1.21±0.11[a]	1.30±0.08[b]	1.40±0.09[c]	1.39±0.11[c]
White kidney bean	8.00±0.59[a]	7.00±0.70[b]	6.64±0.53[c]	6.85±0.51[b]	14.9±1.00[a]	9.41±0.85[b]	9.10±0.73[c]	10.79±0.67[b]	1.32±0.14[a]	1.50±0.10[b]	1.40±0.11[c]	1.50±0.14[b]

Mean values ± SD (on dry basis). Mean values in a row with different superscripts are significantly different at $P < 0.05$.

an increase in NSP content in peas after cooking, which is due to losses of non-fiber material, presumably mainly free sugars.

Cheung and Chau (1998) investigated the effect of cooking on the NSPs of *Phaseolus angularis*, *Phaseolus calcaratus* and *Dolichos lablab* flour. They observed an increase in soluble NSP content with cooking time. The increase in soluble fraction could be due to solubilization of cell wall pectic substances as a result of dissolution of the middle lamellae and some breakdown of pectins through β-elimination during boiling (McDougall et al., 1996; Cheung and Chau, 1998). Redistribution of the insoluble fraction and soluble components of NSP has been reported for peas (Periago et al., 1996), black gram and green gram (Gooneratne et al., 1994a) during thermal treatment. The increase in the solubilization of pectic polysaccharides during the cooking of these pulses was important for their therapeutic values as soup ingredients (Cheung and Chau, 1998). Gooneratne et al. (1994a) observed an increase in soluble NSP fraction in extruded mung bean flour. The increased soluble NSP content relative to the raw flour was due to mainly an increased solubility of arabinose-containing pectic polysaccharides resulting from a slight degradation of pectic polysaccharides. Gooneratne et al. (1994a) stated that the notable increase in glucose-containing soluble NSP was probably due to the presence of some "resistant starch" formed during the extrusion process; this starch was resistant to amylolytic digestion.

6.7 Conclusions

Pulses are potential sources of non-starch polysaccharides. Variations exist in NSP content of pulses which is mainly due to genotypic variation. There is convincing evidence that NSPs play an important role in the human diet and have several physiological benefits including appetite control, slowing down small intestine transit, lowering blood plasma cholesterol and improving bowel function. However, some of these health benefits are still inconclusive. Compared to pulses there is strong evidence that cereal soluble NSPs have benefits in the control of plasma cholesterol (Topping, 2007). However, more work is needed to investigate the function of pulse NSPs in physiological responses such as regulating obesity and diabetes.

References

Aguilera, Y., Martín-Cabrejas, M.A., Benítez, V., Mollá, E., López-Andréu, F.J., Esteban, R.M., 2009. Changes in carbohydrate fraction during dehydration process of common legumes. J. Food Compos. Anal. 22, 678–683.

Anderson, J.W., Bridges, S.R., 1988. Dietary fiber content of selected foods. Am. J. Clin. Nutr. 47, 440–447.

Bailoni, L., Schiavon, S., Pagnin, G., Tagliapietra, F., Bonsembiante, M., 2005. Quanti-qualitative evaluation of pectins in the dietary fibre of 24 foods. Ital. J. Anim. Sci. 4, 49–58.

Bravo, L., 1999. Effect of processing on the non-starch polysaccharides and *in vitro* starch digestibility of legumes. Food Sci. Technol. Int. 5, 415–423.

Bravo, L., Siddhuraju, P., Calixto, F.S., 1999. Composition of underexploited Indian pulses. Comparison with common legumes. Food Chem. 64, 185–192.

Brennan, C.S., 2005. Dietary fibre, glycaemic response, and diabetes. Mol. Nutr. Food Res. 49, 560–570.

Brennan, C.S., Tudorica, C.M., 2008. Evaluation of potential mechanisms by which dietary fibre additions reduce the predicted glycaemic index of fresh pastas. Int. J. Food Sci. Technol. 43, 2151–2162.

Brennan, C.S., Suter, M., Matia-Merino, L., Luethi, T., Ravindran, G., Goh, K., Ovortrup, J., 2006. Gel and pasting behaviour of fenugreek–wheat starch and fenugreek–wheat flour combinations. Starch/Staerke 58, 527–535.

Caffall, K.H., Mohnen, D., 2009. The structure, function, and biosynthesis of plant cell wall pectic polysaccharides. Carbohydr. Res. 344, 1879–1900.

Campos-Vega, R., Reynoso-Camacho, R., Pedraza-Aboytes, G., 2009. Chemical composition and *in vitro* polysaccharide fermentation of different beans (*Phaseolus vulgaris* L.). J. Food Sci. 74, T59–T62.

Champ, M., Langkilde, A.M., Brouns, F., Kettlitz, B., Le Bail-Collet, Y., 2003. Advances in dietary fibre characterisation. 2. Consumption, chemistry, physiology and measurement of resistant starch; implications for health and food labelling. Nutr. Res. Rev. 16, 143–161.

Cheung, P.C.-K., Chau, C.-F., 1998. Changes in the dietary fiber (resistant starch and nonstarch polysaccharides) content of cooked flours prepared from three Chinese indigenous legume seeds. J. Agric. Food Chem. 46, 262–265.

Cobiac, L., McArthur, R., Nestel, P.J., 1990. Can eating baked beans lower plasma-cholesterol? Eur. J. Clin. Nutr. 44, 819–822.

Cosgrove, D.J., 2005. Growth of the plant cell wall. Nat. Rev. Mol. Cell Biol. 6, 850–861.

Cummings, J.H., 1997. The large intestine in nutrition and disease. Institut Danone, Brussels, http://www.danoneinstitute.be/communication/pdf/mono06/mono6full.pdf.

Cummings, J.H., Englyst, H.N., 1995. Gastrointestinal effects of food carbohydrate. Am. J. Clin. Nutr. 61, 938S–945S.

Cummings, J.H., Bingham, S.A., Heaton, K.W., Eastwood, M.A., 1992. Fecal weight, colon cancer risk, and dietary intake of nonstarch polysaccharides (dietary fiber). Gastroenterology 103, 1783–1789.

Englyst, H.N., Cummings, J.H., 1985. Digestion of the polysaccharides of some cereal foods in the human small intestine. Am. J. Clin. Nutr. 42, 778–787.

Englyst, H.N., Cummings, J.H., 1987. Digestion of the polysaccharides of potato in the human small intestine. Am. J. Clin. Nutr. 45, 423–431.

Englyst, H.N., Quigley, M.E., Hudson, G.J., 1994. Determination of dietary fibre as non-starch polysaccharides with gas-liquid chromatographic, high-performance liquid chromatographic or spectrophotometric measurement of constituent sugars. Analyst 119, 1497–1509.

Fernando, W.M.A.D.B., Ranaweera, K.K.D.S., Bamunuarachchi, A., Brennan, C.S., 2008. The influence of rice fibre fractions on the *in vivo* fermentation production of short chain fatty acids using human faecal microflora. Int. J. food Sci. technol. 43, 2237–2244.

Goodlad, J.S., Mathers, J.C., 1991. Digestion by pigs of nonstarch polysaccharides in wheat and raw peas (*Pisum sativum*) fed in mixed diets. Br. J. Nutr. 65, 259–270.

Gooneratne, J., Majsak-Newman, G., Robertson, J.A., Selvendran, R.R., 1994a. Investigation of factors that affect the solubility of dietary fiber, as nonstarch polysaccharides, in seed tissues of mung bean (*Vigna radiata*) and black gram (*Vigna mungo*). J. Agric. Food Chem. 42, 605–611.

Gooneratne, J., Needs, P.W., Ryden, P., Selvendran, R.R., 1994b. Structural features of cell-wall polysaccharides from the cotyledons of mung bean *Vigna radiata*. Carbohydr. Res. 265, 61–77.

Górecka, D., Lampart-Szczapa, E., Janitz, W., Sokolowska, B., 2000. Composition of fractional and functional properties of dietary fiber of lupines (*L. luteus* and *L. albus*). Nahrung 44, 229–232.

Harris, P.J., Ferguson, L.R., 1999. Dietary fibres may protect or enhance carcinogenesis. Mutat. Res. 443, 95–110.

Harris, P.J., Smith, B.G., 2006. Plant cell walls and cell-wall polysaccharides: structures, properties and uses in food products. Int. J. Food Sci. Technol. 41, 129–143.

Jenkins, D.J.A., Taylor, R.H., Goff, D.V., 1981. Scope and specificity of acarbose in slowing carbohydrate-absorption in man. Diabetes 30, 951–954.

Key, F.B., Mathers, J.C., 1993. Complex carbohydrate digestion and large-bowel fermentation in rats given wholemeal bread and cooked haricot beans (*Phaseolus vulgaris*) fed in mixed diets. Br. J. Nutr. 69, 497–509.

Lairon, D., 1996. Dietary fibres: effects on lipid metabolism and mechanisms of action. Eur. J. Clin. Nutr. 50, 125–133.

Lintas, C., Cappelloni, M., Bonmassar, L., Clementi, A., Deltoma, E., Ceccarelli, G., 1995. Dietary fiber, resistant starch and *in-vitro* starch digestibility of cereal meals – glycemic and insulinemic responses in NIDDM patients. Eur. J. Clin. Nutr. 49, S264–S267.

Lundin, L., Golding, M., Wooster, T.J., 2008. Understanding food structure and function in developing food for appetite control. Nutr. Dietet. 65, S79–S85.

Lyly, M., Soini, E., Rauramo, U., Lahteenmaki, L., 2004. Perceived role of fibre in a healthy diet among Finnish consumers. J. Hum. Nutr. Dietet. 17, 231–239.

Marconi, E., Ruggeri, S., Cappelloni, M., Leonardi, D., Carnovale, E., 2000. Physicochemical, nutritional, and microstructural characteristics of chickpeas (*Cicer arietinum* L.) and common beans (*Phaseolus vulgaris* L.) following microwave cooking. J. Agric. Food Chem. 48, 5986–5994.

McDougall, G.J., Morrison, I.M., Stewart, D., Hillman, J.R., 1996. Plant cell walls as dietary fiber: range, structure, processing and function. J. Sci. Food Agric. 70, 133–150.

Perez-Hidalgo, M.A., Guerra-Hernandez, E., Garcia-Villanova, B., 1997. Dietary fiber in three raw legumes and processing effect on chick peas by an enzymatic-gravimetric method. J. Food Compos. Anal. 10, 66–72.

Periago, M.J., Ros, G., Martinez, C., 1996. Relationships between physical-chemical composition of raw peas and sensory attributes of canned peas. J. Food Qual. 19, 91–106.

Periago, M.J., Ros, G., Casas, J.L., 1997. Non-starch polysaccharides and *in vitro* starch digestibility of raw and cooked chick peas. J. Food Sci. 62, 93–96.

Plaami, S.P., 1997. Content of dietary fiber in foods and its physiological effects. Food Rev. Int. 13, 29–76.

Rehman, Z.-u., Shah, W.H., 2004. Domestic processing effects on some insoluble dietary fibre components of various food legumes. Food Chem. 87, 613–617.

Roberfroid, M.B., 2001. Prebiotics: preferential substrates for specific germs? Am. J. Clin. Nutr. 73 (2), 406S–409S.

Rose, J.K.C., Bennett, A.B., 1999. Cooperative disassembly of the cellulose–xyloglucan network of plant cell walls: parallels between cell expansion and fruit ripening. Trends Plant Sci. 4, 176–183.

Sandberg, A.S., Andersson, H., Hallgren, B., Hasselblad, K., Isaksson, B., 1981. Experimental model for *in vivo* determination of dietary fibre and its effect on the absorption of nutrients in the small intestine. Br. J. Nutr. 45, 283–294.

Sasaki, T., Kohyama, K., Yasui, T., 2004. Effect of water-soluble and insoluble non-starch polysaccharides isolated from wheat flour on the rheological properties of wheat starch gel. Carbohydr. Polym. 57, 451–458.

Selvendran, R.R., Robertson, J.A., 1990. The chemistry of dietary fibre – an holistic view of cell wall matrix. In: Southgate, D.A.T. (Ed.), Dietary fibre: chemical and biological aspects. Woodhead Publishing, Cambridge.

Sosulski, F.W., Wu, K.K., 1988. High fiber breads containing field pea hulls, wheat, corn, and wild oat brans. Cereal Chem. 65, 186–191.

Srisuma, N., Ruengsakulrach, S., Uebersax, M.A., Bennink, M.R., Hammerschmidt, R., 1991. Cell-wall polysaccharides of navy beans (*Phaseolus vulgaris*). J. Agric. Food Chem. 39, 855–858.

Stephen, A.M., Dahl, W.J., Sieber, G.M., van Blaricom, J.A., Morgan, D.R., 1995. Effect of green lentils on colonic function, nitrogen balance, and serum lipids in healthy human subjects. Am. J. Clin. Nutr. 62, 1261–1267.

Topping, D., 2007. Cereal complex carbohydrates and their contribution to human health. J. Cereal Sci. 46, 220–229.

Topping, D.L., 1991. Soluble fiber polysaccharides: effects on plasma cholesterol and colonic fermentation. Nutr. Rev. 49, 195–203.

Tudoricã, C.M., Kuri, V., Brennan, C.S., 2002. Nutritional and physico-chemical characteristics of dietary fiber enriched pasta. J. Agric. Food Chem. 50, 347–356.

Vidal-Valverde, C., Frias, J., 1991. Legume processing effects on dietary fiber components. J. Food Sci. 56, 1350–1352.

Willats, W.G.T., McCartney, L., Mackie, W., Knox, J.P., 2001. Pectin: cell biology and prospects for functional analysis. Plant Mol. Biol. 47, 9–27.

Post-harvest technology of pulses

7

Rangarajan Jagan Mohan[1], Arumugam Sangeetha[1],
Hampapur V. Narasimha[2], Brijesh K. Tiwari[3]
[1]Department of Food Product Development, Indian Institute of Crop Processing Technology, Thanjavur, India
[2]Department of Grain Processing, Central Food Technological Research Institute, Mysore, India
[3]Department of Food and Tourism, Manchester Metropolitan University, Manchester, UK

7.1   Introduction

Post-harvest handling of pulses plays a significant role on the level of quantitative and qualitative losses experienced in the grain-food supply chain. The post-harvest requirements of pulses are diverse and include threshing, storage and cleaning, whether by traditional, i.e. manual, or more advanced methods, i.e. mechanized and/or with controlled atmospheric storage. Post-harvest technology combines the use of different unit operations, which taking place after harvesting. The sequence of activities and operations that make up a post-harvest technology can be divided into various stages as shown in Figure 7.1, with each post-harvest stage involving critical factors which affect qualitative and quantitative losses. These losses occur at all stages in the post-harvest system from harvesting, through handling, storage, processing and marketing to final delivery to the consumer.

Pulse Foods: Processing, Quality and Nutraceutical Applications. DOI: 10.1016/B978-0-1238-2018-1.00007-0

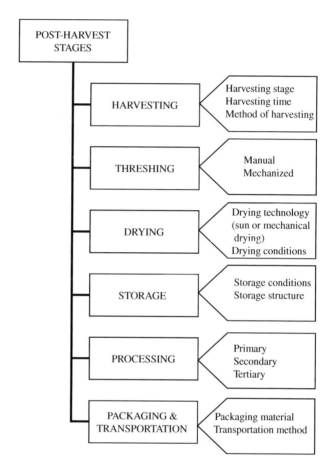

Figure 7.1 Important stages of post-harvest systems for pulses.

The post-harvest system should be thought of as encompassing the delivery of a crop from the time and place of harvest to the time and place of consumption, with minimum loss, maximum efficiency and maximum return for all involved.

Spurgeon

Therefore, post-harvest loss of pulses indicates any measurable reduction in terms of quality or quantity, which could be economic (e.g. reduction in money value), quantitative (e.g. moisture loss), qualitative (e.g. hard-to-cook phenomena, discoloration of pulses due to microbial attack) or nutritional (e.g. protein digestibility). These losses may occur individually or as a combination of premature harvest, poor maturation, poor threshing, insufficient drying and cleaning, and pests (birds, rodents, insects or microorganisms). The remainder of this chapter will look into the effective use of post-harvest technologies with pulses and how they can be used in a manner that will reduce product losses.

7.2 Post-harvest losses

Pulse production systems including pre-harvest and post-harvest techniques are mechanized in developed countries but developing countries still employ traditional techniques which incur losses of about 20–25% post-harvest losses (Maneepun, 2003). Pre-harvest losses can be influenced by the harvesting time of pulses, which itself is determined by the degree of pulse maturity and varies according to the pulse variety. Harvesting prior to maturity leads to lower yield and inferior quality grain whereas delay in harvesting may result in shattering of pods leading to losses. Moisture content of pulses during harvesting is one of the important factors that affect subsequent grain drying and storage. Traditionally, grains are dried in fields which ensures optimum moisture prior to storage but a potential risk is associated quantitative losses due to birds, insect infestation, rodents, etc. Threshing of pulses after harvesting is mainly done to separate edible grains from the pod. Threshing should be performed at optimum moisture. Threshing at high moisture content of pulses may pose difficulty in removal of chaffy material during cleaning and may be the source of microbial cross-contamination. Further, granary or bag storage of moist grains will be susceptible to fungal attack leading to qualitative and quantitative losses.

Even though pre-harvest losses can be significant, post-harvest losses in pulses account for up to 25% depending upon different extrinsic and intrinsic factors. Most of these losses are due to improper post-harvest management; among these grain handling, transportation and storage are the most important factors influencing post-harvest losses in pulses. Improper storage conditions such as temperature, moisture and oxygen content influence the grain quality. For example, the cooking quality of beans is related to post-harvest handling and storage conditions (Yousif et al., 2007). Pulses require long storage duration and losses during storage are mainly due to external factors (moisture, temperature and pests) compared to perishable commodities like fruits and vegetables, where both intrinsic and extrinsic factors are equally responsible for post-harvest losses. Moreover, as will be seen in the following sections, drying and handling prior to storage will all affect the quality of the grain and its suitability for each potential pest (Cox and Collins, 2002).

7.3 Drying of pulses

Drying of pulses takes place immediately after harvesting, before threshing (pre-drying) and during storage and/or primary processing. Drying is an important post-harvest operation for safe storage, processing and grain quality preservation. The sole objective of grain drying is to remove excess moisture. Inadequate drying results in excess grain moisture and coupled with temperature favors microbial growth and enzymatic activity leading to degradation of grain quality. Drying and conditioning of pulses are mostly done by artificial methods; however, the most common drying method for pulses in the world is open sun drying, particularly at farm level. In some tropical and subtropical countries, sun drying is performed, particularly field drying. Most of the pulse cultivars are non-synchronous and pod maturity extends over long periods. Moreover, pods also have a tendency to shatter when left for a long period of time before they reach complete maturity. This non-synchronous maturity of pulses leads to non-uniformity in grain size, making it necessary to spread the grain in thin layers on paved ground and expose it to sun and wind for drying.

Carefully controlled artificial drying, as practiced in temperate regions, is an expensive process and its use in the tropics is limited to some export crops (Aidoo, 1993). The most important advantage of sun drying is the low cost but there are several disadvantages which include prolonged drying time, high labor requirement, weather dependence and potential exposure to environmental contamination. Further, it is not always feasible to reduce the moisture content of grains to safe levels in the field, which necessitates the artificial removal of moisture from grains (Pabis et al., 1998; Jayas and White, 2003). In general, sun-dried products do not fulfil the international quality standards and therefore cannot be sold on international markets (Esper and Miihlbauer, 1998). Drying of pulses is essential because the moisture content at the time of harvesting is on the higher side (18–25%) and, for safe storage, the optimum moisture content is in the range of 9–12% to avoid production of mycotoxins. It is essential that grain is dried to a safe moisture level as quickly as possible regardless of the drying system employed. There are different methods of artificial drying with hot air such as fixed/moving-bed drying, fluidized-bed drying, spouted-bed drying, thin-layer drying, etc. (Kundu et al., 2005). General food grain dryers can also be used for drying of pulses. Jayas and White (2003) classified dryers based on:

1. Flow of grain (batch, recirculating and continuous)
2. Relative motions of grain and drying air (concurrent, counter-current, cross and mixed flow)
3. Source of heat (solar, propane and electrical).

Irrespective of the type of dryer involved, the drying process of pulses, like other grains, includes simultaneous heat and moisture transfer. Drying of grains involves removal of free moisture, i.e. grain is dried until it reaches its equilibrium, known as equilibrium moisture content. The equilibrium moisture content is the final moisture content of the grain at a certain temperature and relative humidity. The important factors influencing drying of grains are temperature and moisture content of the grain and, to some extent, relative humidity and air velocity (Kundu et al., 2005).

Drying kinetics of biological materials have been studied using a number of empirical and semi-empirical equations; among these Lewis (1921) (Equation 7.1), Brooker et al. (1974) (Equation 7.2), Henderson and Henderson (1968) (Equation 7.3), Page (1949) (Equation 7.4) and Overhults et al. (1973) (Equation 7.5) are applicable for studying drying kinetics of grains. These model equations are generally derived by simplifying general series solution of Fick's second law. Equation 7.1 is similar to Newton's law of cooling whereas Equations 7.2 and 7.3 are simplifications of the diffusion model. Equations 7.4 and 7.5 are derived from Levi's model with slight modification.

$$MR = exp(-Kt) \quad \text{where } K = A \, exp(-B/T) \tag{7.1}$$

$$MR = C \, exp(-Kt) \quad \text{where } K = A \, exp(-B/T) \tag{7.2}$$

$$MR = C[exp(-Kt) + (1/9)exp(-9Kt)]$$
$$\text{where } K = A \, exp(-B/T) \tag{7.3}$$

$$MR = exp[-(Kt)d] \quad \text{where } K = exp(-B/T) \tag{7.4}$$

$$MR = exp[-(Kt)d] \quad \text{where } K = exp(A + B/T) \tag{7.5}$$

where $MR = M_t - M_e M_o - M_e$ is a moisture ratio, M_t is moisture content at time t, M_o is initial moisture content, M_e is equilibrium moisture content, K is a drying constant which follows the Arrhenius model except for Equations 7.4 and 7.5, and A, B, C and d are constants.

The drying rate of grains can be calculated using Equation 7.6, which is expressed as rate of change of moisture with respect to time.

$$DR = -\frac{dm}{dt} = K(m_t - m_e) \qquad (7.6)$$

The variation of the drying rate during the drying of grains can be expressed as constant and falling drying rates, and the critical moisture content can be easily obtained by plotting drying rate against moisture content. Statistical models can be developed by using an appropriate equilibrium moisture content based on a multiple regression method for constant drying rate as a function of drying air temperature (Kundu et al., 2005) using the Arrhenius relationship. Knowledge of the drying kinetics of biological materials under various process conditions is extremely important for both the design of dryers and modeling of drying processes (Mathioulakis et al., 1998; Xia and Sun, 2002; Barrozo et al., 2004).

7.4 Storage of pulses

Storage is an integral part of any food grain processing chain. Storage of raw and processed grains, including legumes, has been used by humans since the beginning of history as a prerequisite for insuring food security due to off time availability, and for withholding seed grain for long periods. Grain storage and handling is a major concern for pulse growers and processors worldwide. Pulses can remain in an edible condition for several years if properly stored. However, during ambient storage conditions, grains are subjected to various physical, chemical and biological changes. The basic advantage of good storage is to create environmental conditions which protect the product from the extremes of outside temperature and relative humidity fluctuations, both diurnal and seasonal, and maintain its quality (Cox and Collins, 2002).

7.4.1 Storage systems

A food grain storage system is an artificial ecological system in which deterioration of the stored product results from interactions among physical (e.g. temperature, moisture, storage structure), chemical (e.g. carbon dioxide, oxygen) and biological (e.g. grain characteristics, microorganisms, insects, mites, rodents, birds) factors (Jayas and White, 2003). Both pulse grain quality and quantity are affected by intrinsic and extrinsic factors (Fig. 7.2) and, among these, temperature and moisture content are the most important influences of shelf-life. Grain viability and reproduction of biological agents in grain are dependent to a great extent on the temperature and moisture levels (White, 1995). Improper storage conditions influence the post-harvest storability of pulses. Various pests and microorganisms attack the pulses and their products after harvest, in storage, during transportation to the market, etc. The application of pesticides results in chemical residues in the food that are extremely hazardous to health. Establishment of efficient and effective post-production storage systems is warranted in order to minimize qualitative and quantitative losses.

A storage system is like an ecosystem where quality changes gradually as a result of complex interactions (Multon, 1989; Tipples, 1995; Fleurat-Lessard, 2002), as shown in Figure 7.3. Fleurat-Lessard

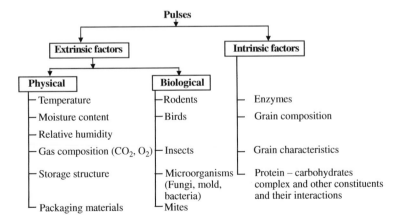

Figure 7.2 Factors influencing grain quality and quantity.

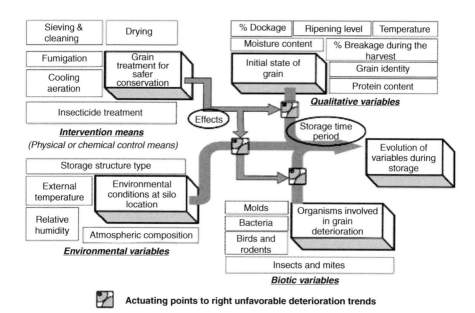

Figure 7.3 Simplified functional diagram of the complex relationships existing in the stored grain ecosystem (Fleurat-Lessard, 2002).

(2002) suggested that parameters involved in quality changes can be classified into the following categories:

1. Initial condition when grain is delivered to the grain store
2. Storage conditions between harvest and first processing
3. Technical operations and treatments of the grain batch, which is called the storage technical route
4. Biological deterioration factors (also called "deteriorative forces"), mainly represented by storage fungi and invertebrate pests (insects and mites).

Storage of pulses is done at several stages of the pulse production chain, mainly on the farm, in warehouses and in other commercial storage facilities at grain processors prior to processing. Both raw whole grains (unprocessed) and processed grains such as dehulled whole, split (dhals) and unhulled splits are stored. The major causes of loss of quality and weight loss in stored grain are the activities of insects, mites, rodents, birds and fungi.

7.4.2 Storage pests

Pulse grain infestation by pests, microbial spoilage or contamination is a significant problem. Infestation of grains during storage
may make grains totally inedible through associated microbial spoilage and contamination (Neethirajan et al., 2007). Insect damage in
pulses may amount to 10–40% in countries where modern storage
technologies have not been introduced (Shaaya et al., 1997). Most
of the qualitative and quantitative losses during storage are due to
Callosobruchus species (Salunke et al., 2005). Several control strategies have been reported for control of insect pests during storage of
pulses. Until recently, several insecticides and fumigation techniques
were used for storage pest management. Currently, commonly used
pesticides (fumigants) for grain storage include aluminum phos-
phide, methyl bromide and phosphine (Tiwari et al., 2010). Among
these, the use of methyl bromide has been nearly phased out as agreed
in the Montreal Protocol. The use of chemicals is now restricted due
to the growing concern over the use of harmful pesticides to kill
storage pests. The Montreal Protocol on substances that deplete the
ozone layer (Fields and White, 2002), increased insect resistance
and increased consumer demand for chemical-free grains have led
grain processors to seek alternatives to control storage pests. The
persistent use of these pesticides has been reported to disrupt biological control systems by natural agents, leading to outbreaks of
insect pests, widespread development of resistance, undesirable effects on non-target organisms, and environmental and human health
concerns (Kells et al., 2001; Collins et al., 2005; Pimentel et al.,
2007, 2009; Islam et al., 2009). The increasing concern about their
adverse effects has highlighted the need for the development of selective insect control alternatives (Fields and White, 2002). Some
studies show that pulses contain a wide range of allelochemicals
with toxic and deterrent effects against insect pests (Harborne et al.,
1971; Bell, 1978; Fields et al., 2001). Fields et al. (2001) reported
repellence behaviors of pea fractions (fibers and proteins) against
storage pests. Similarly, Bodnaryk et al. (1999) observed toxicity,
oviposition deterrency and reproduction inhibition of pea. Several other alternatives to synthetic chemicals, focusing on the use
of natural pesticides such as plant volatile oils, have been used in
the protection of stored grains (Saxena and Mathur, 1976; El-Nahal
et al., 1989; Risha et al., 1990; Gbolade and Adebayo, 1993; Raja
et al., 2001).

Among microorganisms, growth of bacteria is restricted due to the low moisture content whereas insects and molds are major spoilage organisms in stored grain ecosystems. Insects and molds require aerobic conditions for their growth and development. Therefore, creating an anaerobic atmosphere in the stored grain ecosystems has a lethal effect on insects and molds and extends the storageability of grains (Jayas and Jeyamkondan, 2002). Storage of grains including pulses at higher levels of carbon dioxide (>60%) and low levels of oxygen (<10%) with moderate humidity (<50%) and high temperature (>27%) is lethal to most of the storage pests responsible for qualitative and quantitative losses; however, these conditions may vary depending upon the nature and size of the infestation (Jayas and Jeyamkondan, 2002).

Ozone as a fumigant is reported to kill stored grain insects such as *Tribolium castaneum*, *Rhyzopertha dominica*, *Oryzaephilus surinamensis*, *Sitophilus oryzae* and *Ephestia elutella* (Sousa et al., 2008). Laboratory and field studies report the efficacy of ozone in controlling both phosphine-susceptible and phosphine-resistant strains of *Sitophilus zeamais*, *S. oryzae*, *R. dominica* and *T. castaneum* (Qin et al., 2003). Ozone toxicity for insects varies depending on the stage within their life cycle. To date, application of ozone has been extensively investigated for cereals. However, studies involving common storage pests for cereals and pulses indicate that ozone is an alternative method of grain disinfestation which is environmentally benign, resulting in no toxic residues. However, the effectiveness of ozone depends on several factors including the amount of ozone applied, various environmental factors such as grain mass temperature, moisture and the surface characteristics (Tiwari et al., 2010).

7.4.3 Detection of storage pests

Detection and removal of internal insects from grains are important control measures for ensuring storage longevity, seed quality and food safety (Neethirajan et al., 2007). Currently, manual inspection, sieving, cracking-flotation and Berlese funnels are used to detect insects in grain handling facilities (Gowen et al., 2010). However, these methods are not efficient and are time consuming (Neethirajan et al., 2007) with poor accuracy for the developing life stages of pests. Infestation of grains by hidden insects can also be detected by staining of kernels to identify entrance holes for eggs, flotation, radiographic techniques, acoustic techniques, uric acid measurement, nuclear magnetic resonance imaging and immuno-assays (Neethirajan et al., 2007). Detection of pest infestation in food commodities and in their storage premises is essential to ensure wholesome and acceptable produce for human consumption, for regulatory compliance, for diagnosis of incipient infestation and to ascertain the success of control measures such as fumigants (Rajendran, 1999).

The International Standards Organization (ISO 6639-4, 1987) has standardized five methods to determine the hidden insect infestations in pulses, which are also applicable to cereals (Manickavasagan et al., 2008a):

1. Determination of carbon dioxide production method
2. Ninhydrin reaction with amino acids
3. Whole grain flotation method
4. The acoustic method
5. The X-ray method.

Application of thermal imaging for the detection of all insect post-embryonic stages has been investigated in grains. Thermal imaging is based on the temperature difference due to heat production from respiration compared to grain temperature (Emekci et al., 2002, 2004). This temperature difference due to infestation inside a kernel can be detected by thermal imaging. Research to date on the detection of insects in cereal grains (Manickavasagan et al., 2005, 2008a,b) suggests a potential for application in the detection of insects in pulses.

7.5 Effect of post-harvest technology on quality

Pulses are generally subjected to long-term storage. The storage conditions, such as time, temperature and moisture content, induce physicochemical and biological changes and have a significant effect on the nutritional composition, germination and longevity (Burr et al., 1968; Garcia-Vela and Stanley, 1989; Hentges et al., 1991; Menkov, 2000). The presence of insects, fungi and molds raises both temperature and moisture in the storage ecosystem and thereby accelerates the activity of the enzymes, resulting in loss of nutrients.

The cooking quality of pulses is related to post-harvest handling and storage conditions. Storage of beans under unfavorable conditions is reported to influence end-product quality by reducing water uptake during cooking, which leads to poor cell separation, incomplete starch gelatinization, and protein denaturation (Hayakawa and Breene, 1982; Garcia-Vela and Stanley, 1989; Yousif et al., 2007). One of the major causes of deterioration in pulse quality is poor cookability (ability to soften with cooking), followed by deterioration of color, texture and loss of nutritive value (Martin-Cabrejas et al., 1997; Yousic and Deeth, 2002; Yousif et al., 2003; Nasar-Abbas et al., 2008). Though the cooking time of pulses depends strongly on genotype (Singh, 1999) the hard-to-cook phenomenon as a function of extended storage time and high temperature is a common characteristic across various pulses (Yousif et al., 2007; Nasar-Abbas et al., 2008, 2009). The hard-to-cook (HTC) defect that develops in dry beans stored under conditions of high temperature and high humidity has been investigated extensively. Nasar-Abbas et al. (2008) observed a linear increase in the HTC state with storage temperature of faba beans. They observed that faba beans stored for an extended period at high temperature ($\geq 37\,^{\circ}\text{C}$) develop a harder texture after cooking compared to beans stored at $25\,^{\circ}\text{C}$ for the same time period of 12 months. Similarly, Reyes-Moreno et al. (2001) also reported that chickpeas developed the HTC defect during storage at high temperatures ($>25\,^{\circ}\text{C}$) and high relative humidities (RH $> 65\%$). The HTC defect in pulses increases cooking time, thus needing additional energy during cooking, and this is one of the factors responsible for wider under-utilization of pulses (Deshpande et al., 1982). The HTC defect is attributed to several mechanisms; "pectin-phytate" and "lignification" (cell wall thickening) are possibly responsible for the HTC defect (Hincks and Stanley, 1986). Hardening of pulses during high temperature and relative humidity is caused by phytate hydrolysis and strengthening of cell walls and results in increased difficulty in achieving cell separation during cooking (Stanley and Aguilera, 1985; Del Valle and Stanley, 1995). The scanning electron micrographs of cell wall isolates of common bean stored at $5\,^{\circ}\text{C}/40\%$ RH (Fig. 7.4A) and $35\,^{\circ}\text{C}/75\%$ RH (Fig. 7.4B) show a thickening of the walls at the cell junction and confirm the possible lignification of the middle

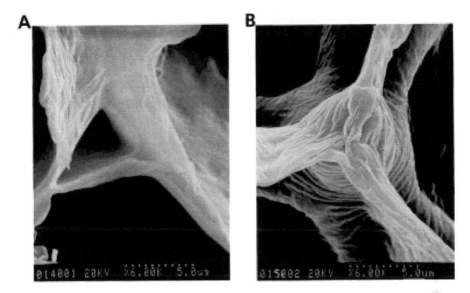

Figure 7.4 Scanning electron micrographs of isolated cell walls of (A) control (5 °C/40% RH) and (B) HTC (35 °C/75% RH) bean cotyledons after storage period of 6.5 months (Garcia et al., 1998).

lamella as an explanation for the HTC of beans (Garcia et al., 1998). Water absorption, soluble solids and electrolyte leaching are also important quality factors associated with the bean hardness defect (Berrios et al., 1999) and are good indicators of the loss in bean quality during storage.

Nasar-Abbas et al. (2008) investigated physical and chemical characteristics of faba beans after 12 months' storage at 5, 15, 25, 37, 45 or 50 °C. They observed a significant decrease in hydration and swelling coefficient with an increase in electric conductivity (μS cm^{-1}), solute leaching (mg g^{-1}) and hardness (N seed^{-1}) (Table 7.1). Similarly, Liu et al. (1992) found an increase in hardness of cowpea during 18 months at 30 °C/64% RH. Research studies show that long duration storage of pulses at elevated temperature and humidity results in a reduction of the hydration capacity, reduction of cooking quality (increase in cooking time), darkening, development of off-flavor and increased lipid acidity (Burr et al., 1968; Antunes and Sgarbrieri, 1979; Kon and Sanshuck, 1981) in common beans. However, bean quality can be maintained at low storage temperature (10–20 °C), moderate relative humidity (50% RH) and safe moisture content (10– 14%) (Yousif et al., 2007).

Table 7.1 Changes in physical properties of faba beans stored at different temperatures for 12 months

Storage temperature (°C)	Hydration coefficient	Swelling coefficient	Electrical conductivity (μS cm^{-1})	Solute leaching (mg g^{-1})	Hardness (N seed^{-1})
5 (Control)	193±3[a]	215±4[a]	827±23[a]	3.8±0.4[a]	3.3±0.2[a]
15	191±2[a]	210±2[ab]	905±19[a]	6.6±0.5[ab]	5.2±0.3[b]
25	188±2[a]	209±2[ab]	1115±55[b]	7.4±0.2[b]	7.1±0.2[c]
37	174±2[b]	201±3[b]	2523±49[c]	18.1±0.7[c]	10.7±0.2[d]
45	170±3[b]	189±4[c]	3216±61[d]	24.8±1.8[d]	13.7±0.3[e]
50	158±2[c]	175±2[d]	3467±60[e]	36.1±1.8[e]	15.2±0.3[f]

Means (±S.E., n=3 except hardness where n=30) sharing the same letter in the column are non-significant (P=0.05) according to Tukey's HSD test.
Source: Nasar-Abbas et al. (2008).

Apart from cooking quality and HTC defect, loss of water-soluble proteins has also been reported during storage at higher temperature and relative humidity. Reduction in water-soluble proteins is attributed to a decrease in tissue pH during storage (Saio et al., 1980; Liu et al., 1992). Hardening of pulses has an unfavorable effect on protein quality and significantly decreases protein efficiency ratio, net protein ratio, *in vivo* and *in vitro* protein digestibilities and availability of essential amino acids such as methionine, lysine and tryptophan (Uma-Reddy and Pushpamma, 1986; Sievwright and Shipe, 1986; Tuan and Phillips, 1991; Reyes-Moreno et al., 2001). There were significant decreases in protein digestibility, protein efficiency ratio and lightness of chickpea, with significant increases in starch gelatinization temperature (T_g, °C) and enthalpy (ΔH_g, Jg^{-1}) during a storage period of 160 days at high temperature and high relative humidity (33–35 °C, RH 75%) (Reyes-Moreno et al., 2001). Table 7.2 shows the effect of storage at high temperature and high relative humidity (33–35 °C, RH 75%, 160 days) on the some of the quality characteristics of chickpea seeds.

Granito et al. (2008) investigated the effect of three storage conditions, i.e. 30 °C/11% RH, 30 °C/80% RH and 50 °C/80% RH for 150 days on total polyphenol content and antioxidant capacity of beans. They observed that the combination of high temperatures and high relative humidities during the post-harvest storage of the beans negatively affects the antioxidant capacity, whereas storage at 30% and 11% relative humidity allows preservation of the antioxidant activity of the beans. They also observed a significant increase in total

Table 7.2 Effect of storage[1] on functional, physicochemical and nutritional characteristics of Blanco Sinaloa chickpea seeds

Characteristic	Fresh	Stored
Water activity	9.50 ± 0.24^b	15.51 ± 0.30^a
Water activity	0.454 ± 0.1^b	0.752 ± 0.01^a
Water absorption capacity (%)		
Whole grains	115 ± 0.9^a	98 ± 1.7^b
Cotyledons	90 ± 0.2^a	79 ± 1.2^b
Seed coats	1072 ± 12^a	902 ± 9.7^b
Cooking time (min)		
Whole grains	142 ± 4.2^b	213 ± 3.9^a
Cotyledons	81 ± 1.4^b	116 ± 4.1^a
Seed coats	104 ± 2.8^b	217 ± 8.5^a
Hardness (N per seed)	5.2 ± 0.2^b	12.1 ± 0.3^a
Color		
Hunter L value	64.1 ± 0.7^a	55.6 ± 0.7^b
Total color difference	37.9 ± 1.0^b	44.4 ± 0.7^a
T_g (°C)[2]	67.8 ± 0.6^b	70.1 ± 0.4^a
ΔH_g (J g^{-1} sample)[2]	3.28 ± 0.2^b	4.03 ± 0.3^a
Protein digestibility (%)		
Apparent	82.7 ± 0.6^a	77.8 ± 0.8^b
True	84.9 ± 0.7^a	78.5 ± 0.8^b
PER[3]	1.79 ± 0.04^a	1.46 ± 0.01^b

[1]33–35 °C, RH 75%, 160 days.
[a,b]Means with the same letters are not significantly different (P ≤ 0.05).
[2]T_g, starch gelatinization temperature; ΔH_g, starch gelatinization enthalpy.
[3]Protein efficiency ratio = weight gain (g)/protein consumed (g); casein PER = 2.5.
Source: Reyes-Moreno et al. (2001).

phenol content during the first 90 days of storage at 30 °C/80% RH and 50 °C/80% RH which decreased beyond 90 days. The increase in polyphenol content might be due to the deamination of the aromatic amino acids L-phenylalanine and L-tyrosine, catalyzed by the phenylalanine and tyrosine ammonium lyase, to produce hydrocinnamic acid (Srisuma et al., 1989). Garcia et al. (1998) also reported an increase in free phenolics during adverse storage conditions (35 °C/75% RH). Shiga et al. (2011) investigated the effect of storage conditions (30 °C/75% RH) for 8 months on discoloration of common beans and observed a visible discoloration of the hulls, turning them a bright dark brown color (Fig. 7.5). Observed discoloration of the hull by darkening may be due to polymerization of phenolic compounds. Shiga et al. (2011) also observed a significant

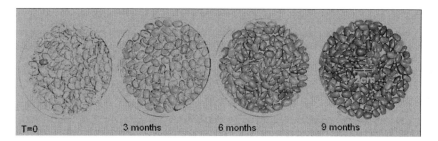

Figure 7.5 Changes in common bean hull color during storage (30 °C/75% RH) (Shiga et al., 2011).

decrease of about 1% in water-soluble polysaccharides during 8 months of storage suggesting that the dietary fiber profile may be altered during adverse storage conditions.

7.6 Conclusions

Post-harvest handling of pulses plays an important role in pulse production systems and in food supply. Both quantitative and qualitative food losses of extremely variable magnitude occur at all stages in the post-harvest system from harvesting, through handling, storage, processing and marketing to final delivery to the consumer. Drying is a particularly important post-harvest operation for safe storage, processing and preservation of pulses. Pulses are harvested at higher moisture, therefore, drying of pulses to a safe moisture level is critical for improving the storage ability of pulses. Improper storage may provide favorable conditions for storage pests, such as insects and microorganisms. Pulses are more difficult to store than cereals and they suffer much greater damage from insects and microorganisms. This results not only in quantitative losses, but also in qualitative reduction of the nutritive value, cooking quality and protein quality. Most of these losses are the outcome of improper post-harvest management such as unsuitable storage conditions, preliminary processing and processing. Post-harvest management of pulses is a complex problem requiring a multidisciplinary approach; there is a need to combine the current and innovative technologies to minimize the post-harvest losses. Post-harvest handling, transportation and storage are the most important factors influencing post-harvest losses. Improper storage conditions, such as temperature, moisture and oxygen content, influence the post-harvest storability of legumes and affect grain quality. Various pests such as insects and microorganisms attack pulses and their legume products after harvest, in storage, during transportation to the market, etc. The application of pesticides results in chemical residues in the food that are extremely hazardous to health. Recent research focus for developing greener and environmentally friendly approaches such as gas storage or using ozone as a fumigant for pulse grain preservation should continue. Establishment of efficient and effective post-production storage systems should complement the integrated sustainable crop production system. An integrated, environmentally friendly, safe pest management system, and proper pre- and post-harvest handling of the crops, would minimize the post-production losses and would improve the quality of pulses for utilization.

References

Aidoo, K.E., 1993. Post-harvest storage and preservation of tropical crops. Int. Biodeterior. Biodegradation 32, 161–173.

Antunes, P.L., Sgarbrieri, V.C., 1979. Influence of time and conditions of storage on technological and nutritional properties of a dry bean (*Phaseolus vulgaris* L.) variety Rosinha G2. J. Food Sci. 44, 1703–1706.

Barrozo, M.A.S., Sartori, D.J.M., Freire, J.T., 2004. A study of the statistical discrimination of the drying kinetics equations. Food Bioprod. Process. 82, 219–225.

Bell, E.A., 1978. Toxins in seeds. In: Harborne, J.B. (Ed.), Biochemical aspects of plant and animal coevolution. Academic Press, New York, pp. 143–161.

Berrios, J.D.J., Swanson, B.G., Cheong, W.A., 1999. Physicochemical characterisation of stored black beans (*Phaseolus vulgaris* L.). Food Res. Int. 32, 669–676.

Bodnaryk, R., Fields, P., Xie, Y., Fulcher, K., 1999. Insecticidal factors from field pea. US Patent 5,955,082.

Brooker, D.B., Bakker-Arkema, F.W., Hall, C.W., 1974. Drying cereal grains. AVI, Westport.

Burr, H.K., Kon, S., Morris, H.J., 1968. Cooking rates of dry beans as influenced by moisture content and temperature and time of storage. Food Technol. 22, 88–90.

Collins, P.J., Daglish, G.J., Pavic, H., Kopittke, R.A., 2005. Response of mixed-age cultures of phosphine-resistant and susceptible strains of lesser grain borer, *Rhyzopertha dominica*, to phosphine at a range of concentrations and exposure periods. J. Stored Prod. Res. 41, 373–385.

Cox, P.D., Collins, L.E., 2002. Factors affecting the behaviour of beetle pests in stored grain, with particular reference to the development of lures. J. Stored Prod. Res. 38, 95–115.

Del Valle, J.M., Stanley, D.W., 1995. Reversible and irreversible component of bean hardening. Food Res. Int. 28, 455–463.

Deshpande, S.S., Sathe, S.K., Salunkhe, D.K., Cornforth, D.P., 1982. Effects of dehulling on phytic acid, polyphenols and enzyme inhibitors of dry beans (*Phaseolus vulgaris*). J. Food Sci. 47, 1846–1850.

El-Nahal, A.K.M., Schmidt, G.H., Risha, E.M., 1989. Vapours of *Acorus calamus* oil: a space treatment for stored product insects. J. Stored Prod. Res. 25, 211–216.

Emekci, M., Navarro, S., Donahaye, E., Rindner, M., Azrieli, A., 2002. Respiration of *Tribolium castaneum* (Herbst) at reduced oxygen concentrations. J. Stored Prod. Res. 38, 413–425.

Emekci, M., Navarro, S., Donahaye, E., Rindner, M., Azrieli, A., 2004. Respiration of *Rhyzopertha dominica* (F.) at reduced oxygen concentrations. J. Stored Prod. Res. 40, 27–38.

Esper, A., Miihlbauer, W., 1998. Solar drying – an effective means of food preservation. Renewable Energy 15, 95–100.

Fields, P.G., White, N.D.G., 2002. Alternatives to methyl bromide treatments for stored-product and quarantine insects. Ann. Rev. Entomol. 47, 331–359.

Fields, P.G., Xie, Y.S., Hou, X., 2001. Repellent effect of pea (*Pisum sativum*) fractions against stored-product insects. J. Stored Prod. Res. 37, 359–370.

Fleurat-Lessard, F., 2002. Qualitative reasoning and integrated management of the quality of stored grain: a promising new approach. J. Stored Prod. Res. 38, 191–218.

Garcia, E., Filisetti, T.M.C.C., Udaeta, J.E.M., Lajolo, F.M., 1998. Hard-to-cook beans (*Phaseolus vulgaris*): involvement of phenolic compounds and ectates. J. Agric. Food Chem. 46, 2110–2116.

Garcia-Vela, L.A., Stanley, D.W., 1989. Protein denaturation and starch gelatinisation in hard-to-cook beans. J. Food Sci. 5495, 1284–1292.

Gbolade, A.A., Adebayo, T.A., 1993. Fumigant effects of some volatile oils on fecundity and adult emergence of *Callosobruchus maculatus* F. Insect Sci. Applic. 14, 631–636.

Gowen, A.A., Tiwari, B.K., Cullen, P.J., McDonnell, K., O'Donnell, C.P., 2010. Applications of thermal imaging in food quality and safety assessment. Trends Food Sci. Technol. 21, 190–200.

Granito, M., Paolini, M., Pérez, S., 2008. Polyphenols and antioxidant capacity of *Phaseolus vulgaris* stored under extreme conditions and processed. LWT – Food Sci. Technol. 41, 994–999.

Harborne, J.B., Boulter, D., Turner, B.L., 1971. Chemotaxonomy of the Leguminosae. Academic Press, London.

Hayakawa, I., Breene, W.M., 1982. A study on the relationship between cooking properties of adzuki bean and storage conditions. J. Fac. Agr. -Kyushu Univ. 27, 83–88.

Henderson, J.M., Henderson, S.M., 1968. A computational procedure for deep-bed drying analysis. J. Agric. Eng. Res. 13, 87–95.

Hentges, D.L., Weaver, C.M., Nielsen, S.S., 1991. Changes of selected physical and chemical components in the development of the hard-to-cook bean defect. J. Food Sci. 56, 436–442.

Hincks, M.J., Stanley, D.W., 1986. Multiple mechanisms of bean hardening. Food Technol. 21, 731–750.

Islam, M.S., Hasan, M.M., Xiong, W., Zhang, S.C., Lei, C.L., 2009. Fumigant and repellent activities of essential oil from *Coriandrum sativum* (L.) (Apiaceae) against red flour beetle *Tribolium castaneum* (Herbst) (Coleoptera: Tenebrionidae). J. Pest Sci. 82, 171–177.

Jayas, D.S., Jeyamkondan, S., 2002. PH – post-harvest technology: modified atmosphere storage of grains meats fruits and vegetables. Biosyst. Eng. 82, 235–251.

Jayas, D.S., White, N.D.G., 2003. Storage and drying of grain in Canada: low cost approaches. Food Control 14, 255–261.

Kells, S.A., Mason, L.J., Maier, D.E., Woloshuk, C.P., 2001. Efficacy and fumigation characteristics of ozone in stored maize. J. Stored Prod. Res. 37, 371–382.

Kon, S., Sanshuck, D.W., 1981. Phytate content and its effect on cooking quality of beans. J. Food Process Preserv. 5, 169–178.

Kundu, K.M., Das, R., Datta, A.B., Chatterjee, P.K., 2005. On the analysis of drying process. Drying Technol. 23, 1093–1105.

Lewis, W.K., 1921. The rate of drying of solid materials. Ind. Eng. Chem. 13, 427.

Liu, K., McWatters, K.H., Phillips, R.D., 1992. Protein insolubilisation and thermal destabilisation during storage as related to hard-to-cook defect in cowpeas. J. Agric. Food Chem. 40, 2483–2487.

Maneepun, S., 2003. Traditional processing and utilization of legumes. In: Shanmugasundaram, S. (Ed.), Report of the APO Seminar on

Processing and Utilization of Legumes, Japan, 9–14 October 2000. Asian Productivity Organization, Tokyo, pp. 53–62.

Manickavasagan, A., Jayas, D.S., White, N.D.G., Paliwal, J., 2005. Applications of thermal imaging in agriculture – a review. Can. Soc. Eng. Agric. Food Biol. Syst., Paper no. 05-002.

Manickavasagan, A., Jayas, D.S., White, N.D.G., 2008a. Thermal imaging to detect infestation by *Cryptolestes ferrugineus* inside wheat kernels. J. Stored Prod. Res. 44, 186–192.

Manickavasagan, A., Jayas, D.S., White, N.D.G., Paliwal, J., 2008b. Wheat class identification using thermal imaging: a potential innovative technique. Trans. ASABE 51, 649–651.

Martin-Cabrejas, M.A., Esteban, R.M., Perez, P., Maina, G., Waldron, K.W., 1997. Changes in physicochemical properties of dry beans (*Phaseolus vulgaris* L.) during long-term storage. J. Agric. Food Chem. 45, 3223–3227.

Mathioulakis, E., Karathanos, V.T., Belessiotis, V.G., 1998. Simulation of air movement in a dryer by computational fluid dynamics: application for the drying of fruits. J. Food Eng. 36, 182–200.

Menkov, N.D., 2000. Moisture sorption isotherms of chickpea seeds at several temperatures. J. Food Eng. 45, 189–194.

Multon, J.L., 1989. Spoilage mechanisms of grains and seeds in the post-harvest ecosystem, the resulting losses and strategies for the defense of stocks. In: Multon, J.L. (Ed.), Preservation and storage of grains, seeds and their by-products cereals, oilseeds, pulses and animal feed. CBS Publishers, New Delhi, pp. 3–11.

Nasar-Abbas, S.M., Plummer, J.A., Siddique, K.H.M., White, P., Harris, D., Dods, K., 2008. Cooking quality of faba bean after storage at high temperature and the role of lignins and other phenolics in bean hardening. LWT – Food Sci. Technol. 41, 1260–1267.

Nasar-Abbas, S.M., Siddique, K.H.M., Plummer, J.A., 2009. Faba bean (*Vicia faba* L.) seeds darken rapidly and phenolic content falls when stored at higher temperature, moisture and light intensity. LWT – Food Sci. Technol. 42, 1703–1711.

Neethirajan, S., Karunakaran, C., Jayas, D.S., White, N.D.G., 2007. Detection techniques for stored-product insects in grain. Food Control 18, 157–162.

Overhults, D.G., White, G.M., Hamilton, H.E., Ross, I.J., 1973. Drying soybeans with heated air. Trans. ASAE 16, 112–113.

Pabis, S., Jayas, D.S., Cenkowski, S., 1998. Grain drying: theory and practice. John Wiley, New York.

Page, G.E., 1949. Factors influencing the maximum rates of air drying shelled corn in thin-layer. Purdue University, West Lafayette.

Pimentel, M.A.G., Faroni, L.R.D., Guedes, R.N.C., Sousa, A.H., Tótola, M.R., 2009. Phosphine resistance in Brazilian populations of *Sitophilus zeamais* Motschulsky (Coleoptera: Curculionidae). J. Stored Prod. Res. 45, 71–74.

Pimentel, M.A.G., Faroni, L.R.D., Tótola, M.R., Guedes, R.N.C., 2007. Phosphine resistance, respiration rate and fitness consequences in stored-product insects. Pest Manage. Sci. 63, 876–881.

Qin, Z.G., Wu, X., Deng, G., 2003. Investigation of the use of ozone fumigation to control several species of stored grain insects. Adv. Stored Prod. Protect., 846–851.

Raja, N., Albert, S., Ignacimuthu, S., Dorn, S., 2001. Effect of plant volatile oils in protecting stored cowpea *Vigna unguiculata* (L.) walpers against *Callosobruchus maculatus* (F.) (Coleoptera: Bruchidae) infestation. J. Stored Prod. Res. 37, 127–132.

Rajendran, S., 1999. Detection of insect infestation in stored food commodities. J. Food Sci. Technol. 36, 283–300.

Reyes-Moreno, C., Rouzaud-Sandez, O., Carrillo, J.M., Tiznado, J.A.G., Hernandez, L.C., 2001. Hard-to-cook tendency of chickpea (*Cicer arietinum* L.) varieties. J. Sci. Food Agric. 81, 1008–1012.

Risha, E.M., El Nahal, K.M., Schmidt, G.H., 1990. Toxicity of vapours of *Acorus calamus* L. oil to the immature stages of some stored product Coleoptera. J. Stored Prod. Res. 26, 133–137.

Saio, K., Nikkuni, I., Ando, Y., Otsuru, M., Terauchi, Y., Kito, M., 1980. Soybean quality changes during model storage studies. Cereal Chem. 57, 77–82.

Salunke, B.K., Kotkar, H.M., Mendki, P.S., Upasani, S.M., Maheshwari, V.L., 2005. Efficacy of flavonoids in controlling *Callosobruchus chinensis* (L.) (Coleoptera: Bruchidae), a post-harvest pest of grain legumes. Crop Protect. 24, 888–893.

Saxena, B.P., Mathur, A.C., 1976. Loss of fecundity in *Dysdercus koenigii* (F.) due to vapours of *Acorus calamus* oil. Experientia 32, 315–316.

Shaaya, E., Kostjukovski, M., Eilberg, J., Sukprakarn, C., 1997. Plant oils as fumigant and contact insecticides for the control of stored product insects. J. Stored Prod. Res. 33, 7–15.

Shiga, T.M., Cordenunsi, B.R., Lajolo, F.M., 2011. The effect of storage on the solubilization pattern of bean hull non-starch polysaccharides. Carbohydr. Polymers 83, 362–367.

Sievwright, C.A., Shipe, W.F., 1986. Effect of storage conditions and chemical treatment on firmness, *in vitro* protein digestibility, condensed tannins, phytic acid divalent cations of cooked black bean (*Phaseolus vulgaris*). J. Food. Sci. 51, 982–987.

Singh, U., 1999. Cooking quality of pulses. J. Food Sci. Technol. Mysore 36, 1–14.

Sousa, A.H., Faroni, L.R.D.A., Guedes, R.N.C., Tótola, M.R., Urruchi, W.I., 2008. Ozone as a management alternative against phosphine-resistant insect pests of stored products. J. Stored prod. Res. 44, 379–385.

Srisuma, N., Hammerschmidt, R., Uebersax, M.A., Ruengssakulrach, S., Bennink, M.R., Hosfield, G.L., 1989. Storage induced changes of phenolic acids and the development of hard-to-cook in dry beans (*Phaseolus vulgaris*, var. Seafarer). J. Food Sci. 54, 311–318.

Stanley, D.W., Aguilera, J.M., 1985. A review of textural defects in cooked reconstituted legumes – the influence of structure and composition. J. Food Biochem. 9, 277–323.

Tipples, K.H., 1995. Quality and nutritional changes in stored grain. In: Jayas, D.S., White, N.D.G., Muir, W.E. (Eds.), Stored grain eco-systems. Marcel Dekker, New York, pp. 325–351.

Tiwari, B.K., Brennan, C.S., Curran, T., Gallagher, E., Cullen, P.J., O' Donnell, C.P., 2010. Application of ozone in grain processing. J. Cereal Sci. 51, 248–255.

Tuan, Y.H., Phillips, R.D., 1991. Effect of hard-to-cook defect and processing on the protein and starch digestibility of cowpea. Cereal Chem. 68, 413–419.

Uma-Reddy, M., Pushpamma, P., 1986. Effect of storage on amino acid and biological quality of protein in different varieties of pigeon pea, green gram and chickpea. Nutr. Rep. Int. 33, 1020–1029.

White, N.D.G., 1995. Insects, mites, and insecticides in stored-grain ecosystems. In: Jayas, D.S., White, N.D.G., Muir, W.E. (Eds.), Stored-grain ecosystems. Marcel Dekker, New York, pp. 123–167.

Xia, B., Sun, D.W., 2002. Application of computational fluid dynamics (CFD) in the food industry: a review. Comput. Electron. Agric. 34, 5–24.

Yousif, A.M., Deeth, H.C., 2003. Effect of storage time and conditions on the cotyledon cell wall of the adzuki bean (*Vigna angularis*). Food Chem. 81, 169–174.

Yousif, A.M., Deeth, H.C., Caffin, N., Lisle, A., 2002. The effect of storage time and conditions on the hardness and cooking quality of adzuki (*Vigna angularis* L.) beans. LWT – Food Sci. Technol. 3494, 338–343.

Yousif, A.M., Kato, J., Deeth, H.C., 2007. Effect of storage on the biochemical structure and processing quality of adzuki bean (*Vigna angularis*). Food Rev. Int. 23, 1–33.

Pulse milling technologies

8

Jennifer A. Wood[1], Linda J. Malcolmson[2]
[1]Tamworth Agricultural Institute, Industry & Investment NSW, Calala, NSW, Australia
[2]Canadian International Grains Institute, Winnipeg, MB, Canada

8.1   Introduction

Milling, by definition, is a process by which materials are reduced from a larger size to a smaller size. In the case of pulses there are several processes that can be defined in this way:

1. Dehulling or decortication – loosening and removal of the seed coat to produce *polished* seed, "*footballs*" or "*gota*".
2. Splitting – loosening and cleavage of the two cotyledons to produce *splits* or "*dhal*". Dehulling and splitting often occur simultaneously in the same "milling" process for many pulses.
3. Flour milling or grinding – reducing whole seed or cotyledons to a flour.

Pulse crops were first domesticated at the beginning of agriculture more than 8000 years ago in a fertile band of land stretching from Turkey to Iran (Zohary and Hopf, 1973). Milling of pulse seeds also dates back to ancient times (Bar-Yosef, 1998). Tools for milling seeds (e.g. quern stones) date back to the Neolithic period, c. 5600 BC (Encyclopædia Britannica, 2010) and were used to produce either *dhal*

Pulse Foods: Processing, Quality and Nutraceutical Applications. DOI: 10.1016/B978-0-1238-2018-1.00007-0

or flour. *Dhal* (also spelled *dahl*, *dal* or *daal*) or *splits* refers to the two cotyledons that remain after dehulling and splitting a pulse seed. The word *dhal* is derived from the Sanskrit word *dal* which means "to split" (Monier-Williams Sanskrit-English Dictionary, 2010). The Sanskrit language dates back to about 1500 BC. Chickpea (*Cicer arietinum*) and pigeon pea (*Cajanus cajan*) *dhal* were mentioned specifically by Charaka, c. 700 BC (Achaya, 1998). Charaka was sometimes referred to as the "Father of Anatomy" as he was one of the founders of medicine and lifestyle advice promoting health and well-being in ancient India. Chickpea flour was also mentioned by Susruta in 400 BC (Nene, 2006). Susruta was recognized as the "Father of Indian Surgery" and wrote texts on general medicine and surgical procedures. Pulses were considered an important part of a healthy diet. Considering the dates of pulse cultivation, Sanskrit manuscripts and the age of early milling equipment, the origins of pulse milling would date somewhere between 3500 and 7600 years ago.

India is the largest producer and consumer of pulses. Pulses consumed in the form of *dhal* or flour account for around 80% of all pulse consumption in India (Mangaraj et al., 2005). Hence, milling is the most important process for human consumption of pulses, followed by various methods of cooking and food preparation. Traditional milling in India yields a low 65–70% *dhal* (Planning Department of the Government of Uttar Pradesh, 2000) compared to a theoretical yield of 80–90% (deduced from the seed coat comprising 10–20% of the seed). These two facts explain the importance of scientific research in pulse milling related areas.

8.2 Traditional methods of dehulling, splitting and grinding

Pulses are traditionally milled in South-East Asian countries where they constitute staple foods in the diet. India alone consumed over 12 million tonnes of pulses in 2005 (FAO, 2010). Pakistan, Turkey and Bangladesh are also very large consumers.

Most traditional methods are performed at home by families or by small communities. The original method used in ancient times was hand pounding with stones or mortar and pestle, followed by the use of quern stones. The earliest form was a saddle quern that consisted of a large gritty base stone and a small, fist-sized smooth

Figure 8.1 Stone quern mill. (Source: Sole of the Warrior LLC, ©2001.)

stone. Seed could be fractured into rough splits, or with continued pounding reduced to a flour. The saddle quern later evolved into rotary and oscillatory querns (also known as *chakki*). These consist of two large abrasive stones that fit together, the bottom one slightly convex and the upper one slightly concave. The top stone has a central hole in which seed is slowly fed while the top stone is rotated by a wooden handle (Fig. 8.1). As the top stone is rotated, the seed tumbles between the two abrasive stones, resulting in dehulling and splitting of the seed. Continued turning of the top stone further reduces the split seed to flour. *Chakki*-type mills are still often used by many households in South-East Asia (Singh et al., 1992, 2000).

After milling, an edible oil (such as linseed oil) and/or water may be added to the dehulled seed or *dhal* to impart shine and improve appearance (although oil addition may also decrease shelf-life by increasing the rate of rancidity). Alternatively, *dhal* may be polished without oil/water addition using a cone-type polisher (similar to a rice polisher) or a buffing machine consisting of leather straps on a rotating paddle.

Milling techniques are essentially founded on trial and error over many hundreds of years and these secrets have been passed from one generation to the next as tradition.

8.2.1 Dehulling without splitting

Removal of the seed coat is beneficial for the following reasons:

- Reduces antinutritional factors, such as tannins and insoluble fiber (non-nutrients that can bind protein and other nutrients), thereby improving nutritional quality, protein digestibility, texture and palatability.

- Removes astringent taste caused by tannins.
- Allows the production of higher quality flours, without browning/speckling (also increases leavening ability).

The dehulling process, without splitting, is only performed on pulses whose cotyledons are very tightly held together, such as lentil (*Lens culinaris*), green mung bean (*Vigna radiata*), black mung bean (*Vigna mungo*) and pigeon pea (*Cajanus cajan*). The methods used are similar to those for dehulling and splitting but more care is required to avoid seed splitting that lowers yields and therefore profitability. The resultant product is often a specialty and traded at a premium price.

8.2.2 Dehulling and splitting

This process is used to produce *dhal*. In addition to the above benefits of seed coat removal, the resulting *dhal* takes less time to cook than whole seeds and does not require pre-soaking. The technique can be performed on any of the pulses, with the exception of some beans. Dry pea (*Pisum sativum*), pigeon pea, desi chickpea, lentil, mung bean and faba bean (*Vicia faba*) are the most commonly split species. Variation in the ease of milling between the different pulse species is large and primarily explains the wide variation in methodologies and pre-treatments used to optimize yields (Kurien, 1984; Ehiwe and Reichert, 1987; Singh et al., 1992).

8.3 Modern/industrial methods of dehulling and splitting

The traditional stone *chakki* design was used as a template for the attrition-type mills as commercial-scale dehulling and splitting of pulses emerged. The two-stone principle was retained and the much larger stones were rotated using the energy of harnessed animals (such as bullocks) or running water (such as in the flour mills of Europe). The mills were adapted as electricity became an available power source, and automation increased.

Now, the stones are artificial and coated with carborundum (derived from silicon carbide) of various abrasive grades (grit size). These new improved attrition-type mills are often called under

Figure 8.2 Commercial *chakki* mill. (Source: Government of Punjab.)

runner disk shellers (URD Shellers). The orientation of the stones can be either horizontal (as in the original *chakkis*) or vertical, and the gap between the stones can be adjusted to the seed size to optimize dehulled seed and/or *dhal* yields (Fig. 8.2).

The other modern mill type is a carborundum roller mill (Fig. 8.3). It has a cylindrical carborundum stone that is tapered and rotates inside a perforated metal casing so that the gap between the roller and casing decreases from the inlet to the outlet. The stone or casing can

Figure 8.3 Carborundum roller mill: CIAE mill. (Source: Central Institute of Agricultural Engineering.)

sometimes be moved to adjust the gap depending on the seed size. These mills are often mounted horizontally or on a slight downward angle to facilitate passage of the seed.

Since milling performance (yield and resulting product quality) is now considered an important quality trait for pulse breeders, processors and exporters, a range of laboratory-scale mills is available and is currently used in pulse milling research (Fig. 8.4). These include:

- Barley pearlers, consisting of a single rotating carborundum stone positioned vertically a set distance from a fixed metal plate, hence suitable for small pulses only (Singh et al., 1992)

A

B

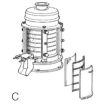
C

D

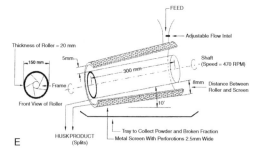
E

Figure 8.4 (A) Barley pearler (Source: Seedburo Equipment Company); (B) TADD (Source: Venables Machine Works Ltd); (C) Satake mill (Source: Satake Australia Pty Ltd); (D) SKE mill, "sheller" component (Source: Industry & Investment NSW); (E) AGT mill (Source: Dr L. Iyer, PhD Thesis, Australia, 1998).

- The tangential abrasive dehulling device (TADD), consisting of a single abrasive stone (or emery paper) on the horizontal plane (Ehiwe and Reichert, 1987; Singh et al., 1992)
- The Satake mill, consisting of a vertical stack of several carborundum stones (Wang, 2005)
- The SK Engineering mill, the "sheller" component modeled on the URD sheller commercial mills in India (Wood et al., 2008)
- The SK Engineering mill modeled on full capability Indian commercial mills and comprising a "pitter" (carborundum roller mill), "sheller" (URD sheller), "conditioner" (bin-dryer) and aspirator (to remove seed coat and flour from milled product). It allows great flexibility for researching a range of pulse species using any selection of milling techniques (Burridge et al., 2001; Wood et al., 2008)
- Carborundum roller mills, such as the AGT mill (Iyer, 1998), the "pitter" part of the SK Engineering mill (Burridge et al., 2001), the CIAE mill (Mangaraj and Singh, 2009) and others (Sahay and Bisht, 1988).

The abrasiveness of stones (grit), speed of stones/rollers, gap size, feed rates and time seed is retained within the mill will all influence milling performance. The effect of these parameters has been demonstrated using a range of mills on various pulse species (Singh et al., 2004; Wang, 2005; Goyal et al., 2009; Mangaraj and Singh, 2009). The ultimate aim of researchers is to maximize milling yields without increasing wastage (broken cotyledons and flour).

8.3.1 Milling calculations

Researchers have used several common calculations to determine the effectiveness of a milling test, although the descriptive names of the calculations often vary:

- Dehulling efficiency (DE; from APQ Method 104.1, Burridge et al., 2001): sometimes called degree of dehulling or dehulling index, is essentially the yield of dehulled product material (dehulled whole seed and *dhal*) as a percentage of original seed weight:

$$DE\ (\%) = \frac{(DWS + D) \times 100}{Wt}$$

where:

DWS=mass of dehulled whole seeds
D=mass of dehulled *dhal*
Wt=original weight

- Dehulling index (TADD; from Ikebudu et al., 2000), sometimes called pearling index: Maximum +1 (100% seed is dehulled with no fines or undehulled grain) and minimum −1 (indicates 100% unsuccessful result with all seed undehulled or broken into fines):

$$n = \frac{(M_c + M_h) - (M_{uh} + M_f)}{M_g}$$

where:

M_c=mass of cotyledons and broken cotyledons
M_h=mass of removed hulls
M_{uh}=mass of kernels that remained undehulled
M_f=mass of fines in the final product
M_g=total mass of original grain fed into the dehuller

- Dehulling index (TADD; from Ehiwe and Reichert, 1987):

$$DI = \frac{100SR}{AF}$$

where:

AF (abraded fines)$= 100(W1 - W2)/W1$
SR (seed coat removed)$= 100 - (100W3/W4)$
$W1$=weight of original sample placed in each TADD cup
$W2$=weight of partially dehulled seed
$W3$=weight of seed coat (dried)
$W4$=weight of fully dehulled seed (dried)

- Splits yield (SY; from APQ Method 104.1, Burridge et al., 2001): also known as *dhal* yield, is essentially the yield of dehulled and split product material (*dhal*) as a percentage of original seed weight:

$$SY\,(\%) = \frac{D \times 100}{Wt}$$

where:

D=mass of dehulled splits. This calculation can be adapted to calculate the percentage of other desired milled products such as dehulled whole seeds.
Wt=original weight

- Percentage kibble or brokens (from APQ Method 104.1, Burridge et al., 2001): are essentially the yield of kibble or broken product material as a percentage of the original seed weight:

$$\text{Kibble (\%)} = \frac{K \times 100}{Wt}$$

where:

K = mass of kibble or brokens (kibble or brokens are defined as cotyledon fragments retained on a 0.85 mm sieve, but this aperture may differ depending on pulse species and researcher preference). Percentage seed coat (hull, husk, testa), undehulled seed and percentage flour loss (powder, fines; passes though the 0.85 mm sieve) may also be calculated using a similar calculation.

Wt = original weight

- Percentage loss: is essentially the sum of all defective products as a percentage of the original seed sample. The constituents that may or may not be included as defective will vary depending on the pulse species, the milled product of interest and individual researcher preferences. Losses may include undehulled whole seeds, dehulled whole seeds, undehulled splits (or caps), *dhal*, seed coat, brokens, kibble and flour.

8.4 Dry versus wet processes of dehulling and splitting

Wet milling methods differ from dry methods mainly by the insertion of a step where seeds are submerged in water. Methods have evolved reflecting the pulse species and/or equipment used.

8.4.1 Wet milling

A generalized flowchart of a wet milling process is shown in Figure 8.5. Seed is cleaned of all foreign matter and usually graded into specific sizes to optimize yields. It is then soaked for 3–14 h (depending on pulse species and individual milling preferences),

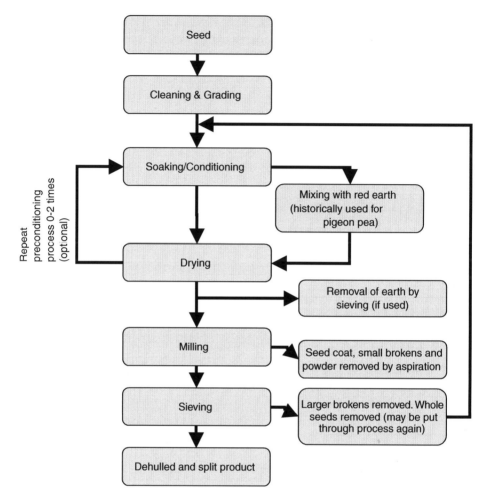

Figure 8.5 Flow diagram of wet milling (only used to produce dehulled and split product, not dehulled whole seeds).

sun-dried for 3–4 days (in developing countries) or mechanically dried (in Western countries) and finally milled.

This type of process is common among rural-based processors in South-East Asian countries, such as India. The method is simple and often a single pass through the mill will be sufficient to produce *dhal*. A long period of soaking will increase the ease of milling but also increase the percentage of broken cotyledons; the resulting *dhal* can also be unacceptably firm after cooking (Singh et al., 2000). Long soaking can also cause hollow-heart resulting in concave adaxial cotyledon surfaces (Kurien and Parpia, 1968) which are not desirable.

A variation of this method, traditionally used for pigeon pea, involves mixing seed with red earth (5%) after soaking (3–12 h) followed by conditioning overnight and alternate sun-drying and conditioning for 2–6 days. The red earth is then removed by sieving and the seeds are milled. It is believed that the red earth increases the rate of drying which facilitates loosening of the seed coat of difficult-to-mill species (Kurien and Parpia, 1968). The adhesive and abrasive properties of the red earth also appear to help with the dehulling. Disadvantages of this method are that seeds can be squashed by workers' feet during the red earth mixing and that a fine dust often remains adhered to the resulting *dhal* (Reddy, 2006). This method was adopted around 50 years ago and, despite appearing in recent publications, the last known use of this method was around 15 years ago (A. Gupta, personal communication, 2010). It has now been superseded by a dry milling method using mustard oil (300–400 g 100 kg^{-1} seed) and heating to 80–85 °C.

8.4.2 Dry milling

A generalized flowchart of a dry milling process is shown in Figure 8.6. Seed is cleaned of all foreign matter and may be graded into specific sizes to optimize yields. Some easy-to-mill species can be milled directly without prior conditioning/tempering or drying processes. This is sometimes the routine for some desi chickpea and field pea varieties that are easy to dehull and split.

For difficult-to-mill species, however, seed is usually pitted (scarification and/or breakage of the seed coat) to facilitate absorption of the following agents:

1. Water – added at a rate of ≈10% (w:w) prior to tempering for 4–8 h. In developing countries, the tempered seeds are sun-dried for 16–20 hours before milling. This method is commonly used for desi chickpea and field pea.
2. Oil – added at a rate of ≈0.4% (w:w) prior to conditioning for 2–3 days. In developing countries, the conditioned seed is then heated in hot sand or ash before milling. This method is often used for pigeon pea, mung bean and black mung bean.

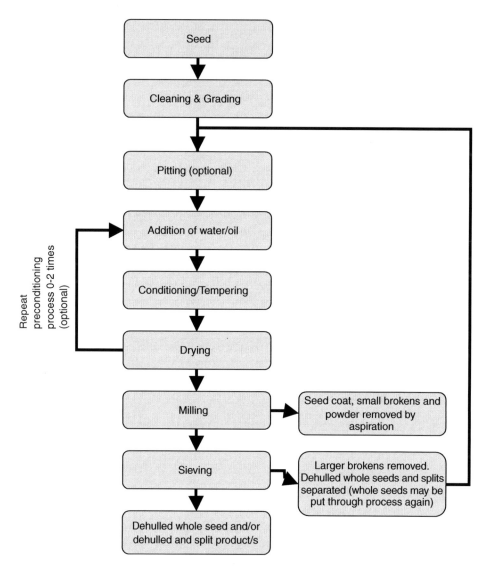

Figure 8.6 Flow diagram of dry milling (used to produce either dehulled whole seeds and/or dehulled and split product).

Rates, conditioning times and oil type may differ with pulse species and individual milling preferences and sometimes both water and oil pre-treatments may be used. Dry milling methods usually produce better quality *dhal* than wet milling procedures; however, the seed may need to be passed through the process more than once to achieve acceptable *dhal* yields.

Sun-drying, hot sand and ash treatments are not practical for most Western countries that mill pulses (mainly Canada, Australia, USA and Europe). Labor is expensive there, so mechanical methods have been devised to replicate the traditional dry process described above. Western processors apply oil and/or water treatments to seed as it travels along a screw conveyor. The screw conveyor mixes seed as it is transported to a holding container. Seed may then be tempered/conditioned for up to 12 h followed by drying. Typically, hot air is mechanically blown through the tempered/conditioned seed (typically using a batch-bin drying technique) until seed reaches the desired moisture content for milling (about 7–11%). Excessively hot air or prolonged seed heating can adversely affect splitting yields and quality; in the extreme, overheated *dhal* will diminish appearance, taste and cooking quality.

8.5 Current and future trends – improving dehulling and splitting performance

8.5.1 Seed quality attributes

Research has shown that milling yield is an important parameter for pulses. The main seed attributes influencing milling yields are:

1. Seed coat content. A lower seed coat content (or thinner seed coat) increases the theoretical milling yields from all pulse species, assuming all other attributes are equal.
2. Moisture content. Seeds that are too dry or too moist can decrease milling yields due to increased breakage and/or powder (Wang, 2005; Goyal et al., 2009). Hence, the drying process is crucial to the final milling yields.
3. Seed size and uniformity of size. Seeds that are too large or too small are often not suitable for milling, mainly due to milling equipment limitations (Erskine et al., 1991b; Wood et al., 2008). The more uniform the seed size, the higher the milling yields: less seed is shattered/broken (larger seeds) and less seed passes through untouched (smaller seeds).
4. Seed shape. Rounder seeds are more likely to have larger milling yields due to more of the seed surface coming into contact with the milling stones. Unevenly shaped seed is more likely to shatter

or produce a higher percentage of brokens and *dhal* with abrasions on the most exposed surfaces. This has been demonstrated in two isogenic lines of desi chickpea (Wood, 2010).

5. Seed hardness. Seeds that are too hard or too soft can decrease milling yields due to increased breakage and/or percentage powder (Reichert et al., 1984).

Hence, it is not surprising that large genetic differences in milling performance exist within species (Ramakrishnaiah and Kurien, 1983; Erskine et al., 1991a; Swamy et al., 1991; Singh et al., 1992; Black et al., 1998; Wood et al., 2008). Improved milling yields are now being sought by screening genotypes within breeding programs (Agrawal and Singh, 2003; Wang, 2008; Wood et al., 2008).

8.5.2 Manipulation of milling variables

Higher milling yields are also being sought through experimentation with processing methods. Researchers have demonstrated significant effects of pre-treatment and drying methods on the milling performance of a range of pulse species (Mandhyan and Jain, 1993; Singh et al., 2004; Wang, 2005; Tiwari et al., 2007, 2010; Mangaraj and Singh, 2009). However, it is difficult to compare results as the species, milling equipment and methodologies varied widely.

Pre-treatments commonly examined have included water, various oils (palm, vegetable, coconut, castor, cottonseed, mustard, sunflower, groundnut, peanut, sesame) and chemical treatments (salt solutions, sodium carbonate, sodium bicarbonate, sodium hydroxide, acetic acid, vinegar and urea solutions) with variable success (Saxena et al., 1989; Erskine et al., 1991b; Phirke and Bhole, 2000; Mangaraj et al., 2005; Anton et al., 2008a; Tiwari et al., 2008). More recently, researchers have found promise in hydrothermal pre-treatment (Tiwari et al., 2010) and pre-treating with various enzymes (Phirke and Bhole, 2000; Sreerama et al., 2009; Wood, 2010).

The effect of pre-treatments on the quality of end-products is limited. Soaking (wet method) in most pre-treatments will result in nutrient losses due to leaching. Soaking in sodium bicarbonate was shown to result in loss of protein, starch and vitamins (Srivastava et al., 1988). Tempering (dry method) is likely to result in smaller losses, although research on this is scant. Wet methods (water) resulted in more undesirable beany odor in flour from white cowpeas than dry methods, but the dry methods undesirably reduced water absorption of the flour; there was little effect on particle size distribution, protein solubility, foam volume, foam stability or least gelation capacity (Amonsou et al., 2009). Sreerama et al. (2009) found no difference in cookability (cooking time or solids dispersed) of *dhal* (pigeon pea, mung bean and horse gram) after oil (peanut) and enzyme (xylanase and protease) pre-treatments and milling. The effect of different pre-treatments on the nutritional value, cooking quality and sensory aspects requires further investigation to ensure that the quality of end-products is not compromised.

8.5.3 Advances in drying techniques

Western countries are disadvantaged by expensive labor and suboptimal temperature and sunshine climates relative to South-East Asian countries. To counteract this, mills are more automated and use mechanical driers. These driers are typically batch-bin types that consist of a vessel confining the seeds and hot air blown through the sample (typically around 60–75 °C). Some Western processors are experimenting with new techniques that increase speed and throughput to save on heating costs. These include the addition of a stirring device in a batch-bin drier, recirculating bin driers and continuous flow bin driers. The next logical step is continuous flow drying with cross-flow heated air.

Other technologies may be suitable for drying tempered/conditioned seed:

- Flash drying is a continuous process where superhot steam (up to 300 °C) is blown through the tempered/conditioned seeds causing rapid evaporation of seed moisture. It is exponentially faster than bin drying methods. An added benefit is that latent heat can be recovered by recompression leading to energy savings.
- Dielectric drying is a process whereby the tempered/conditioned seeds are dried by the absorption of electromagnetic radiation (radio waves, microwaves or infrared radiation). Infrared technology is currently being used commercially to dehull rice and corn.

While the application of flash drying and dielectric drying to pulse milling is yet to be proven, these methods may prove useful on their own or in combination with other air drying techniques to reduce drying times, increase throughput and lower energy costs.

8.6 Distribution of pulse dehulling and splitting

Pulse milling is one of the largest agricultural processing industries in India with an estimated 14 000 pulse mills in operation (Planning Department of the Government of Uttar Pradesh, 2000). In contrast, there is only a handful of pulse mills in each of the major Western pulse milling countries, Canada, Australia, USA and in Europe.

Most pulse mills in India are traditional types, achieving low milling yields of 65–70% *dhal* (Planning Department of the Government of Uttar Pradesh, 2000). In comparison, many of the Indian agricultural universities and research institutes have developed modern improved mills (carborundum roller mills and URD shellers, similar to those used in Western countries), achieving better milling yields of 70–85% depending on the species/variety and the techniques used.

Traditional household milling of pulses will probably continue in many South-East Asian households, especially in the lower socio-economic demographic where people cannot afford commercially milled *dhal*. Similarly, the commercial milling industry in South-East Asia will continue to provide domestic product.

In India, the industry is beset by high capital requirements and a large number of intermediaries involved in both pulse procurement and marketing of the milled product (Planning Department of the Government of Uttar Pradesh, 2000). As developing countries, such as India, become more affluent, commercial pulse mills in the larger cities may find workers more difficult to retain and labor rates will start to increase. There is also a risk that increasing affluence will lead to increasing consumer preference for premium *dhal* and flour products imported from Western countries where hygiene is monitored more closely. These factors have the potential to cause a significant shift in commercial pulse milling from South-East Asia to Western countries.

8.7 Modern/industrial methods of milling pulse flours

Pulse flours are traditionally used in the preparation of pastes, batters and doughs which are baked (e.g. *roti, kitta, dabo, boondi*), steamed (e.g. *dhokla, moin-moin*) or fried (e.g. falafel, *pakora, shiro, bhujia, akara*).

Figure 8.7 Impact hammer mill. (Source: Industry & Investment NSW.)

The production of pulse flours involves grinding whole or dehulled seeds/*dhal* into small particles. To obtain a flour of more uniform particle size, ground particles are passed through one or more screens. The grinding and screening process may be done as separate processes, or may be simultaneously performed in an automated mill, such as an impact mill or roller mill (Figs 8.7 and 8.8, respectively). The resulting flour differs markedly depending on whether the seed coat is removed prior to flour milling (more common) or included to produce a wholemeal flour. Wholemeal flour use in pulses is less common than cereals, but is sometimes prepared from pulses with very thin and opaque/white seed coats that are difficult to remove, such as kabuli chickpeas or beans where the seed coat is difficult to remove. Wholemeal flours contain a substantially higher fiber content that will affect the functionality and processing properties compared to flour prepared from dehulled seeds/*dhal*.

Impact mills, such as hammer mills, are commonly used to produce pulse flours from dry seeds (without pre-conditioning). The principle behind a hammer mill is simple. The mill is composed of

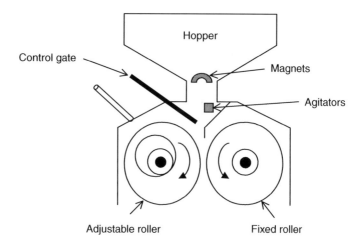

Figure 8.8 Flour roller mill. (Source: © 2008, Government of Saskatchewan.)

a steel drum containing either a vertical or horizontal rotating shaft. The shaft is fitted with hammer bars and is spun inside the drum as seeds are fed into the mill. Upon impact with the hammer bars, the seeds are reduced until the particles are small enough to exit the mill through the metal screen. Thus, the process involves reduction in size by a series of impacts against a rigid housing. The metal screen is often changed to obtain the particle size desired.

Research examined the effect of milling technology on the functional properties of pulse flours. These studies have used a number of different experimental mills to produce pulse flours based on three main types; *chakki* mills, impact mills and roller mills.

Ravi and Harte (2009) used a vertical *chakki* grinder to produce flours from desi and kabuli chickpea with minor differences in functional and thermal properties.

Impact mills used for pulse flour research include: hammer mills (Indira and Bhattacharya, 2006; Siddiq et al., 2010), turbo impact mills (Sanjeewa et al., 2010) and cyclone mills (Deshpande et al., 1983; Shahzadi et al., 2005). Differences among pulse species show inherent properties that influence grinding behavior. In addition, the flours produced from different species have different physiochemical, thermal and functional characteristics, and varietal differences have been observed.

In impact milling, the stress before rupture is primarily concentrated at a single point, in contrast to a roller mill where the energy is more evenly dispersed. Roller mills consist of a series of paired cylindrical rollers with a gap that can be manually adjusted. This gap decreases with each set of subsequent rollers. The first set of rollers is designed to crush the seed and remove the seed coat, while the latter rollers are designed to allow particular flour particle sizes to pass through screens. Roller mills are routinely used for wheat milling, after grain is tempered to a particular moisture content. This pre-conditioning is less suitable for many pulses which have higher oil and water-soluble non-starch polysaccharide contents. Such pulses are better milled dry to avoid potential residues being deposited in the mills. This may explain why only a few researchers have examined milling pulses with roller mills: green mung beans (Thompson et al., 1976); peas (Kosson et al., 1994; Otto et al., 1997) and kabuli chickpea (Otto et al., 1997). Research by Kosson et al. (1994) showed that the milling fractions obtained from the roller mill (shorts, reduction flours and break flours) are derived from the outer, intermediate and inner layers of the cotyledon, respectively.

Clearly, a number of factors affect the physiochemical and functional properties of pulse flours. These include the mill selected for preparing the flours, milling conditions and selection of screen and/ or sieves for separation of the ground material. Flour specifications are needed for pulse flours and it is highly probable that the specifications will vary with intended end-use, similar to wheat flour specifications.

8.7.1 Particle size

The greatest variable in flour quality, apart from the pulse species and varietal differences, is the particle size distribution resulting from the milling process. Much work has been done to examine the effects of flour particle size on cowpeas (*Vigna unguiculata*). Studies by Kerr et al. (2000) found differences in the functional and physiochemical properties of cowpea flours hammer milled to varying particle sizes (Thomas-Wiley laboratory mill, model 4, Arthur Thomas Co., Philadelphia, USA). Differences in milling and separation procedures significantly affected water absorption, solids lost and protein solubility. Finely milled flour had lower initial gelatinization temperatures compared to more coarsely milled flour. The flour with small particle size had relatively constant diameters (19–21 μm) and was associated with starch granules. The flour with large particle size was associated with aggregated flour particles. In a subsequent study, Kerr et al. (2001) found chips made from fine cowpea flour had higher snapping force than chips made from coarser flour. In addition, finely milled flours had higher starch content than coarsely milled flours.

Indira and Bhattacharya (2006) used a hammer mill (GEO Adlam & Sons Engineers, Bristol, London) to grind dehulled lentil, cowpea, black and green mung bean and chickpea, and found differences in the particle size, surface area and number of particles in the flours.

Particle size is also known to affect the product quality of *akara*, a fried cowpea paste (McWatters, 1983; Ngoddy et al., 1986). *Akara* made from fine cowpea flour (400 mesh) had decreased hydration and reduced air incorporation resulting in an unacceptably drier, denser and less spongy texture than that made from coarser flour (100 mesh). They recommended that flours used for *akura* and *moin-moin* (a steamed cowpea paste) be milled with 65–75% of the particles in the size range of 45–150 μm (350–100 mesh). Chickpea flour (*besan*) is traditionally milled to different particle sizes depending on the end-product for which it will be used; a finer flour for desserts and a coarser flour for savory foods.

8.8 Substitution of pulse flours into other products

Recent studies have shown that pulse flours can partially replace wheat flour in bakery products, pasta, noodles, crackers and snack foods (Sabanis et al., 2006; Anton et al., 2008b; Gómez et al., 2008; Wood, 2009; Hood-Niefer and Tyler, 2010; Petitot et al., 2010). Other potential uses of pulse flours include processed and prepared meat systems (Verma et al., 1984; Pietrasik and Janz, 2010; Sanjeewa et al., 2010), batters and coatings (Chidanandaiah and Keshri, 2007), beverages, yogurt, salad dressings and desserts. Pulse-based foods offer a number of health and nutritional benefits including high fiber and high protein, and can be gluten-free (Wood and Grusak, 2007).

Deshpande et al. (1983) showed that whole bean (*Phaseolus vulgaris*) flours of 60 mesh (0.018 mm) particle size had different functional properties depending on the ratio of substitution with wheat flour and the subspecies of bean. Overall, bean flour addition increased water and oil absorption, and foaming and emulsification capabilities compared to 100% wheat flour. Pasting and mixing properties were also altered with the addition of bean flours. The researchers noted that the pronounced dough weakening observed in the composite flours may have been a result of proteolytic activity in the bean flours (as they were not inactivated prior to use).

Shahzadi et al. (2005) blended wheat flour with dehulled chickpea and lentil flours and observed decreased water absorption and increased dough development time compared to 100% wheat flour. Rheological behavior of the composite flours was also significantly affected by flour storage.

The functional properties of pulse flours are critical to the quality characteristics of the final food product. Therefore, care must be taken in the production of the flour to ensure that the desired functional properties (such as water and fat absorption, protein solubility and starch gelatinization) are achieved, otherwise processing and end-product quality will be compromised.

8.9 Production of roasted pulse flours

Raw pulse flours can develop undesirable odors and flavors during storage. Lipoxidase has been implicated in off-flavor development by catalyzing the formation of peroxides from unsaturated fatty acids (Kon, 1979). Treatment of the flour with dry heat for 6–8 min at 104–105 °C has been shown to inactivate the enzyme (Smith and Circle, 1972). With dry beans (*Phaseolus vulgaris*), heat treatment of the beans prior to milling to flour may be important to inactivate hemagglutinin compounds. Conventional air heating does not provide a high rate of heat transfer and a number of alternative heating methods have been developed. Carvalho et al. (1977) roasted dry beans in a bed of salt prior to grinding into flour. Peterson and Harper (1978) developed a granular bed roaster. Based on this work, Aguilera et al. (1982) developed a particle heat exchanger for roasting dry beans. The roasted products showed low trypsin inhibitors and hemagglutinin activities, good gel-forming capacity and increased water-holding capacity and cold paste viscosities.

8.10 Production of precooked pulse flours and powders

Instant precooked pulse flours can be prepared by soaking, cooking and slurrying, followed by drum or spray drying. Precooked or partially cooked pulse flours and flakes can also be prepared by micronization using infrared heating. Precooked pulse flours have increased color and shelf-life stability due to inactivation of enzymes. Precooking also reduces antinutritional compounds and alters physiochemical and functional properties of the flour. According to Siegel and Fawcett (1976) precooked pulse flours prepared in a way that preserves cellular integrity will retain their original flavor.

Nagmani and Prakash (1997) compared the functional properties of raw and thermally treated pulse flours made from chickpea, black and green mung beans and lentils. Thermal treatment lowered nitrogen solubility profiles of all flours and increased water-absorption capacity in chickpea, black mung bean and lentil flours. Fat absorption and foam capacities decreased in thermally treated chickpea and black mung bean flours.

Gomes et al. (2006) examined different processing treatments in the development of a precooked dry bean flour. Factors such as cooking method and the addition of sodium carbonate were found to influence the physicochemical and functional properties of the flour.

Jangchud and Bunnag (2001) examined the effects of soaking and cooking time on oligosaccharide content (responsible for flatulence) in precooked red kidney bean (*Phaseolus vulgaris*) flour. Increasing the soaking time significantly reduced raffinose and stachyose content but only stachyose reduced with increased cooking time. Recent evidence has suggested that oligosaccharides have prebiotic effects, encouraging the growth of beneficial bacteria in the large intestine. Oligosaccharides also slow digestion, increase satiety and help maintain blood glucose levels, thereby being beneficial in fighting obesity and diabetes. Thus, their removal from the flour may not be desirable.

8.11 Production of germinated pulse powders

Germination of pulse seeds involves soaking the whole seed for 24 h and then spreading the soaked seeds on a damp cloth for up to 48 h. Depending on the species used, germination is allowed to take place until sprouts of 1–2 cm or 6 cm appear. Sprouted seeds are usually eaten raw, fried or boiled, but some are dried and ground into a modified pulse flour. The advantages of germinated pulse seeds include a reduction in phytic acid and trypsin inhibitors, and an increase in vitamin levels. Germination also reduces levels of oligosaccharides. Kadlec et al. (2001) treated wet germinated yellow pea seeds with microwave energy followed by drying in hot air at 80 °C. Microwave treatment shortened the time required to dry germinated seeds.

8.12 Production of pulse fractions

Pulse flour can be fractionated to produce ingredients such as fiber, protein and starch. Seed coats can be removed by dehulling to produce a concentrated source of fiber. More importantly, various protein and starch fractions can be produced by dry milling into flour followed by air classification. In this process, pin-milled flour is classified in a spiral air stream. Concentration of the protein and starch is possible because of the differences in size, shape and density of starch granules and protein-containing particles. The starch fraction is present in the heavier particles whereas the protein fraction is contained in the fine particles. Protein fractionation takes place when opposing centrifugal and centripetal forces are employed such that the heavier particles (starch) move in the opposite direction to the fine protein-rich particles. Since pulse flours contain relatively large starch granules, protein concentration by air classification is practical (Siegel and Fawcett, 1976).

Tyler and Panchuk (1982) found seed moisture affected starch separation efficiency, starch fraction yields and protein content of the resulting starch and protein fractions of pea and faba bean. Wu and Nichols (2005) showed that more intense grinding prior to air classification increased the yield and purity of both the protein and starch fractions for pea. The resulting high protein fractions from pea, faba bean and lentil generally contain twice the original whole seed protein contents (Uebersax and Ruengsakulrach, 1989).

Pulse fractions can also be produced by wet fractionation technology based on the original process developed by Schoch and Maywald (1968). Wet fractionation involves the preparation of a slurry in alkaline solution followed by centrifugation and drying. Alkali neutralization and salt removal are also required. A recent study by Naguleswaran and Vasanthan (2010) reported on a wet fractionation method that incorporated sonication and protease treatment to maximize yield and purity of pea starch.

Acknowledgment

Thank you to Mr Adarsh Gupta, Canny Overseas P/L, Delhi, India, for his assistance in ensuring the accuracy of this chapter.

References

Achaya, K.T., 1998. A historical dictionary of Indian food. Oxford University Press, Delhi.

Agrawal, K., Singh, G., 2003. Physico-chemical and milling quality of some improved varieties of chickpea (*Cicer arietinum*). J. Food Sci. Technol. (Mysore) 40, 439–442.

Aguilera, J.M., Lusas, E.W., Uebersax, M.A., Zabik, M.E., 1982. Roasting of navy beans (*Phaseolus vulgaris*). J. Food Sci. 47, 996–1000.

Anton, A.A., Ross, K.A., Beta, T., Fulcher, R.G., Arntfield, S.D., 2008a. Effect of pre-dehulling treatments on some nutritional and physical properties of navy and pinto beans (*Phaseolus vulgaris* L.). LWT – Food Sci. Technol. 41, 771–778.

Anton, A.A., Ross, K.A., Lukow, O.M., Fulcher, R.G., Arntfield, S.D., 2008b. Influence of added bean flour (*Phaseolus vulgaris* L.) on some physical and nutritional properties of wheat flour tortillas. Food Chem. 109, 33–41.

Bar-Yosef, O., 1998. The Natufian culture in the Levant, threshold to the origins of agriculture. Evolutionary anthropology. Issues, News Rev. 6, 159–177.

Black, R.G., Singh, U., Meares, C., 1998. Effect of genotype and pretreatment of field peas (*Pisum sativum*) on their dehulling and cooking quality. J. Sci. Food Agric. 77, 251–258.

Burridge, P., Hensing, A., Petterson, D., 2001. Australian pulse quality laboratory manual. SARDI Grain Laboratory for GRDC, Urrabrae.

Carvalho, C.C.C., Jansen, G.R., Harper, J.M., 1977. Protein quality evaluation of an instant bean powder produced by dry heat processing. J. Food Sci. 42, 553–554.

Chidanandaiah, Keshri, R.C., 2007. Quality of buffalo meat patties enrobed with batter mix containing Bengal gram flour, finger millet flour and/or corn flour. J. Food Sci. Technol. (Mysore) 44, 307–309.

Deshpande, S.S., Rangnekar, P.D., Sathe, S.K., Salunkhe, D.K., 1983. Functional properties of wheat–bean composite flours. J. Food Sci. 48, 1659–1662.

Ehiwe, A.O.F., Reichert, R.D., 1987. Variability in dehulling quality of cowpea, pigeon pea, and mung bean cultivars determined with the tangential abrasive dehulling device. Cereal Chem. 64, 86–90.

Encyclopædia Britannica, 2010. Quern. Encyclopædia Britannica Online. 21 April 2010. http://www.britannica.com/EBchecked/topic/487081/quern.

Erskine, W., Williams, P.C., Nakkoul, H., 1991a. Splitting and dehulling lentil (*Lens culinaris*): effects of genotype and location. J. Sci. Food Agric. 57, 85–92.

Erskine, W., Williams, P.C., Nakkoul, H., 1991b. Splitting and dehulling lentil (*Lens culinaris*): effects of seed size and different pretreatments. J. Sci. Food Agric. 57, 77–84.

FAO, 2010. FAOSTAT Database. Food and Agriculture Organization of the United Nations, Rome.

Gomes, J.C., da Silva, C.O., Costa, N.M.B., Pirozi, M.R., 2006. Development and characterization of common bean flour. Rev. Ceres 53, 548–558.

Gómez, M., Oliete, B., Rosell, C.M., Pando, V., Fernández, E., 2008. Studies on cake quality made of wheat–chickpea flour blends. LWT – Food Sci. Technol. 41, 1701–1709.

Goyal, R., Vishwakarma, R., Wanjari, O., 2009. Effect of moisture content on pitting and milling efficiency of pigeon pea grain. Food Bioprocess Technol. 3, 146–149.

Hood-Niefer, S.D., Tyler, R.T., 2010. Effect of protein, moisture content and barrel temperature on the physicochemical characteristics of pea flour extrudates. Food Res. Int. 43, 659–663.

Ikebudu, J.A., Sokhansanj, S., Tyler, R.T., Milne, B.J., Thakor, N.S., 2000. Grain conditioning for dehulling of canola. Can. Agric. Eng. 42, 4.1–4.13.

Indira, T.N., Bhattacharya, S., 2006. Grinding characteristics of some legumes. J. Food Eng. 76, 113–118.

Iyer, L., 1998. Physical properties, chemical composition and food processing applications of Australian chickpea. PhD thesis, Faculty of Engineering and Science, School of Life Sciences and Technology, Victoria University of Technology, Melbourne.

Jangchud, K., Bunnag, N., 2001. Effect of soaking time and cooking time on qualities of red bean flour. Kasetsart J. Nat. Sci. 35, 409–415.

Kadlec, P., Rubecova, A., Hinkova, A., Kaasova, J., Bubnik, Z., Pour, V., 2001. Processing of yellow pea by germination, microwave treatment and drying. Innov. Food Sci. Emerg. Technol. 2, 133–137.

Kerr, W.L., Ward, C.D.W., McWatters, K.H., Resurreccion, A.V.A., 2000. Effect of milling and particle size on functionality and physiochemical properties of cowpea flour. Cereal Chem. 77, 213–219.

Kerr, W.L., Ward, C.D.W., McWatters, K.H., Resurreccion, A.V.A., 2001. Milling and particle size of cowpea flour and snack chip quality. Food Res. Int. 34, 39–45.

Kon, S., 1979. Effect of soaking temperature on cooking and nutritional quality of beans. J. Food Sci. 44, 1329–1335.

Kosson, R., Czuchajowska, Z., Pomeranz, Y., 1994. Smooth and wrinkled peas. II. Distribution of protein, lipid, and fatty acids in seed and milling fractions. J. Agric. Food Chem. 42, 96–99.

Kurien, P.P., 1984. Dehulling technology of pulses. Res. Ind. 29, 207–214.

Kurien, P.P., Parpia, H.A.B., 1968. Pulse milling in India I – Processing and milling of Tur, Arhar (*Cajanus cajan* Linn). J. Food Sci. Technol. 5, 203–207.

Mandhyan, B.L., Jain, S.K., 1993. Optimisation of machine conditions for milling of pigeon pea. J. Food Eng. 18, 91–96.

Mangaraj, S., Singh, K., 2009. Optimization of machine parameters for milling of pigeon pea using RSM. Food Bioprocess. Technol., doi: 10.1007/s11947-009-0215-x.

Mangaraj, S., Agrawal, S., Kulkarni, S.D., Kapur, T., 2005. Studies on physical properties and effect of pre-milling treatment on cooking quality of pulses. J. Food Sci. Technol. 42, 258–262.

McWatters, K.H., 1983. Compositional, physical, and sensory characteristics of akara processed from cowpea paste and Nigerian cowpea flour. Cereal Chem. 60, 333–336.

Monier-Williams Sanskrit-English Dictionary, 2010. 21 April 2010. http://webapps.uni-koeln.de/tamil/.

Nagmani, B., Prakash, J., 1997. Functional properties of thermally treated legume flours. Int. J. Food Sci. Nutr. 48, 205–214.

Naguleswaran, S., Vasanthan, T., 2010. Dry milling of field pea (*Pisum sativum* L.) groats prior to wet fractionation influences the starch yield and purity. Food Chem. 118, 627–633.

Nene, Y.L., 2006. Indian pulses through the millennia. Asian Agri-Hist. 10, 179–202.

Ngoddy, P.O., Enwere, N.J., Onurah, V.I., 1986. Cowpea flour performance in akara and moin-moin preparations. Trop. Sci. 26, 101–119.

Otto, T., Baik, B.K., Czuchajowska, Z., 1997. Microstructure of seeds, flours, and starches of legumes. Cereal Chem. 74, 445–451.

Petitot, M., Boyer, L., Minier, C., Micard, V., 2010. Fortification of pasta with split pea and faba bean flours: pasta processing and quality evaluation. Food Res. Int. 43, 634–641.

Peterson, D.G., Harper, J.M., 1978. Granular bed roaster construction. US Patent 4,094,633.

Phirke, P.S., Bhole, N.G., 2000. Pretreatment of pigeonpea grain for improvement of dehulling characteristics. Int. J. Food Sci. Technol. 35, 305–313.

Pietrasik, Z., Janz, J.A.M., 2010. Utilization of pea flour, starch-rich and fiber-rich fractions in low fat bologna. Food Res. Int. 43, 602–608.

Planning Department of the Government of Uttar Pradesh, 2000. Model project on improved dal mill. Government of Uttar Pradesh website. 21 April 2010. http://planning.up.nic.in/innovations/inno3/ae/dal.htm.

Ramakrishnaiah, N., Kurien, P.P., 1983. Variabilities in the dehulling characteristics of pigeon pea (*Cajanus cajan* L.) cultivars. J. Food Sci. Technol. India 20, 287–291.

Ravi, R., Harte, J.B., 2009. Milling and physiochemical properties of chickpeas (*Cicer arietinum* L.) varieties. J. Sci. Food Agric. 89, 258–266.

Reddy, B.S., 2006. Indigenous technical knowledge on pulses storage and processing practices in Andhra Pradesh. Ind. J. Tradit. Knowl. 5, 87–94.

Reichert, R.D., Oomah, B.D., Youngs, C.G., 1984. Factors affecting the efficiency of abrasive-type dehulling of grain legumes investigated with a new intermediate-sized, batch dehuller. J. Food Sci. 49, 267–272.

Sabanis, D., Makri, E., Doxastakis, G., 2006. Effect of durum flour enrichment with chickpea flour on the characteristics of dough and lasagne. J. Sci. Food Agric. 86, 1938–1944.

Sahay, K.M., Bisht, B.S., 1988. Development of a small abrasive cylindrical mill for milling pulses. Int. J. Food Sci. Technol. 23, 17–22.

Sanjeewa, T.W.G., Wanasundara, J.P.D., Pietrasik, Z., Shand, P.J., 2010. Characterization of chickpea (*Cicer arietinum* L.) flours and application in low-fat pork bologna as a model system. Food Res. Int. 43, 617–626.

Saxena, R.P., Laxmi, C., Garg, G.K., Singh, B.P.N., 1989. Effect of bicarbonate soaking on dehusking efficiency and composition of pigeon pea (*Cajanus cajan* L.) seeds and dhal. Int. J. Food Sci. Technol. 24, 237–241.

Schoch, T.J., Maywald, E.C., 1968. Preparation and properties of various legume starches. Cereal Chem. 45, 564–573.

Shahzadi, N., Butt, M.S., Rehman, S.U., Sharif, K., 2005. Rheological and baking performance of composite flours. Int. J. Agric. Biol. 7, 100–104.

Siddiq, M., Ravi, R., Harte, J.B., Dolan, K.D., 2010. Physical and functional characteristics of selected dry bean (*Phaseolus vulgaris* L.) flours. LWT – Food Sci. Technol. 43, 232–237.

Siegel, A., Fawcett, B., 1976. Food legume processing and utilization. International Development Research Centre, Ottawa, Canada.

Singh, S.K., Agrawal, U.S., Saxena, R.P., 2004. Optimization of processing parameters for milling of green gram (*Phaseolus aureus*). J. Food Sci. Technol. 41, 124–130.

Singh, U., Santosa, B.A.S., Rao, P.V., 1992. Effect of dehulling methods and physical characteristics of grains on dhal yield of pigeon-pea (*Cajanus cajan* L.) genotypes. J. Food Sci. Technol. India 29, 350–353.

Singh, U., Williams, P.C., Petterson, D.S., 2000. Processing and grain quality to meet market demands. In: Knight, R. (Ed.), Linking research and marketing opportunities for pulses in the 21st century. Kluwer Academic Publishers, Dordrecht, pp. 155–166.

Smith, A.K., Circle, S.J., 1972. Soybeans: chemistry and technology. AVI Publishing Company, Westport.

Sreerama, Y.N., Sashikala, V.B., Pratape, V.M., 2009. Effect of enzyme pre-dehulling treatments on dehulling and cooking properties of legumes. J. Food Eng. 92, 389–395.

Srivastava, V., Mishra, D.P., Chand, L., Gupta, R.K., Singh, B.P.N., 1988. Influence of soaking on various biochemical changes and dehusking efficiency in pigeon pea (*Cajanus cajon*) seeds. J. Food Sci. Technol. 25, 267–271.

Swamy, N.R., Ramakrishnaiah, N., Kurien, P.P., Salimath, P.V., 1991. Studies on carbohydrates of red gram (*Cajanus cajan*) in relation to milling. J. Sci. Food Agric. 57, 379–390.

Thompson, L.U., Hung, L., Wang, N., Rasper, V., Gade, H., 1976. Preparation of mung bean flour and its application in bread baking. J. Can. Inst. Food Technol. 9, 1–5.

Tiwari, B.K., Jagan Mohan, R., Vasan, B.S., 2007. Effect of heat processing on milling of black gram and its end product quality. J. Food Eng. 78, 356–360.

Tiwari, B.K., Tiwari, U., Jagan Mohan, R., Alagusundaram, K., 2008. Effect of various pre-treatments on functional, physiochemical, and cooking properties of pigeon pea (*Cajanus cajan* L). Food Sci. Technol. Int. 14, 487–495.

Tiwari, B.K., Jagan Mohan, R., Venkatachalapathy, N., Tito Anand, M., Surabi, A., Alagusundaram, K., 2010. Optimisation of hydrothermal treatment for dehulling pigeon pea. Food Res. Int. 43, 496–500.

Tyler, R.T., Panchuk, B.D., 1982. Effect of seed moisture content on the air classification of field peas and faba beans. Cereal Chem. 53, 928–936.

Uebersax, M.A., Ruengsakulrach, S., 1989. Utilization of dry field beans, peas, and lentils. In: Applewhite, T.H. (Ed.), Proceedings of the World Congress on vegetable protein utilization in human foods and animal feedstuff. American Oil Chemists' Society, Champaign, pp. 123–130.

Verma, M.M., Ledward, D.A., Lawrie, R.A., 1984. Utilization of chickpea flour in sausages. Meat Sci. 11, 109–121.

Wang, N., 2005. Optimization of a laboratory dehulling process for lentil (*Lens culinaris*). Cereal Chem. 82, 671–676.

Wang, N., 2008. Effect of variety and crude protein content on dehulling quality and on the resulting chemical composition of red lentil (*Lens culinaris*). J. Sci. Food Agric. 88, 885–890.

Wood, J.A., 2009. The texture, processing and organoleptic properties of chickpea fortified spaghetti with insights to the underlying mechanisms of traditional durum pasta quality. J. Cereal Sci. 49, 128–133.

Wood, J.A., 2010. The influence of seed morphology, ultra-structure and chemistry on the milling quality of chickpea. Phd thesis, School of Environment and Rural Science, University of New England, Armidale.

Wood, J.A., Grusak, M.A., 2007. Nutritional value of chickpea. In: Yadav, S.S., Redden, B., Chen, W., Sharma, B. (Eds.), Chickpea breeding and management. CAB International, Wallingford, pp. 101–142.

Wood, J.A., Knights, E.J., Harden, S., 2008. Milling performance in desi-type chickpea (*Cicer arietinum* L.): effects of genotype, environment and seed size. J. Sci. Food Agric. 88, 108–115.

Wu, Y.V., Nichols, N.N., 2005. Fine grinding and air classification of field peas. Cereal Chem. 82, 341–344.

Zohary, D., Hopf, M., 1973. Domestication of pulses in the Old World: legumes were companions of wheat and barley when agriculture began in the Near East. Science 182, 887–894.

Emerging technologies for pulse processing

9

Jasim Ahmed
Polymer Source Inc., Dorval (Montreal), QC, Canada

9.1   Introduction

Thermal processing remains the key processing technology for food products. The major limitations for thermally processed products are in sensory and nutritional quality. In conventional thermal processing, slow heat conduction from the heating medium to the cold-spot often results in treatment of the material at the periphery of the container that is far more severe than that required to achieve commercial sterility (Meredith, 1998). Furthermore, today's consumers are looking for fresh-like products that are free of chemical preservatives, yet have a longer shelf-life while retaining their natural appearance and flavor. Thus, food processors are looking for alternative technologies which can replace conventional thermal processing while keeping food safety in mind. Recent interests in novel or emerging processing technologies are very much in demand. These newly developed non-thermal processing technologies preserve the food effectively and also modify the functional properties of food components by retaining sensory quality attributes.

Pulse Foods: Processing, Quality and Nutraceutical Applications. DOI: 10.1016/B978-0-1238-2018-1.00007-0

Emerging technologies can be divided into two categories:

1. Electrotechnologies, which include pulsed electric field, radio-frequency heating, microwave heating, infrared heating and ohmic heating.
2. Non-thermal technologies, which include high hydrostatic pressure, oscillating magnetic field, pulse-light, ultrasound, irradiation and ozonization (Ahmed et al., 2009).

Most of the electrotechnologies focus on novel approaches to produce heat and follow conventional thermal mechanisms for achieving preservation and processing. Non-thermal processing technologies can inactivate enzymes and microorganisms, and modify functional properties of food by alternate means without an abrupt increase in product temperature. A combination of thermal and non-thermal processing is also a promising technology which can be used to make a microbiologically safe and quality product.

Pulses are an excellent source of protein, carbohydrates and fiber and, in addition, provide many essential vitamins and minerals. Their highly nutritional properties have been associated with many beneficial health-promoting properties, such as managing high cholesterol and type 2 diabetes and in the prevention of various forms of cancer (Roy et al., 2010). However, these crops contain various antinutritional proteins, e.g. lectins, protease inhibitors and the non-antinutritional compound, angiotensin I-converting enzyme (ACE) inhibitor, and cause various diseases and ingestion problems including hemagglutination, bloating, vomiting and pancreatic enlargement. Thus, care has to be taken to remove or inactivate antinutritional components from pulses to make them more acceptable to consumers. Because of their growing importance as food ingredients, proteins from several pulses are receiving increased attention from biochemists and food scientists who are gradually unraveling their complex properties and trying to relate them to problems encountered in food applications. This chapter will focus on applications of emerging technologies to storage, processing and functionality of pulses.

9.2 Brief description of emerging technologies

In this section, a brief discussion of emerging food technologies is given to provide easy understanding for readers. These emerging technologies offer quality enhancements and energy efficiency, and some are practically waste free. However, it is worth mentioning that not all emerging technologies are suitable or relevant for all types of food processing. The selection of emerging process technologies depends upon the nature of the food material under development, potential market, available technical resources and, above all, investment.

9.2.1 Dielectric heating

Electrotechnologies like microwave (MW) and radiofrequency (RF) dielectric heating are used in thermal processing of non-conductive materials with industrial applications for drying of food and tempering of frozen foods. Microwave and radiofrequency heating refers to the use of electromagnetic waves of certain frequencies to generate heat in a material. There are a number of frequencies allocated by the Federal Communications Commission (FCC) for heating purposes. However, two frequencies, 2450 and 915 MHz, are used for MW heating applications (Decareau, 1985). The most commonly used frequency for RF food processing is 27 MHz.

The rate of heat generation per unit volume, Q, at a particular location in the food during microwave and radiofrequency heating can be characterized by (Datta, 2001):

$$Q = 2\pi f \varepsilon_0 \varepsilon'' E^2 \tag{9.1}$$

where E is the strength of electric field of the wave at that location, f is the frequency of the microwaves or the radiofrequency waves, ε_0 is the permittivity of free space (a physical constant), and ε'' is the dielectric loss factor.

MW and RF heating is a result of interaction between alternating electromagnetic field and dielectric material. Compared with conventional heating using water or steam as the heating medium for package foods, MW/RF energy has the potential to provide more uniform and rapid volumetric heating. There are two mechanisms involved in MW and RF heating: dielectric and ionic. Water in the food is often the primary component responsible for dielectric heating. Due to their dipolar nature, water molecules try to follow the electric field associated with electromagnetic radiation as it oscillates at the very high frequencies. Such oscillations of the water molecules generate heat. The second mechanism of heating with MW and RF is through the oscillatory migration of ions in the food which produces heat under the influence of the oscillating electric field.

The basic difference between MW and RF heating is in penetration. Since RF waves have a longer wavelength than MW, the electromagnetic waves in the RF system can penetrate deeper into the product so there is less surface overheating, and fewer hot or cold spots – a common problem with MW heating (Ahmed et al., 2007a). In addition, RF heating offers simple uniform field patterns as opposed to the more complex and characteristically non-uniform standing wave patterns in a microwave oven.

RF heaters are similar to microwave ovens where food products are passed through a heater and subjected to a direct or volumetric heating process in the form of a radiofrequency energy source. In the simplest form of RF heating, the material to be heated is placed between two metal plates which form an electrical capacitor (Fig. 9.1). The material becomes a "lossy" dielectric and absorbs energy from the RF generator which is connected across the two plates.

Both MW and RF are promising for many heating and drying processes; however, there is still a need for more readily available data related to the design of the applicators, targeted product formulations with specific dielectric properties and control systems for the development of MW/RF heating process applications.

9.2.2 Other electrotechnologies

Pulsed electric field (PEF) treatment is one of the emerging non-thermal food preservation technologies which is exclusively used for liquid foods. PEF treatment of liquid foods is based on a pulsing power delivered to the product placed between two electrodes that confine the treatment gap of the PEF chamber. Electric field pulses of short duration ($1-100\,\mu s$) and high intensity ($10-50\,kV\,cm^{-1}$) are applied

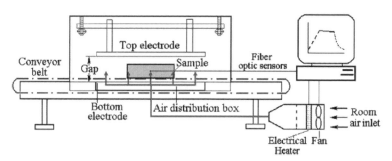

Figure 9.1 Schematic view of a pilot plant RF unit with two pair plate electrodes, conveyor belt and the hot air system. (Source: Wang et al., 2010.)

to a food sample placed between the electrodes. This technology is based on the phenomenon that biological membranes are punctured when an external electrical impulse is applied and inactivates microbial cells. An electrical breakdown of the cell can be expected when a critical transmembrane potential larger than approximately 1 volt is provided (Zimmermann et al., 1974). This process is often referred to as non-thermal, since structural damage to membranes is realized at significantly lower energy levels when compared to the process of heating. As a consequence, the required minimal electric field strength necessary for proper inactivation depends on the cell size and geometry (Heinz et al., 2002). The generation of a PEF requires a system for generating high-voltage pulses and a treatment chamber that converts the pulsed voltage into PEFs.

PEF treatment better preserves the sensory, nutritional and functional properties of foods as compared to thermal treatment (Grahl and Märkl, 1996; Manas and Pagan, 2005), which is compatible with the increasing consumer demand for "fresh-like" food. Several studies have proven that a sufficient reduction of microorganisms can be achieved by the application of PEF (Wouters et al., 1999). Most successful was the treatment of liquid foods in a continuous process (Ho and Mittal, 2000; Heinz et al., 2003). However, the PEF technology has not reached yet a stage of commercial usability, which shows the demand for further intensive research to complete the understanding and to minimize the drawbacks. Limited knowledge is available about the electric field distribution, temperature profile and flow dynamics during the treatment in a continuous flow-through PEF treatment chamber. Similar to conventional thermal processes, a homogeneous treatment intensity distribution is crucial to ensure product safety and to avoid overprocessing and related detrimental effects on product quality.

Ohmic heating, also referred to as Joule heating, electrical resistance heating or electroconductive heating, is another emerging processing technique which is used for food processing (Palaniappan and Sastry, 1990; Marcotte, 1999). In ohmic heating, food is placed between two electrodes serving as an electrical resistor and an alternating electric current is passed through the circuit. Heat is generated throughout the food matrix due to electrical resistance. The electrical energy is directly converted into heat, causing a rise in temperature, and depends mostly on the electrical conductivity of the product. When compared to conventional heating, where heat is conducted from the outside in using a hot surface, ohmic heating conducts heat throughout the entire mass of the food uniformly. Using electric current, food can be pasteurized, fermented or sterilized in a manner that is comparable to, if not better than the current methods of processing.

As a non-thermal technology, ultrasound is attracting considerable interest in the food industry. Ultrasound refers to sound waves, mechanical vibrations, which propagate through solids, liquids or gases with a frequency greater than the upper limit of human hearing. High frequencies in the range of 0.1–20 MHz, pulsed operation and low-power levels (100 mW) are used for non-destructive testing (Gunasekaran and Chiyung, 1994). By means of mechanical vibrations of high enough intensity, power ultrasound can produce changes in food by either disrupting its structure or promoting certain chemical reactions. The range of applications of power ultrasound in the food industry is vast and is growing rapidly. Ultrasonic excitation has been tested for non-destructive evaluation of the internal quality and latent defects of whole fruits and vegetables in a manner similar to the use of ultrasound for viewing the developing fetus in a mother's womb (Mizrach et al., 1994). Some examples of ultrasound in food processing include emulsification (one of the earliest applications), drying, degassing and inactivation of microorganisms. The use of this technology provides some valuable benefits such as reduced processing and maintenance costs, and shorter processing times. However, there is very little information about the effects of this technology on sensory and nutritional food properties.

Ionizing radiation (IR) was one of the first ever emerging technologies used in food products for the enhancement of shelf-life. IR provides numerous technological advantages by minimizing food losses and improving food quality. Irradiation is generally applied in the range of 1.5–7.0 kGy to food materials; this is similar to pasteurization and termed radurization. It has been confirmed that IR is safe for food use within prescribed doses and IR is permitted by all specialized agencies of the United Nations (WHO, FAO, IAEA).

The use of ozone in food processing has recently come to the forefront as a result of the recent approval by the US Food and Drug Administration (FDA) for the use of ozone as an antimicrobial agent for food treatment, storage and processing. The FDA approval marks a watershed event for the food industry. Relatively small quantities of ozone and short contact times are sufficient for the desired antimicrobial effect as it rapidly decomposes into oxygen leaving no toxic residues in food (Muthukumarappan et al., 2000).

The interest in ozone as an antimicrobial agent is based on its high biocidal efficacy and wide antimicrobial spectrum that is active against bacteria, fungi, viruses, protozoa, and bacterial and fungal spores (Khadre et al., 2001).

Supercritical fluid extrusion (SCFX) is another emerging technology which is a blend of supercritical fluid and conventional extrusion technology. This new process utilizes supercritical carbon dioxide ($SC\text{-}CO_2$) as a blowing agent, a nutrient carrier and, if necessary, an in-line process modifier. The technology has been patented by Rizvi and Mulvaney (1992, 1995) for the production of highly expanded starch foams. The technology has tremendous benefits over conventional extrusion processing. Low-shear and low-temperature processing, formation of highly expanded products with a unique microcellular structure comprising non-porous skin and homogeneous cell size distribution, and use of $SC\text{-}CO_2$ as a carrier for soluble flavors and an in-line process modifier are some of the defining characteristics of the SCFX process.

9.2.3 High-pressure (HP) processing

High hydrostatic pressure offers an alternative potential non-thermal preservation method by inactivating microorganisms and producing a safe product. There are two principles that well describe the effects of high pressure. The first is the principle of le Chatelier, which describes that any phenomenon (phase transition, chemical reactivity and reaction, change in molecular configuration) accompanied by a decrease in volume will be increased by pressure. It is expected that increase in temperature results in a volume increase. In addition, the reaction rate increases with increasing temperature (Arrhenius' law). The second is that pressure is instantaneously and uniformly transmitted independent of the size and the geometry of the food – what is known as the isostatic principle.

Research in the area of high-pressure (HP) food processing was further advanced following the work carried out in Japan in the 1980s (Gomes et al., 1998). HP processing has been commercialized to a limited extent in Japan, France, the USA and some other countries. The first commercial products produced by HP appeared on the market in 1991 in Japan. Currently, HP processing is one of the most studied emerging technologies (Fig. 9.2) and successfully used for products like fruit juices, jams, sauces, rice, cakes and desserts.

Figure 9.2 A pilot-scale high-pressure unit.

HP acts instantaneously and uniformly throughout the volume of the food. Generally, water has been used as the transmitting medium while the pressure is applied uniformly and instantaneously. The food is preserved evenly throughout due to the uniform isostatic application of the pressure. The compression process during pressurization generally increases the temperature of foods by approximately $3\,°C\,100\,MPa^{-1}$. The increase depends on food composition being higher for fatty foods. The process temperature during pressurization can be varied from below $0\,°C$ (to minimize any effects of adiabatic heat) to above $100\,°C$. Vessels are uniquely designed to withstand these pressures safely over many cycles. The main drawback of HP processing is the high capital cost of equipment.

HP has been applied and tested for many types of food products, with the majority of research being focused on the processing of fruit juices and inactivation of pectin methyl esterase or microbial inactivation of meat and dairy products. Recently, HP applications at elevated temperatures (initial temperatures $>70\,°C$), commonly referred to as high-pressure sterilization (HPS), have been introduced and suggested as a food preservation technology for spore inactivation to reach commercially sterile conditions (Meyer, 2000; Wilson and Baker, 2001; Van Schepdael et al., 2002). Furthermore,

HP combined with other treatments (combination of mild heat and pressure treatment) may increase its effectiveness and diminish the detrimental effects caused by excessive pressure on foods and food ingredients (Estrada-Giron et al., 2005). HP is also used for food structure modification and improves emulsification (Ahmed et al., 2003; Ahmed and Ramaswamy, 2004).

9.3   Applications of emerging technologies to pulses and processing

9.3.1   Dielectric heating: radiofrequency (RF) heating

The biggest challenge in the pulse industry is infestation by insect pests during storage, transport and distribution. These pests reduce the quality of products by direct damage through feeding and by the production of webbing and feces (Wang et al., 2010). Most of the importing countries have specific conditions concerning pests and their control. These may include phytosanitary disinfestation treatments, such as fumigation or certification that the shipment is free of certain insects. Methyl bromide and hot air treatment of pulses are commonly used techniques for fumigation. However, these techniques have several limitations including product quality degradation. Radiofrequency energy offers the possibility of rapidly increasing temperatures within bulk materials. Thus, there has been an increasing interest in developing advanced thermal treatments for post-harvest insect control in pulses using this method (Nelson, 1996; Wang et al., 2007).

Wang et al. (2010) studied post-harvest disinfestation of selected pulses (chickpea, green pea and lentil) by RF heating. A pilot-scale 27 MHz, 6 kW RF unit was used for the heating. A time period of only 5–7 min was sufficient to raise the temperature of 3 kg legume samples to 60 °C using RF energy, compared to more than 275 min when using forced hot air at 60 °C. RF heating uniformity in legume samples was improved by adding forced hot air, and back and forth movements on the conveyor at 0.56 m min^{-1}. The final temperatures exceeded 55.8 °C in the interior of the sample container and 57.3 °C on the surface for all three legumes, resulting in low uniformity index values of 0.014–0.016 (ratio of standard deviation to the average temperature rise) for the interior temperature distributions and 0.061–0.078 for surface temperature distributions. RF treatments combined with forced hot air at 60 °C to maintain the target treatment temperature for 10 min followed by forced room air cooling through a 1 cm product layer provided good product quality. No significant differences in weight loss, moisture content, color or germination were observed between RF-treated and control samples.

9.3.1.1 Dielectric properties of pulses

The dielectric properties of food materials describe how those materials interact with electromagnetic radiation (Ryynanen, 1995). The interaction between dielectric energy and food products at selected frequency range provides better understanding pertaining to processing by either microwave or radiofrequency heating. These properties of foods are important in the design of RF and MW processing equipment and in the design of foods and meals intended to make use of MW ovens (Mudgett, 1995). Limited information on dielectric properties of pulses is available.

Dielectric properties of three pulse flour samples (chickpea, green pea and lentil) at four different moisture contents (8–22% w/w) were recently reported by Guo et al. (2008, 2010). The measurements were carried out with an open-ended coaxial probe and impedance analyzer at frequencies of 10–1800 MHz and temperatures of 20–90 °C. The dielectric constant (ε_0) and loss factor (ε') of the pulse samples decreased with increasing frequency, and increased with increasing temperature and moisture content.

The log–log plot of ε_0 and ε'' at different frequencies is shown in Figure 9.3 for compressed lentil flour samples over the temperature range from 20 to 90 °C at 21.5% moisture content. It was found that both ε_0 and ε'' decreased with increasing frequency, especially at high temperatures. Furthermore, ε_0 and ε'' increased with increasing temperature, especially at RF frequencies. The loss factor had a negative relationship with frequency, especially at high temperatures and the low frequency range (e.g. below 200 MHz).

At the highest moisture content (20% w/w) lentil had the highest loss factor at 27 MHz and 20 °C, followed by green pea and chickpea. The difference in loss factor among the pulses did not show definite trends at 915 MHz. Deep penetration depths at 27 MHz could help in developing large-scale industrial RF treatments for post-harvest

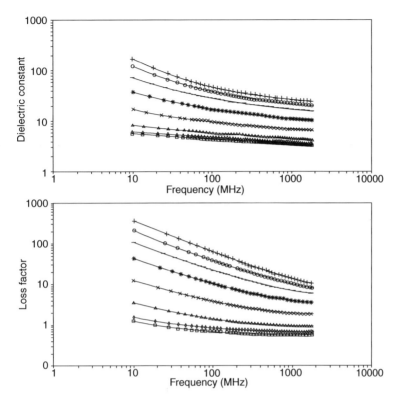

Figure 9.3 Frequency dependence of the dielectric properties of compressed lentil flour samples at temperature of 20($\square$), 30 ($\lozenge$), 40 (Δ), 50 ($\times$), 60 ($*$), 70 ($-$), 80 ($\circ$) and 90 °C ($+$) at maximum moisture content. (Source: Guo et al., 2010.)

insect control or other applications that require bulk heating in pulses with acceptable heating uniformity and throughputs.

9.3.2 Microwave (MW) heating

Microwave cooking of pulses, such as chickpeas and common beans, was evaluated by assessing the cooking quality (cooking time, firmness, cooking losses and water uptake) and the physicochemical, nutritional and microstructural modifications in starch and non-starch polysaccharides (Marconi et al., 2000). It was observed that MW cooking with sealed vessels enabled a significant reduction in cooking time, from 110 to 11 min for chickpeas and from 55 to 9 min for common beans compared to conventional cooking. The solid losses, released in the cooking water, were significantly less after MW cooking than after conventional cooking. Both cooking procedures produced a redistribution of the insoluble non-starch polysaccharides to the soluble fraction, although the total non-starch polysaccharides were not affected. Increases in *in vitro* starch digestibility were similar after both cooking processes. Scanning electron microscopy (SEM) studies revealed that the cotyledons maintained a regular structure; however, most of the cell wall was disrupted and shattered by both cooking methods. With regard to the heating treatments, it is interesting to note that despite the considerable differences between both heating methods (heating principle, pressure, temperature and speed), MW and traditional heating had similar effects on *in vitro* starch digestibility.

Microwave irradiation and extrusion cooking individually caused several changes in the functional, rheological and morphological characteristics of lentil starch (Gonzalez and Perez, 2002). Both treatments caused a decrease ($P < 0.05$) of moisture, crude protein, crude fiber, water absorption, solubility and swelling power, and an increase ($P < 0.05$) of ash, reducing sugars and absolute density. Granular morphology was significantly affected by microwave irradiation. It is worth mentioning that both methods reduced the retrogradation tendency of lentil starch, which could be interesting from an industrial point of view, because this is one reason that has limited its commercial use.

Khatoon and Prakash (2006) studied the effects of germination on the cooking quality and nutrient retention in microwave-cooked Bengal gram, green gram and horse gram. Ungerminated (UGL) and germinated legumes (GL) were cooked in a microwave oven and under pressure, followed by analysis of macro- and micronutrients (protein, ash, iron, thiamine and ascorbic acid), *in vitro* protein digestibility (IVPD) and starch digestibility (IVSD) and bioavailable iron. Results indicated that MW cooking required more water and time than pressure cooking. MW cooking caused 36–57% reduction of ascorbic acid while pressure cooking caused 10–30% loss. Cooking of UGL and GL, by both methods, increased the starch digestibility threefold. The IVPD of raw UGL ranged from 64.6% to 66.2% and that of GL was 72.4–73.9%. In cooked UGL, the IVPD ranged from 70.9% to 82.3% and, in GL, from 78.4% to 84.2%, showing a significant difference in cooking methods only in UGL. The iron bioavailability ranged from 11.5% to 18.7% in raw UGL, whereas it was 18.3–20.6% in GL. GL showed a higher content of thiamine and ascorbic acid, higher protein and starch digestibility and higher bioavailable iron, even after cooking. Cooking by both methods decreased the starch content, ranging from 7.5% to 25.1%, with the MW sample exhibiting a higher decrease.

9.3.3 Irradiation (IR)

Radiation treatment at low and moderate doses has been recommended for disinfestation of legumes (International Project in the Field of Food Irradiation News, 1980; Tilton and Burditt, 1983). The treatment has also been found to be effective for the reduction of flatulence-causing oligosaccharides (Rao and Vakil, 1983) and also of trypsin and chymotrypsin inhibitors (Iyer et al., 1980). Further, the nutritive quality of irradiated beans is improved (Reddy et al., 1979).

A study of gamma radiation (1–10 kGy) on selected functional properties in green gram, lentil, horse-bean and Bengal gram revealed that the water-absorption capacity of irradiated samples increased, although pasting temperature was not significantly altered (Rao and Vakil, 1985). Maximum gelatinization viscosity dropped systematically in all samples with increasing radiation dose. A significant reduction in cooking time of all pulses was recorded at 2.5–10.0 kGy compared to control samples. The force required to compress irradiated cooked samples was reduced. Sensory evaluation of cooked control and irradiated (5 kGy) samples revealed that there was no significant difference in acceptability.

The effects of low dose gamma radiation processing for insect disinfestation on functionally important sugars in commonly used pulses (mung, Bengal gram, horse beans, horse gram, cowpeas and rajma) were studied by Machaiah and Pednekar (2002). The separation profiles of legume carbohydrates were qualitatively comparable; distinct legume-specific quantitative changes were observed. The main flatulence-producing raffinose family oligosaccharides (RFO), stachyose and verbascose, which constituted 55–65% of soluble carbohydrates in these pulses, were degraded at different rates during 0–4 days of germination, with concomitant accumulation of easily metabolizable sucrose and reducing sugars. Radiation processing enhanced this in a legume-specific manner. Subtle differences in degradation of these oligosaccharides, between control and irradiated samples (0.25 and 0.75 kGy), were observed in the dry seeds of Bengal gram, horse beans and cowpeas; these were highly significant in mung and horse gram on the second day of germination and no change was noticed in rajma. Degradation of flatulence factors reflected an accumulation of sucrose in Bengal gram, cowpeas and rajma rather than reducing sugars, which were more prominent in mung and horse gram. These results conclusively indicate that radiation processing of the six legumes, at disinfestation dose (0.25 kGy) and germination (0–2 days), results in rapid degradation of flatulence factors without affecting their sprout lengths; this improves their nutritional acceptability, though subtle varietal differences are noticed. At higher dose (0.75 kGy), considerable reductions in their sprout lengths compared to the control were observed; however, their sensory attributes were unaffected.

A study of IR on amino acid profiles of pulses showed some interesting results. Khattak and Klopfenstein (1989) compared the amino acid profile of irradiated (0.5–5 kGy) and control samples of chickpea and mung bean seeds and found that sulfur-containing amino acids were radiation labile. Significant losses were observed for lysine, tyrosine and isoleucine. However, the percent available lysine was higher in the irradiated seeds (Table 9.1).

Celik et al. (2004) studied the effects of gamma irradiation (1 kGy, 5 kGy and 10 kGy) on water absorption properties, cooking quality and electrophoretic patterns of insoluble proteins of red and green lentil samples. The densitometric analysis indicated that the effects of irradiation on sodium dodecyl sulfate–polycrylamide gel electrophoresis (SDS-PAGE) patterns of red and green lentil proteins were not different. Generally, a 1 kGy irradiation dose did not significantly affect the water absorption properties of the lentil samples while significant increases were observed at the 5 kGy level. The dry and wet cooking times were found to be significantly decreased, as the irradiation level increased in all red and green lentil samples.

Cunha et al. (1993) studied the effects of gamma radiation on physical, chemical and sensorial properties of dry beans during

Table 9.1 Effect of γ-irradiation on percent available lysine content of mung bean and chickpea seeds

Dose (kGy)	Mung bean (%)	Chickpea (%)
0	84[b]	82[b]
0.5	85[ab]	83.5[b]
1.0	85[ab]	85.5[a]
2.5	86[ab]	86[a]
5.0	87[b]	86[a]

Means in the same column with the same letter are not significantly different ($P < 0.05$).
Adapted from Khattack and Klopfenstein (1989).

storage. The hydration capacity of IR-treated samples stored at 4–5 °C for 6 months was about 60% that of samples stored at 30 °C, 75% relative humidity (RH). IR of beans increased the hydration rate and decreased cooking time and hardness of the seeds. Sensorial attributes were preserved in all samples kept under refrigeration (4–5 °C) but deteriorated considerably at 30 °C, 75% RH. Overall, the sensory properties of the irradiated samples did not differ significantly from those of the controls. Storage for 6 months at 30 °C, 75% RH reduced phytate to 50% of the original values in the control and irradiated samples. For the same storage condition and time, methionine was reduced to 72 and 75% of the original value, in the control and irradiated samples, respectively.

9.3.4 High-pressure (HP) processing

High-pressure (HP) processing of biological macromolecules has received tremendous research interest for the potential benefits in functionality leading to the development of new products with desirable structures (Ledward et al., 1995). However, limited research works concerning the application of HP in processing of pulse grains have been reported.

At present, cereal and legume protein isolates have been commercially utilized as functional ingredients in food formulations for their high nutritional value and minimum cost. Protein isolates are also of special interest to processors and consumers due to their low fat content and potential use as a substitute for meat, fish and dairy products. The major functional properties of legume proteins are: hydration, gelation, emulsification, foaming and adhesion (Morr, 1990). Application of HP (200–600 MPa) on viccilin-rich red kidney bean protein isolate (KPI) resulted in gradual unfolding of the protein structure, as evidenced by gradual increases in fluorescence strength and disulfide bond (SS) formation from free sulfhydryl (SH) groups, and a decrease in denaturation enthalpy change (Yin et al., 2008). Differential scanning calorimetry (DSC) thermograms showed a prominent endothermic peak, clearly attributed to thermal denaturation of vicilin component (7S). The onset temperature of denaturation (T_o) and thermal denaturation temperature (T_d) of the vicilin component of KPI were unaltered by HP treatment (Table 9.2), suggesting that the structure of this protein component was very compact and not easily susceptible to HP treatment.

The protein solubility of KPI was significantly improved at pressures of 400 MPa or higher, possibly due to formation of soluble

Table 9.2 Effect of high-pressure treatment on thermal properties of kidney bean protein isolate

Sample (MPa)	T_o (°C)	T_d (°C)	ΔH (J g^{-1})
0.101	88.3	93.9	16.2
200	88.1	93.8	13.8
400	88.7	93.9	12.4
600	88.5	93.7	11.2

Adapted from Yin et al. (2008).

aggregate from insoluble precipitate. HP treatment (200–400 MPa) significantly increased the emulsifying activity index (EAI) and emulsion stability index (ESI); however, EAI was significantly decreased at 600 MPa (relative to untreated KPI). The protein solubility (PS) of KPI was significantly improved by HP treatment at a pressure of 400 MPa or higher ($P < 0.05$) and the observation was duly supported by SEC analysis equipped with an RI detector. The PS improvement at pressures above 200 MPa was believed to be due to the transformation of insoluble aggregates to soluble ones with lower molecular weight. The formation of stable soluble aggregate seems to be closely related to the newly formed SS bonds after HP treatment at higher pressure levels (above 200 MPa).

High-pressure treatment of lupin proteins prior to emulsifying provided an interesting modification of the properties of emulsions (Chapleau and de Lamballerie-Anton, 2003). The modification of protein structure appeared from treatment at 400 MPa by aggregation of α-conglutin (11S globulin) and denaturation of β-conglutin (7S globulin). These modifications improved the emulsifying properties as indicated by a decrease of droplet size, flocculation and creaming index. After protein pressurization the adsorption at the interfacial film of protein increased and a selective adsorption was observed leading to modified rheological properties of emulsion.

Digestibility of protein treatment at 500 MPa presents a very beneficial effect on lupin protein digestibility, whereas treatment at 200 MPa is inefficient for lupin proteins (Chapleau and de Lamballerie-Anton, 2003). This study shows that HP treatment of lupin proteins induces modifications of *in vitro* digestibility: HP treatment cannot be regarded as neutral regarding digestibility of proteins.

HP enhances the effectiveness of disinfectant agents to inactivate natural microbial populations in mung bean sprouts. The effects of several combinations of pressure, temperature and two antimicrobial agents, hypochlorite and carvacrol, applied separately on mung bean seeds, on their germination capability and on native microbial loads of sprouts developed from treated seeds, were studied by Peñas et al. (2010). The pressurization at 250 MPa of mung bean seeds treated with hypochlorite (18 000 ppm) or carvacrol (1500 ppm) led to a significant inactivation of all natural microbial populations (>5 log units) in sprouts. Seed viability decreased as pressure increased, at all concentrations of both hypochlorite and carvacrol. Enhanced reductions of total aerobic mesophilic bacteria, total and fecal coliforms and yeast and mold populations were observed as pressure and hypochlorite/carvacrol concentrations increased. The optimal treatment at 250 MPa of seeds soaked in 18 000 ppm and 1500 ppm of calcium hypochlorite and carvacrol, respectively, maintained an acceptable germination rate (80% and 60%, respectively) and improved the microbial quality of the respective sprouts with reductions of more than 5 log cfu g^{-1}. These reduction levels can be considered a preservative goal for industrial mung bean sprout production. Similar results (5–6 D reduction) were also observed for pressure-treated germinated legume seeds (pea, chickpea, lentil, green gram) (Dostalova et al., 2007). The pressurization of germinated grain legume seeds is an effective method for prolonging of their shelf-life during storage.

The effects of high pressure (350–650 MPa) on thermal and rheological characteristics of lentil flour dispersion were studied by Ahmed et al. (2007b). The moisture contents of the lentil slurry were varied between 14 and 58% (w/w). DSC measurements of pressure-treated lentil dispersions indicated incomplete denaturation of lentil proteins. The peak (T_d) ranged between 135 and 110 °C as moisture content increased from 6.5 to 57%. The endothermic peak (T_d) of dispersions is most likely attributed to the thermal denaturation of globulin. A higher T_d for lentil could be associated with variation in legume, nature of protein, and presence of starch and fat.

The protein dentauration temperature (T_d) changed non-systematically with applied pressure and moisture content. No starch gelatinization peak was detected during thermal scanning of lentil slurries before or after pressure treatment. Furthermore, the retrogradation behavior of lentil gel diminished compared to thermally treated gel under similar conditions.

Dynamic rheological measurement of pressure-treated lentil slurry confirmed the gradual transformation of solid-like behavior to liquid-like behavior. The pressure-treated sample behaved as a

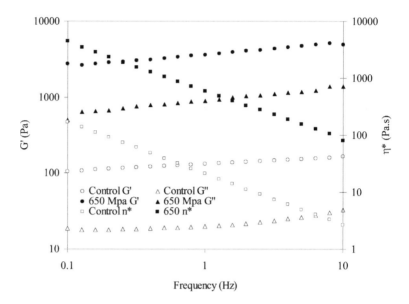

Figure 9.4 Typical rheogram on pressure effect of lentil flour dispersion.

true viscoelastic fluid. A typical rheogram of pressure-treated lentil dispersions at 20 °C is illustrated in Figure 9.4. It demonstrates that G' predominates over G'' in the frequency range of 0.1 to 10 Hz for control and pressure-treated samples, indicating a gel with more solid-like properties. Over the frequency range measured, G' and G'' were relatively frequency independent, due to extensive molecular entanglement. The complex viscosity increased with increasing pressure and supporting more viscoelastic characteristics of pressurized gel.

The viscoelasticity of the lentil dispersion was adequately supported by the calculated slopes of the linear regression of the power-type relationship of ln ω vs ln G' and ln G'' (Equation 9.2) (Table 9.3). The magnitude of the slope clearly indicates a loss of solid-like character and an increase in liquid-like character with an increase in moisture content. In addition, samples behaved as true viscoelastic fluids with equal magnitude of slope. High pressure converted lentil flour dispersion into a viscous gel by gelatinization and/or denaturation of starch and protein components. The slopes increased with increasing pressure level. For example, the slope for the 57% moisture-containing sample increased from 0.10 to 0.15 and from 0.15 to 0.18 for solid-like and liquid-like properties, respectively, after pressure treatment at 650 MPa for 15 min. The effect of high

Table 9.3 Effect of high-pressure treatment on viscoelastic properties of lentil starch gel

Moisture content (g 100 g⁻¹)	Pressure (MPa)	Slope for G' vs ω	Slope for G″ vs ω
37	0.101	0.09	0.11
	450	0.12	0.15
46	0.101	0.10	0.15
	450	0.13	0.17
58	0.101	0.11	0.15
	450	0.13	0.17
	650	0.15	0.18

Adapted from Ahmed et al. (2007b).

pressure on complex viscosity clearly demonstrates an increase in viscoelasticity with increasing pressure intensity and consequently the viscoelastic solid converts to a viscoelastic fluid.

$$G' \text{ or } G'' = k\omega^n \quad \text{ or } \ln G' \text{ or } \ln G'' = \ln k + n \ln \omega \qquad (9.2)$$

Fourier transform infrared (FTIR) spectroscopy confirmed insignificant changes in the amide band of pressure-treated lentil slurry (Ahmed et al., 2007b). The study has provided complementary information on pressure-induced structural changes on both the molecular and the submolecular level of lentil protein. The spectral intense bands for legumes are around 1520 (amide II; N–H bending) and 1660 (amide-I; C = O stretching) cm⁻¹ (Carbanaro et al., 2008). The spectra of control and pressurized samples in the spectral range of amide I' band are shown in Figure 9.5. The spectral shape at 1652 cm⁻¹ (amide I) remained unaffected after pressure treatment whereas an insignificant change was noticed at 1545 cm⁻¹.

Quagila et al. (1996) studied ultra-high isostatic pressure treatment of green peas in order to evaluate the possibility of using this process as an alternative to blanching. Different operative conditions (pressure range 400 to 900 MPa and treatment time 5 and 10 min) combined with a mild heat treatment at 60 °C were examined. The level of peroxidase inactivation by pressurization was comparable with that obtained by blanching treatment. The peroxidase residual activity for the pressure-treated sample (900 MPa for 5 min) was 12% versus 10% found for the water blanching treatment. In addition, the pressure-treated sample showed a higher retention of ascorbic acid (82% at 900 MPa) in comparison with water blanching systems (10%). These results could be

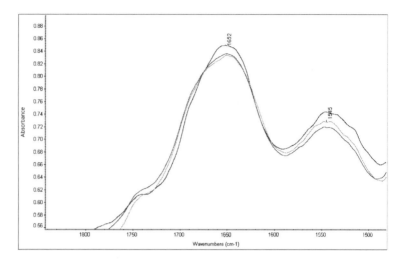

Figure 9.5 Effect of high pressure on Fourier transform infrared (FTIR) spectra of lentil proteins.

Table 9.4 Effect of high pressure and temperature combination on firmness of green pea

Pressure (MPa)	Temperature (°C)	Time (min)	Firmness (kN)
600	39	5	2.74
600	60	10	2.58
700	40.9	5	2.50

Adapted from Quagila et al. (1996).

improved if vegetables were packaged without water, which causes loss of some water-soluble components. Pressurized samples did not show differences in texture, although a significant reduction in firmness was observed with respect to fresh peas (Table 9.4). Increase of pressurization time did not cause any particular effect on the parameters measured, indicating that 900 MPa pressure for 5 min is sufficient.

9.4 Conclusions

This chapter summarizes the principles of emerging technologies and application of these technologies to the pulse-processing industry for the improvement of quality, functionality and safety. Among various emerging technologies, radiofrequency (RF), microwave (MW), irradiation (IR) and high-pressure (HP) processing have found potential application for storage and processing of pulses. Both MW/RF could be used in drying of pulses by minimizing infestation and drying of processed-based products based on pulses. HP studies have shown that the technique has potential for influencing the rheological and structural properties of pulse proteins. The rheological study of various pulse proteins treated with high pressure indicated gradual transformation of sol to gel and a predominant viscoelastic nature of the treated gel. Increased pressure intensity and heating have strong effects on the structural properties of pulse proteins and led to an increase in gel strength. SCFX technique is also a promising technology for temperature-sensitive formulation in extruded pulse foods. Although emerging technologies are promising, the cost of the equipment remains a challenge before food processors can adopt the technology.

References

Ahmed, J., Ramaswamy, H.S., 2004. Effect of high-hydrostatic pressure and concentration on rheological characteristics of xanthan gum. Food Hydrocolloid. 18, 367–373.

Ahmed, J., Ramaswamy, H.S., Alli, I., Ngadi, M., 2003. Effect of high-pressure on rheological characteristics of liquid egg. Lebensm. -Wiss. Technol. 36, 517–524.

Ahmed, J., Ramaswamy, H.S., Alli, I., Raghavan, V.G.S., 2007a. Protein denaturation, rheology, and gelation characteristics of radio-frequency heated egg white dispersions. Int. J. Food Prop. 10, 145–161.

Ahmed, J., Ramaswamy, H.S., Ayad, A., Alli, I., Alvarez, P., 2007b. Effect of high pressure treatment on rheological thermal and structural changes in Basmati rice flour slurry. J. Cereal Sci. 46, 148–156.

Ahmed, J., Ramaswamy, H.S., Kasapis, S., Boye, J.I., 2009. Introduction and plan of the book. In: Ahmed, J., Ramaswamy, H.S., Kasapis, S., Boye, J.I. (Eds.), Novel food processing: effects on rheological and functional properties. CRC Press, Boca Raton, pp. 1–6.

Carbonaro, M., Maselli, P., Dore, P., Nucara, A., 2008. Application of Fourier transform infrared spectroscopy to legume seed flour analysis. Food Chem. 108, 361–368.

Çelik, S., Yalçin, E., Başman, A., Köksel, H., 2004. Effects of irradiation on protein electrophoretic properties, water absorption and cooking quality of lentils. Int. J. Food Sci. Nutr. 55, 641–648.

Chapleau, N., de Lamballerie-Anton, M., 2003. Improvement of emulsifying properties of lupin proteins by high pressure induced aggregation. Food Hydrocoll. 17, 273–280.

Cunha, M.F., Sgarbieri, V.C., Damhio, M.H., 1993. Effects of pretreatment with gamma rays or microwaves on storage stability of dry beans. J. Agric. Food Chem. 4, 1710–1715.

Datta, A.K., 2001. Fundamentals of heat and moisture transport for microwaveable food product and process development. In: Datta, A.K., Anatheswaran, R.C. (Eds.), Handbook of microwave technology for food applications. Marcel Dekker, New York, pp. 115–166.

Decareau, R.V., 1985. Microwaves in the food processing industry. Academic Press, New York, pp. 100-105.

Dostalova, J., Kadlec, P., Strohalm, J., Culkova, J., Houska, M., 2007. Application of high-pressure processing for preservation of germinated legumes. High Pressure Res. 27, 139–142.

Estrada-Giron, Y., Swanson, B.G., Barbosa-Canovas, G.V., 2005. Advances in the use of high hydrostatic pressure for processing cereal grains and legumes. Trends Food Sci. Technol. 16, 194–203.

Gomes, M.R.A., Clark, R., Ledward, D.A., 1998. Effects of high pressure on amylases and starch in wheat and barley flours. Food Chem. 63, 363–372.

Gonzalez, Z., Perez, E., 2002. Evaluation of lentil starches modified by microwave irradiation and extrusion cooking. Food Res. Int. 35, 415–420.

Grahl, T., Märkl, H., 1996. Killing of microorganisms by pulsed electric fields. Appl. Microbiol. Biotechnol. 45, 148–157.

Gunasekaran, S., Chiyung, A., 1994. Evaluating milk coagulation with ultrasonics. Food Technol. 48 (12), 74–78.

Guo, W., Tiwari, G., Tang, J., Wang, S., 2008. Frequency, moisture and temperature-dependent dielectric properties of chickpea flour. Biosyst. Eng. 101, 217–224.

Guo, W., Wang, S., Tiwari, G., Johnson, J.A., Tang, J., 2010. Temperature and moisture dependent dielectric properties of legume flour associated with dielectric heating. LWT – Food Sci. Technol. 43, 193–201.

Heinz, V., Alvarez, I., Angersbach, A., Knorr, D., 2002. Preservation of food by high intensity pulsed electric fields – basic concepts for process design. Trends Food Sci. Technol. 12, 103–111.

Heinz, V., Toepfl, S., Knorr, D., 2003. Impact of temperature on lethality and energy efficiency of apple juice pasteurization by pulsed electric fields treatment. Innovat. Food Sci. Emerg. Technol. 4, 167–175.

Ho, S., Mittal, G.S., 2000. High voltage pulsed electric field for liquid food pasteurization. Food Rev. Int. 16, 395–434.

International Project in the Field of Food Irradiation News, 1980. In: Potter, W.T., Elias, P.S., Gottschalk, H.M. (Eds.), Food irradiation information, Vol. 10. Karlsruhe, Germany, p. 46.

Iyer, V., Salunkhe, D.K., Sathe, S.K., Rockiand, L.B., 1980. Quick cooking beans (*Phaseolus vulgaris* L.). 1. Investigations on quality. Qual. Plant. Foods Hum. Nutr. 30, 27.

Khadre, M.A., Yousef, A.E., Kim, J., 2001. Microbiological aspects of ozone applications in food: a review. J. Food Sci. 6, 1242–1252.

Khatoon, N., Prakash, J., 2006. Nutrient retention in microwave cooked germinated legumes. Food Chem. 97, 115–121.

Khattak, A.B., Klopfenstein, C.F., 1989. Effect of gamma irradiation on the nutritional quality of grains and legumes. II. Changes in amino acid profiles and available lysine. Cereal Chem. 66, 171–172.

Ledward, D.A., Johnston, D.A., Earnshaw, R.G., Hasting, A.P.M. (Eds.), 1995. High pressure processing of foods. Nottingham University Press, Nottingham.

Machaiah, J.P., Pednekar, M.D., 2002. Carbohydrate composition of low dose radiation-processed legumes and reduction in flatulence factors. Food Chem. 79, 293–301.

Manas, P., Pagan, R., 2005. A review: microbial inactivation by new technologies of food preservation. J. Appl. Microbiol. 98, 1387–1399.

Marconi, E., Ruggeri, S., Cappelloni, M., Leonardi, D., Carnovale, E., 2000. Physicochemical, nutritional, and microstructural characteristics of chickpeas (*Cicer arietinum* L.) and common beans (*Phaseolus vulgaris* L.) following microwave cooking. J. Agric. Food Chem. 48, 5986–5994.

Marcotte, M., 1999. Ohmic heating of viscous liquid foods (PhD thesis). Department of Food Science and Agricultural Chemistry, McGill University, Montreal.

Meredith, R., 1998. Engineers' handbook of industrial microwave heating. Institution of Electrical Engineers, London, pp. 1–18.

Meyer, R.S., 2000. Ultra high pressure, high temperature food preservation process. US Patent, 6,017,572.

Mizrach, A., Galili, N., Teitel, D.C., Rosenhouse, G., 1994. Ultrasonic evaluation of some ripening parameters of autumn and winter-grown "Galia" melons. Sci. Horticult. 56, 291–297.

Morr, C.V., 1990. Current status of soy protein functionality in food systems. J. Am. Oil Chem. Soc. 67, 265–271.

Mudgett, R.E., 1995. Electrical properties of foods. In: Rao, M.A., Rizvi, S.S.H. (Eds.), Engineering properties of foods. Marcel Dekker, New York, pp. 389–455.

Muthukumarappan, K., Halaweish, F., Naidu, A.S., 2000. Ozone. In: Naidu, A.S. (Ed.), Natural food anti-microbial systems. CRC Press, Boca Raton, pp. 783–800.

Nelson, S.O., 1996. Review and assessment of radio-frequency and microwave energy for stored-grain insect control. Trans. ASAE 39, 1475–1484.

Palaniappan, S., Sastry, S.K., 1990. Effects of electricity on microorganisms: a review. J. Food Process. Preserv. 14, 393–414.

Peñas, E., Gómez, R., Frías, J., Vidal-Valverde, C., 2010. Effects of combined treatments of high pressure, temperature and antimicrobial products on germination of mung bean seeds and microbial quality of sprouts. Food Control 21, 82–88.

Quaglia, G.B., Gravina, R., Paperi, R., Paoletti, F., 1996. Effect of high pressure treatments on peroxidase activity, ascorbic acid content and texture in green peas. Lebensm. -Wiss. Technol. 29, 552–555.

Rao, V.S., Vakil, V.K., 1983. Effects of gamma-irradiation on flatulance-causing oligosaccharides in green gram (*Phaseolus areus*). J. Food Sci. 48, 1791–1795.

Rao, V.S., Vakil, V.K., 1985. Effects of gamma-radiation on cooking quality and sensory attributes of four legumes. J. Food Sci. 50, 372–375.

Reddy, S.J., Pubols, M.H., McGinnis, J., 1979. Effect of gammairradiation on nutritional value of dry field beans (*Phaseolus vulgaris*) for chicks. J. Nutr. 109, 1307–1312.

Rizvi, S.S.H., Mulvaney, S., 1992. Extrusion processing with supercritical fluids. US Patent 5,120,559.

Rizvi, S.S.H., Mulvaney, S., 1995. Supercritical fluid extrusion process and apparatus. US Patent 5,417,992.

Roy, F., Boye, J.I., Simpson, B.K., 2010. Bioactive proteins and peptides in pulse crops: pea, chickpea and lentil. Food Res. Int. 43, 432–442.

Ryynanen, S., 1995. The electromagnetic properties of food materials: a review of the basic principles. J. Food Eng. 26, 409–429.

Tilton, E.W., Burditt, A.K., 1983. Insect disinfestation of grain and fruit. In: Josephson, E.S., Peterson, M.S. (Eds.), Preservation of food by ionizing radiation. CRC Press, Boca Raton, p. 215.

Van Schepdael, L.J.M.M., De Heij, W.B.C., Hoogland, H., 2002. Method for high pressure preservation. European Patent WO 02/45528.

Wang, S., Monzon, M., Johnson, J.A., Mitcham, E.J., Tang, J., 2007. Industrial-scale radio frequency treatments for insect control in walnuts: I. Heating uniformity and energy efficiency. Postharvest Biol. Technol. 45, 240–246.

Wang, S., Tiwari, G., Jiao, S., Johnson, J.A., Tang, J., 2010. Developing postharvest disinfestation treatments for legumes using radio frequency energy. Biosyst. Eng. 105, 341–349.

Wilson, M.J., Baker, R., 2001. High pressure/ultra-high pressure sterilization of foods. US Patent 6,207,215.

Wouters, P.C., Dutreux, N., Smelt, J.P.P.M., Lelieveld, H.L.M., 1999. Effects of pulsed electric fields on inactivation kinetics of *Listeria innocua*. Appl. Environ. Microbiol. 65, 5364–5371.

Yin, S.W., Tang, C.H., Wen, Q.B., Yang, X.Q., Li, L., 2008. Functional properties and *in vitro* trypsin digestibility of red kidney bean (*Phaseolus vulgaris* L.) protein isolate: effect of high-pressure treatment. Food Chem. 110, 938–945.

Zimmermann, U., Pilwat, G., Riemann, F., 1974. Dielectric breakdown of cell membranes. Biophys. J. 14, 881–899.

Pulse-based food products

10

Nissreen Abu-Ghannam[1], Aoife Gowen[2]
[1]School of Food Science and Environmental Health, Dublin Institute of Technology, Dublin, Ireland
[2]UCD School of Agriculture, Food Science and Veterinary Medicine, University College Dublin, Dublin, Ireland

10.1 Introduction

Interest in the utilization of pulses in the developed world is on the increase. Factors contributing to this include their reported nutritional and health benefits, changes in consumer preferences, increasing demand for variety/balance, changes in demographics (age, racial diversity), rise in the incidence of food allergies and ongoing research on production and processing technologies (Boye et al., 2010). In many developing countries, pulses have had a long history of use as a human food. In India, dhal (split chickpea) and food products prepared from dhal flour are quite popular. In Pakistan and Bangladesh, a range of staple products are based on pulses including "phutna" (roasted grains), "pakora" (oil fried), "kadi" (boiled in buttermilk), "roti" (chickpea flour in combination with wheat flour), "dholka" (fermented product) and "satu" (roasted chickpea flour with cereal flours). Cowpea is widely utilized in Thailand in a range of main, snack and dessert products. In many African countries, cowpea is cooked with vegetables and spices to make a gruel which is eaten with other staples such as cassava, yam or cereal. In

Pulse Foods: Processing, Quality and Nutraceutical Applications. DOI: 10.1016/B978-0-1238-2018-1.00007-0

Turkey, chickpea is a major staple and it is consumed in many forms including soup, meals, desserts, baby foods and snacks (Williams and Singh, 1988).

Despite the increasing interest in pulse production and considering the general trend towards healthy eating in contemporary culture, it may be surprising to learn that pulses are underutilized as a food source in Europe (Schneider, 2002) and the USA (Messina, 1999). In addition, it has been reported that pulse consumption in developing countries is in decline (FAO, 2005). Widespread use of pulses as a primary staple food has been limited by the presence of antinutritional factors including enzyme inhibitors, lectins, phytates, cyanoglycosides and phenolics. Polyphenols have contradicting positive effects on human health and it has been reported that they have anticarcinogenic and antioxidant properties (Gamez et al., 1998). Public health recommendations and nutritional research advocating the consumption of pulses (Leterme, 2002; Bazzano et al., 2009; Tupe and Chiplonkar, 2010) coupled with developments in convenient pulse products that encourage their increased utilization may reverse the downward trend in pulse consumption.

10.2 Common forms of pulse foods

Soybeans are the most widely produced of all legumes, comprising over 80% of world production. Figure 10.1 shows the world production of various pulse products (excluding soybean) in 2007 (FAO, 2009). The major pulse products, viewed in terms of world production, are dry beans, chickpeas, dry peas and green beans. The following section attempts to give an overview of the most common pulse types used for human consumption.

10.2.1 Green beans

The immature or "green" bean is the unripe form of the legume. China is the main producer of green beans worldwide with a production value exceeding 1000 million dollars in 2007 (FAO, 2009). Consumption of green beans comprises a major part of the Western diet due the convenience and healthy perception associated with this type of pulse product. Many fresh varieties such as petit pois, mangetout, sugar snap peas, green beans and string beans are popular

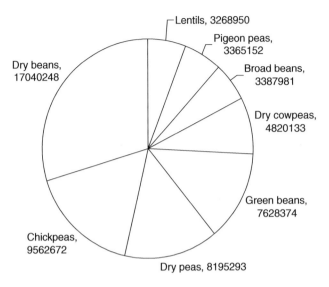

Figure 10.1 World production (MT) of major pulses in 2007 (faostat.org) not including soybeans (http://faostat.fao.org/site/339/default.aspx).

in Europe and the USA. Edamame, a variety of soybean popular in Japanese cuisine, is grown for human consumption and can be eaten in the pod. Blanching by immersion in boiling water for a short period of time (≈ 3 min) is commonly applied to green beans to render them edible. Some factors limiting the consumption of green beans are a short growing season and relatively short shelf-life. Freezing is commonly applied to green beans in order to extend their shelf-life. Many pulse crops are allowed to develop to the mature "dry" form in order to lengthen their shelf-life and to make them available for consumption throughout the year.

Box 10.1 Pulse focus: Cowpeas

The cowpea (Fig. 10.2) is a hardy, multipurpose crop, grown widely in the African continent due to its ability to grow in sandy soil and its tolerance against drought. Its importance in farming systems is well recognized, as it plays a major role in crop rotations, enhancing soil nitrogen levels. It may be eaten green or dried for future use. The Republic of Niger produced the highest value of dry cowpeas in 2007, with a production value exceeding

(*Continued*)

Box 10.1 (Continued)

Figure 10.2 Dry, wild and cultivated cowpea.

$200 million in 2007 (FAO, 2009). The network for genetic improvement of cowpea for Africa (NGICA, 2009) has published a report on the economic impact implications of improved cowpea varieties and storage technology in the peanut basin region of Senegal. "Operation Cowpea" began in 1985 in order to provide Senegalese farmers with a cash crop alternative to the peanut crop. Introduction of metallic drums for grain storage has drastically reduced grain losses which previously occurred due to insufficient drying, leading to mold and/or premature sprouting. Collaborative research carried out between the Senegal Institute for Agricultural Research and the Bean/Cowpea Collaborative Research Support Program has resulted in the development of two new cowpea varieties, Melakh and Mouride, which have relatively short growth cycles, good yields, and are typically consumed as green pods rather than dried, due to taste preferences among consumers.

10.2.2 Dry pulses

Pulses are typically dried post-harvest, to lengthen their shelf-life. After a drying period that can last up to 5 weeks, pulse moisture content is decreased to a level below 15% (d.b.) which inhibits microbial growth and deterioration, generally rendering them shelf stable at room temperature (Hnatowich, 2000). Dry legumes are not edible, as they contain harmful antinutritional factors, such as phytates, polyphenols, enzyme inhibitors and hemaglutinins (Abd El-Hady and Habiba, 2002). In order to inactivate antinutritional factors, dry legumes require hydration and thermal processing (Estevez and Luh, 1985; Abd El-Hady and Habiba, 2002). This is usually achieved by the application of hydrothermal treatment which also facilitates chemical reactions, such as starch gelatinization and protein denaturation.

Dry legumes are somewhat cumbersome to prepare, as they require long soaking (up to 16 h) and cooking (up to 1 h) times. Eihusen and Albrecht (2003) carried out a survey to investigate the consumption of dry beans by women in the age group 18–35. The long cooking time required for their preparation was cited as one of the main reasons for not eating dry beans.

The dry common bean (*Phaseolus* spp.) accounts for around one-third of EU pulse consumption; other dry pulses, e.g. split peas, "marrowfats" and whole green peas, are also an important sector of the EU pulse market. Dry pulses are often further processed for use as ingredients in other products, e.g. milled to make flour or soaked and cooked for canning. Boateng et al. (2008) examined bioactive components of dry beans, reporting that beans with darker seed coats, such as kidney beans and pinto beans, had significantly higher total phenolic content than beans with lighter colored seed coats (black-eyed peas and soybeans).

10.2.3 Canned pulses

Production of canned pulses usually requires the following processing steps: blanching, soaking, canning and pasteurization. Soaked or blanched beans are deposited into cans, into which tomato-sauce (for baked beans) or salted-water (for other bean varieties, e.g. chickpeas) is added. Cans are then heat-sealed and the canned beans are cooked under pressure. Canning is typically applied at 115 °C for 60 min (Van der Merwe et al., 2006). Canned baked beans, consisting of navy beans in tomato sauce, are one of the most popular bean products in the EU and USA. Although canned foods are convenient, people generally regard chilled and frozen foods to be healthier (Mintel, 2005). Uncertainty regarding the health effects of tin and additives, such as copper sulfate (used in some countries to enhance the green color of canned peas), has not improved the perception of canned pulses. However, recent research has shown that daily consumption of baked beans can reduce serum cholesterol in hypercholesterolemic adults (Winham and Hutchins, 2007).

Box 10.2 Pulse focus: Chickpeas

The chickpea (Fig. 10.3) was one of the first pulses to be domesticated, in the region near modern-day Turkey, over 8000 years ago (Singh, 1997). By far the majority of chickpeas consumed worldwide today come from the Indian subcontinent, which produced more than 6 million MT in 2007 (FAOstat, 2009). Kabuli and desi are the main cultivars of chickpea and are used for human consumption. Kabuli chickpeas, also known as Garbanzos (Spanish name), are relatively large in size, light yellow in color, have a thin seed coat and are grown in temperate regions. This type of chickpea is usually sold in either dry or canned format. Desi chickpeas are smaller than the kabuli variety, light tan to black in color, have a thick seed coat and are grown in semi-arid tropic regions (Johnson and Jimmerson, 2003). They require dehulling (removal of skin) prior to processing, and are used to make various food products which are popular in East Asia. In this region, desi type chickpeas are milled into a fine flour called Besan which is blended with wheat flour to make various baked products such as roti and chapatti (Malhotra et al., 1987; Muehlbauer and Singh, 1987).

Figure 10.3 Dry and green forms of chickpea.

10.3 Sprouted pulses

Sprouted pulses are typically produced from dry pulses after soaking followed by storage in high humidity and moderate temperature environments to encourage the process of germination. Mung bean and soybeans are two pulse varieties particularly popular for production of sprouts for human consumption in the USA (Robertson et al., 2002). Recently, both the kabuli and desi cultivars of chickpea have been investigated for the production of a sprouted pulse product. Sensory panel studies indicated preference of desi-based sprouts over kabuli ones; however, the nutritional characteristics of the kabuli type improved more than those of the desi variety after sprouting (Khalil et al., 2007).

It has been reported that sprouting improves the nutritional quality of pulses, increasing vitamin concentration, bioavailability of trace compounds and minerals, while reducing antinutritional factors such as stachyose and raffinose (El Adaway et al., 2002). Rodruiguez et al. (2008) studied the effect of germination conditions (time and total presence or absence of light) on the nutritional quality of a variety of pulses (peas, beans and lentils), concluding that germination time was overall more influential than the presence or absence of light. However, the moderate temperature and high humidity conditions in which germination occurs are ideal environments for microbial growth. Following several outbreaks of food-borne illness associated with their consumption, sprouts have been classified as a source of food-borne illness such as salmonella and *Escherichia coli* O157. This has led to the application of processing technologies, such as antimicrobial treatments to decontaminate sprouted products (USDA, 2001).

10.4 Traditional fermented products

Traditional processes of fermentation are applied to many pulse varieties resulting in products with improved nutritional quality and taste. Roy et al. (2007) investigated the microbiological quality of traditional pulse-based fermented foods in India including wadi and dhokla. Wadi is made from desi type chickpeas. After washing and soaking for 6–12 h, the chickpeas are ground to form a dough, mixed with spices and left at room temperature (20–27 °C) for 1–3 days. Following this they are molded into balls, smeared with oil and allowed to dry in the sun for 4–8 days. Dhokla is made from a blend of soaked rice and chickpeas which is left at room temperature for 12–15 h then steamed in an open cooker for around 10 min.

Young Kwon et al. (2009) provide an overview of fermented soybean foods consumed in China, Japan and Korea. Those originating in China include doubanjiang, douchi, tauchu, yellow soybean paste and dajiang. Natto and miso are two traditional fermented soybean products extremely popular in Japan. Their Korean counterparts, respectively chungkookjang and doenjang, along with kochujang and soy sauce are the most common fermented soybean foods found in Korea. The processing steps involved in the traditional manufacture of doenjang, kochujang and soy sauce are shown in Figure 10.4. After forming them into blocks, cooked soybeans are fermented outside for a period of 20–60 days by naturally present microorganisms. These fermented blocks, known as meju, are used to prepare soy pastes and sauce. In traditional production of meju, fermentation is dominated by *Bacillus* species during early stages of fermentation, after which *Aspergillus* species dominate, with *Aspergillus oryzae* being the major microorganism found in the final product. Chungkookjang and natto are short-term fermented soybean products, while tauchu, miso doenjang, kochujang and soy sauce undergo long-term fermentation.

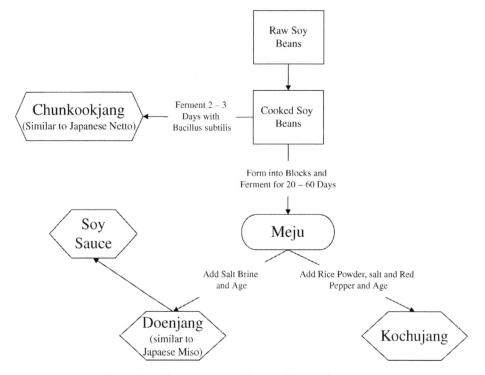

Figure 10.4 The preparation of fermented Korean foods made from soybeans (Young Kwon et al., 2009).

10.5 Developments in pulse products

Motivated by the growing consumer demand for healthy, convenient foods, there is growing research towards innovation in pulse production, particularly in the development of quick-cook pulses. Important characteristics of such products include short preparation times, microbial safety and high quality. Due to the highly competitive nature of the food industry, there is little information in the literature on pulse product development. However, a number of patents have been granted concerning the production of novel pulse products. One patent describes the production of a pulse-based snack food (US Patent 6090433) which retains the whole shape of the pulse from which it was made (e.g. pinto beans, red beans, black beans, navy beans, kidney beans, lentils and peas). The processes involved in the production of this product include soaking, cooking, chilling, dehydrating and enrobing with spices. Another more recent patent (US Patent 7022369) concerns the production of a pulse product by soaking, followed by sprouting, rinsing, regermination and dry roasting. The following sections focus on developments in dehydrated, extruded and fermented pulse products.

10.5.1 Quick-cook dehydrated pulses

Dry legumes generally require 12 h soaking, followed by 1 h cooking. Gowen et al. (2006a, 2008) have investigated the application of combined microwave convective hot air drying to pre-cooked chickpeas and soybeans for the production of a long shelf-life quick cooking pulse product. Combination drying of whole, pre-cooked chickpeas was investigated for three microwave power (210, 300, 560 W) and three air temperature (23, 160, 250 °C) settings. Combination of high air temperatures with high levels of microwave power was an efficient method of dehydration, leading to faster drying and lower dried product moisture content, due to internal fluxes of water heated by microwaves and fast removal of surface water by high-temperature convective currents.

Drying of foods usually results in shrinkage, due to water loss. Soybeans shrank during drying, to reach a final apparent volume less than 50% of their original volume. Chickpeas shrank to a lesser extent, reaching a final apparent volume of around 60% of their original apparent volume. The water activity (a_w) of dry samples

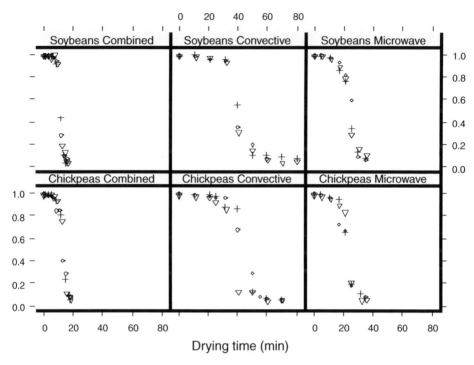

Figure 10.5 Water activity (a_w) as a function of drying method and time for dehydration of cooked chickpeas and soybeans.

(measured prior to blanching, soaking and cooking) was 0.54 ± 0.01 for chickpeas, and 0.67 ± 0.01 for soybeans. After soaking and cooking, the water activity of both chickpeas and soybeans was measured to be 0.99 ± 0.01. During the early stages of drying, water activity was almost constant, close to 0.99 (Fig. 10.5). Towards the end of drying, at a certain time (dependent on drying method), a rapid decrease in a_w to a value between 0.2 and 0.4 was observed. Further drying caused a_w to decrease slowly to a minimum value less than 0.2. In all cases, the time at which water activity became lower than 0.4 corresponded to the time at which volume changes became minimal. Drying chickpeas and soybeans to a water activity of 0.35 would produce shelf-stable dry products.

After dehydration by a combination of microwave power (210 W) and convective hot air (160 °C) for 14 min, dry chickpeas and soybeans were rehydrated in boiling water. Chickpea moisture content did not change significantly after 15 min rehydration, reaching a final moisture content of 60 ± 1 % (w.b.), which corresponds to 95% of their original, cooked moisture content (which is 63 ± 1% (w.b.)).

Similarly, soybean moisture content did not change significantly after 15 min rehydration, reaching a final moisture content of $60 \pm 1\%$ (w.b.), which corresponds to 91% of their original, cooked moisture content (which is $66 \pm 1\%$ (w.b.)).

After 15 min rehydration, chickpea hardness was measured to be $15 \pm 4\,N$, which is lower than the hardness of freshly cooked chickpeas ($22 \pm 4\,N$). Soybean hardness, after 15 min rehydration, was measured to be $10 \pm 1\,N$, which is lower than the hardness of cooked soybeans ($13 \pm 3\,N$). This demonstrates the softening effect of high temperature drying followed by rehydration.

Surface color of samples was measured after 15 min rehydration in terms of lightness (L^*), redness (a^*) and yellowness (b^*). In the case of soybeans, rehydrated samples ($L^* = 61 \pm 3$, $a^* = 7 \pm 2$, $b^* = 20 \pm 1$) were darker, redder and less yellow than cooked samples ($L^* = 68 \pm 1$, $a^* = 5 \pm 2$, $b^* = 23 \pm 2$). In the case of chickpeas, rehydrated samples ($L^* = 58 \pm 3$, $a^* = 9 \pm 2$, $b^* = 19 \pm 2$) were darker, slightly redder and less yellow than cooked samples ($L^* = 63 \pm 2$, $a^* = 8.8 \pm 2$, $b^* = 24 \pm 2$).

Sample appearance, after combined dehydration (MW = 210 W, T = 160 °C) for 14 min, and subsequent rehydration for 15 min in boiling water, is shown in Figure 10.6. Dehydration caused samples

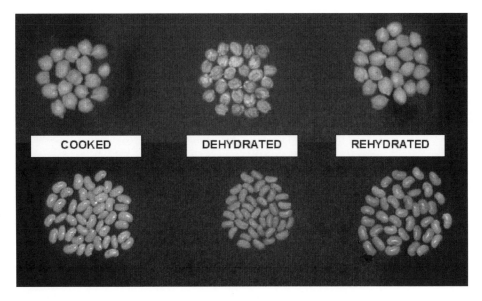

Figure 10.6 Cooked, dehydrated (by combination of microwave power (200 W) and convective hot-air (160 °C) for 13 min) and rehydrated (by immersion in boiling water for 15 min) chickpea (upper half of picture) and soybean (lower half of picture) samples.

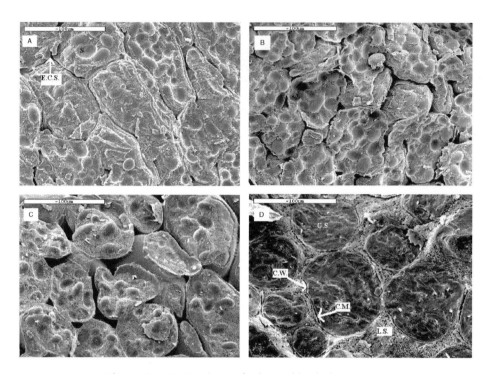

Figure 10.7 CryoSEM images for dry (A), blanched (B), soaked (C) and cooked (D) chickpeas. S = starch globule, G.S. = gelatinized starch, E.C.S. = extracellular space, C.W. = cell wall, C.M. = cell membrane, L.S. = leached solutes.

to become somewhat darker in color. Nonetheless, the appearance of rehydrated samples was quite similar to cooked ones.

The authors also examined cold stage scanning electron microscopy (cryo-SEM) micrographs of chickpeas subjected to convective hot air dehydration (100 °C) combined with two levels of microwave energy (100 W and 200 W) (Gowen et al., 2006b). Micrographs of dry chickpea (prior to soaking and cooking) displayed tightly packed cells, containing embedded (non-gelatinized) starch globules (Fig. 10.7A). Upon blanching, the space between cells increased, due to rapid intake of water and expansion caused by high-temperature blanching (Fig. 10.7B). After soaking, the space between cells was observed to increase again, due to imbibing of water: starch granules were still in the non-gelatinized state, and cell walls were visibly intact (Fig. 10.7C). Major microstructural changes were evident in the chickpea after cooking (Fig. 10.7D). Starch fractions within cells became gelatinized, (due to high temperatures experienced during cooking) and evidence of some cell wall separation was apparent. Extracellular spaces appeared to be filled with solutes that had leaked from cells during cooking, probably as a consequence of cell wall damage.

Dehydration of cooked chickpeas caused their cells to shrivel and shrink, becoming amorphous glassy structures (Fig. 10.8A–C). Leached solutes that were visible in the extracellular spaces of cooked

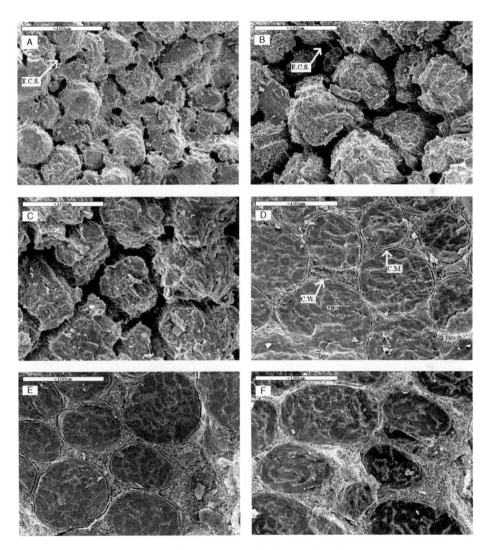

Figure 10.8 CryoSEM images for dehydrated ((A) air dried, (B) combination dried at 100 W, (C) combination dried at 200 W) and rehydrated ((D) air dried, (E) combination dried at 100 W, (F) combination dried at 200 W) chickpeas. C = cell, G.S. = gelatinized starch, E.C.S. = extracellular space, C.W. = cell wall, C.M. = cell membrane.

chickpea (Fig. 10.8D) were no longer present, having adhered to the cell surface during the drying process. Dehydrated structure was totally different to that of the dry chickpea before processing (Fig. 10.8A). For dehydrated samples (Fig. 10.8A–C), non-gelatinized starch was not visible and cell structure was amorphous.

Air-dried chickpeas (Fig. 10.8A) displayed smaller cells that were more tightly packed than samples dried by microwave–air combination. Larger extracellular spaces and cell sizes observed in combination-dried samples may have been due to high internal temperatures reached during microwave application: volumetric heating causes vaporization of internal water, which in turn leads to expansion of the internal cell structure. Cell size and spacing did not seem to be substantially different when samples dried at 200 W (Fig. 10.8C) and those dried at 100 W (Fig. 10.8B) were compared.

Rehydrated chickpeas that were dried using hot air alone (Fig. 10.8D) and at the lower (100 W) level of combination drying (Fig. 10.8E) displayed microstructure strikingly similar to that of cooked chickpea (Fig. 10.8D). Cell walls were clearly visible, gelatinized starch was contained within the cells and intracellular spaces were filled with leached solutes. However, for samples dried at the higher (200 W) level of combination drying (Fig. 10.8F), some cell wall damage was apparent. The space between the cell wall and membrane (for those samples dried at 200 W) was not clearly visible, as it had been for cooked chickpeas (see Fig. 10.7D), rehydrated chickpeas that were dried using hot air alone (Fig. 10.8D), and at the lower (100 W) level of combination drying (Fig. 10.8E). It is hypothesized that the cell wall was damaged due to large internal forces exerted on the cell during vaporization of water, caused by the application of the higher (200 W) level of microwave power. This could have created cracks through which intracellular material leaked, filling the space between the cell wall and membrane and therefore adhering to the cell membrane.

10.5.2 Extruded pulse products

One approach to enhancing the utilization of legumes has been the employment of extrusion cooking, which is characterized as a high temperature, short time process in which moistened, expansive, starchy and/or proteinaceous food materials are plasticized and cooked in a tube by a combination of moisture, pressure, temperature and mechanical shear. This results in fully cooked, shelf-stable food products with enhanced textural attributes and flavor (Havck and Huber, 1989).

Extrusion cooking has advantages including versatility, high productivity, low operating costs, energy efficiency and shorter cooking times. Legume extrusion has developed quickly in the last two decades. One of the benefits derived from pulse extrusion is enhancement of total dietary fiber (Berrios et al., 2010) or destruction of potentially antinutritional factors, especially protease inhibitors, hemagglutinins, tannins and phytates, which limit utilization of nutrients in legume seeds (Alonso et al., 1998, 2000).

Pulse extrusion resulted in the development of a range of value-added nutritious products such as pulse-based snacks or the fortification of cereal-based products such as pasta or tortilla with extruded pulse flours in order to enhance their amino acid profile. Investigation and optimization of the viability of pulse extrusion have been examined and reported.

Abd El-Hady and Habiba (2003) studied the effects of soaking seeds of peas, chickpeas, faba and kidney beans prior to extrusion. They concluded that a pre-soaking step followed by extrusion at barrel temperature and moisture content of 140 °C and 22% or 180 °C and 22% would improve the nutritive value of the studied legumes (Table 10.1). In this work, the *in vitro* protein digestibility of kidney

Table 10.1 Effect of soaking, feed moisture and extrusion temperature on trypsin inhibitor (U mg^{-1})

Treatment	Feed moisture (%)	Barrel temp. (°C)	Trypsin inhibitor (U mg^{-1})			
			Faba beans	Peas	Chickpeas	Kidney beans
Unsoaked	Raw (control)	Raw (control)	2.31	0.78	15.08	19.50
	18	140	n.d. [100]	n.d. [100]	n.d. [100]	n.d. [100]
	18	180	n.d. [100]	n.d. [100]	n.d. [100]	n.d. [100]
	22	140	n.d. [100]	n.d. [100]	n.d. [100]	n.d. [100]
	22	180	n.d. [100]	n.d. [100]	n.d. [100]	n.d. [100]
Soaked	Raw (control)	Raw (control)	1.85 [19.9]	0.66 [15.4]	13.69 [9.2]	19.20 [1.5]
	18	140	n.d. [100]	n.d. [100]	n.d. [100]	n.d. [100]
	18	180	n.d. [100]	n.d. [100]	n.d. [100]	n.d. [100]
	22	140	n.d. [100]	n.d. [100]	n.d. [100]	n.d. [100]
	22	180	n.d. [100]	n.d. [100]	n.d. [100]	n.d. [100]

n.d. = not detected. Values in the brackets indicate the percentage decrease over the control values.
Source: Abd El-Hady and Habiba (2003).

beans increased from 70.59% in unsoaked raw seeds to 79.26% in soaked extruded seeds at the same conditions. Also, this study reported that combination of soaking followed by extrusion treatments completely abolished the activity of α-amylase inhibitor.

Berrios et al. (2010) reported an increase in dietary fiber content and a decrease in the concentration of the raffinose family of oligosaccharides upon extruding lentil, dry pea and chickpea flours. The fabrication of value-added, pulse-based products with high dietary fiber content presents a viable opportunity for providing the population with beneficial food components that have been associated with lowering blood cholesterol, reducing risk of heart disease and improving glucose tolerance.

Extrusion cooking has also been applied in the production of instant type protein-rich porridge products by the incorporation of sorghum with cowpea as reported by Pelembe et al. (2002). Sorghum has a low protein content quality, particularly with respect to lysine. The addition of cowpea, a pulse which is slightly limiting in the sulfur-containing amino acids but rich in lysine, would complement the sorghum cereal and provide an opportunity for the production of a healthier extruded product. Pelembe et al. (2002) reported that addition of cowpea resulted in reduced viscosity of the developed instant porridge which could impact negatively on the acceptability of these products as food for adults. However, reduced viscosity could be very beneficial for infant feeding. It has been reported that cereal weaning porridges are a major cause of infant malnutrition in Africa because of their low nutrient density (Da et al., 1982). The addition of cowpeas not only improves the protein content and quality of the extruded porridge, but the reduced viscosity will make it possible to feed this product to infants and thus enhance food energy intake.

An important attribute of mixing pulse and cereal flours is to enhance the quality of the protein content. However, extrusion parameters could affect both the final nutritional content and sensory properties of the extruded product. Singh et al. (2007) provided a comprehensive review on the nutritional aspects of food extrusion and have reported that among the extrusion process variables, the feed ratio has the maximum effect on protein digestibility and that extrusion, even at 140 °C, does not have any adverse effect on protein digestibility which could be attributed to lesser residence time of food dough within the extruder. The same previous review highlighted that key variables for the destruction of trypsin inhibitors (which are considered one of the main factors for limiting pulse consumption) are high extrusion temperature, longer residence time and lower feed moisture content.

10.5.3 Snack-based products

The world snack food market was valued at US $66 billion in 2003, and typical products that come under this sector include potato chips, corn chips, popped corn, pretzels, seed snacks, mixed nuts and peanuts.

The majority of current extruded snack products are cereal based and the application of extrusion cooking to cereal flours has proven to be a commercially successful process.

Consumer acceptance of extruded foods is mainly due to the convenience, value and attractive appearance that such products offer. However, the production of extruded snack products that are mainly legume based has not yet been a viable application, mainly due to textural defects of the produced extrudates including low expansion ratio and harder and denser structure as compared to cereal-based extrudates.

In order to overcome this shortcoming, there have been numerous reported research works on the incorporation of legume flours with cereal flours for the extrusion of a new type of snack product with improved textural and nutritional characteristics. The addition of legume flour to cereal-based formulations has proven positively to impact their essential amino acid balance (Tharanathan and Mahadevamma, 2003). The addition of high-protein lysine material is known to affect positively the protein quality of cereal foods, since cereal grains are deficient in this amino acid. For example, corn starch provides all the features for production of a highly acceptable extruded snack food, but its nutritional value is far from satisfying the needs of health conscious consumers (Rampersad et al., 2003). The incorporation of navy and red bean flours with corn flour at levels of 15, 30 and 45% is one such attempt to produce fortified puffed snacks (Anton et al., 2009). In this reported study, the process variables (screw speed, moisture and temperature of the final zones) of a twin screw extruder were kept constant at 150 rpm, 22% and 160 °C. Using 30% bean flour, extrudates were produced with a level of crispness that was comparable to extrudates made from corn starch alone. At this level, crude protein was found to increase 12-fold, and red bean fortification yielded extrudates with higher levels of phenols and both 2,2-diphenyl-1-picrylhydrazyl radical (DPPH·)

and oxygen radical absorbance capacity (ORAC) compared to navy beans (Table 10.2). Phytic acid and trypsin inhibitors levels were reduced by nearly 50% and 100% in all bean extrudates compared to raw mixtures. Over 30% concentration of bean flour resulted in denser, less expanded and harder extrudates which is due to the high fiber content of bean flour that ruptures cell walls and prevents air bubbles from expanding to their maximum. This textural defect has been addressed by Berrios et al. (2004) where they reported adding increased levels of sodium bicarbonate to black bean flour to produce more expanded bean extrudates. This observation was attributed to the release of CO_2 from $NaHCO_3$ facilitated through the heat and moisture provided by the extrusion process.

10.6 Value-added pulse-based products

There is a growing market for the development of value-added bean-based products mainly targeting opportunities in the functional and nutraceutical sector. One such example is the use of isolated bean hulls as an ingredient for novel food products featuring high dietary fiber and high antioxidant levels. Antioxidant activity was reported in extracts, condensed tannins and pure flavonoids from colored genotypes of common seed coats (Madhujith and Shahidi, 2005). Anton et al. (2008) studied the effects of pre-dehulling treatments on some nutritional and physical properties of navy and pinto beans and concluded that, depending on the temperature and moisture conditions, important nutritional and physical properties of beans can be developed such as antioxidant activity and pasting properties, thus leading to the development of high quality bean-based products.

The fortification of bakery goods with flours of peas, beans and chickpeas has been already employed and appears to be promising in particular for the functional food market (Dalgetty and Baik, 2006). Pulses have a low glycemic index which benefits people with diabetes and cardiovascular disease. One of the practical applications of the common bean (*Phaseolus vulgaris* L.) is lowering the glycemic index of wheat flours. Anton et al. (2009) studied the shelf-stability and sensory properties of flour tortillas when fortified with pinto bean flour and reported a promising sensory acceptability (Table 10.3) of

Table 10.2 Effect of extrusion cooking on some nutritional properties of corn starch added to bean flour

	%	Protein (g100g⁻¹)	TP (mg FAE 100g⁻¹)		AOX¹ (µmol TE 100g⁻¹)		AOX² (µmol TE 100g⁻¹)	
			Raw	Cooked	Raw	Cooked	Raw	Cooked
Starch	100	$0.3 \pm 0.2^{*}$	$22.99 \pm 2.8^{*}$	$20.61 \pm 0.7^{*}$	$34.04 \pm 1.3^{*}$	$25.60 \pm 1.9^{**}$	$240 \pm 18.1^{*}$	$160.82 \pm 7.2^{**}$
Navy	15	3.89 ± 0.1^{a}	31.96 ± 2.9^{a}	$28.27 \pm 0.3^{a**}$	62.98 ± 7.1^{a}	$41.42 \pm 3.1^{a**}$	289.95 ± 20.4^{a}	$254.36 \pm 4.5^{a**}$
	30	6.88 ± 0.3^{b}	38.04 ± 0.9^{b}	$36.45 \pm 0.9^{b**}$	83.11 ± 3.7^{b}	$71.60 \pm 6.2^{b**}$	569.96 ± 55.4^{b}	508.22 ± 38.2^{b}
	45	10.09 ± 0.3^{c}	52.83 ± 1.3^{c}	$45.96 \pm 1.7^{c*}$	140.6 ± 7.5^{c}	$126.23 \pm 6.3^{c**}$	617.99 ± 12.9^{c}	584.46 ± 7.9^{cx}
Small red	15	3.03 ± 0.05^{a}	119.38 ± 1.3^{c}	$40.94 \pm 0.01^{a**}$	528.76 ± 4.1^{a}	$213.93 \pm 2^{a**}$	$615.28 \pm 8.3a$	$388.69 \pm 24.2^{a**}$
	30	6.23 ± 0.5^{b}	217.62 ± 9.4^{b}	$67.09 \pm 2.7^{b**}$	1131.12 ± 99.6^{b}	$399.38 \pm 4.1^{b**}$	1173.68 ± 84.7^{b}	1164.93 ± 42.2^{b}
	45	9.14 ± 0.6^{c}	361.56 ± 4.01^{c}	$94.82 \pm 1.8^{c**}$	1632.84 ± 20.3^{c}	$642.33 \pm 6.1^{c**}$	1750.36 ± 56.4^{c}	$1527.27 \pm 121.6^{c**}$

TP, total phenol content; FAE, ferrulic acid equivalents; AOX, DPPH. antioxidant activity; TE, trolox equivalent; AOX², ORAC antioxidant activity.

All the values are mean ± SD of three/four determinations adjusted to dry matter.

Data followed by the same character in the same column, within the same bean flour, are not significantly different ($P > 0.05$) using a Tukey test comparing all pairs of columns. *Significantly different compared to added bean flours in the same column using a two-tailed t-test ($P < 0.05$).

Source: Anton et al. (2009).

Table 10.3 Sensory scores of selected tortillas

Sensory attributes	Population (n)	Wheat control	Bean control[a]	Bean GG[b]
Overall acceptability	Total (55)	5.80 ± 1.71^a	7.18 ± 1.29^b	6.96 ± 1.36^b
	HT[c] (41)	5.77 ± 1.89^a	7.34 ± 1.16^b	$7.14 \pm 1.23^{b*}$
	RT[d] (14)	5.73 ± 2.28^a	6.55 ± 1.63^a	6.27 ± 1.68^a
Overall flavor	Total	5.73 ± 1.84^a	6.95 ± 1.41^b	7.00 ± 1.47^b
	HT	5.77 ± 1.89^a	6.98 ± 1.37^b	$7.20 \pm 1.37^{b*}$
	RT	5.55 ± 1.69^a	6.82 ± 1.6^a	6.18 ± 1.6^a
Texture	Total	6.11 ± 1.96^a	7.36 ± 1.37^b	7.24 ± 1.29^b
	HT	6.23 ± 1.92^a	7.45 ± 1.3^b	$7.52 \pm 1.07^{b***}$
	RT	5.64 ± 2.16^a	7.00 ± 1.61^a	6.09 ± 1.51^a

Sensory scores: 9 = most liked; 1 = most disliked. Data followed by the same character in the same row are not significantly different ($P > 0.05$).
*Significantly different compared to scores of the RT population, considering the same type of tortilla for the same sensory attribute, using a two-tailed t-test ($*P < 0.05$, $***P < 0.001$).
[a]Tortilla made out of composite flour containing $25\,g\,100\,g^{-1}$ pinto bean flour and $75\,g\,100\,g^{-1}$ Canada Western Red Spring wheat flour.
[b]Tortilla made out of composite flour containing $25\,g\,100\,g^{-1}$ pinto bean flour and $75\,g\,100\,g^{-1}$ Canada Western Red Spring wheat flour with $0.5\,g\,100\,g^{-1}$ guar gum.
[c]HT: Participants who declared that they usually consumed healthy bread and tortillas (multigrain, whole wheat, high-fiber and flax).
[d]RT: Participants who declared that they usually consumed regular white flour bread and tortillas.
Source: Anton et al. (2009).

the end-product, in addition to enhancing the antioxidant activity and increasing the levels of crude proteins.

Pulse-fortified pasta offers consumers the possibility to improve the nutritional quality of their diet by offering enrichment in proteins and fibers, vitamins, minerals and amino acid complementarity. The World Health Organization (WHO) and Food and Drug Administration (FDA) consider pasta a good vehicle for the addition of nutrients and pasta was one of the first foods for which the FDA permitted vitamin and iron enrichment in the 1940s (Marconi and Carcea, 2001). Petitot et al. (2010) investigated the addition of 35% (d.b.) of split pea or faba bean to durum wheat semolina to develop a nutritionally enhanced spaghetti (Table 10.4) The production of a fortified pasta required an adaptation in the pasta-making extrusion process including higher hydration levels and mixing speed to limit agglomeration of particles during mixing. Moreover, addition of legume flours induced a decrease in some pasta quality attributes such as higher cooking loss and lower breaking energy due to absence of gluten proteins which are responsible for the textural and cooking properties of durum wheat pasta. Durum wheat semolina

Table 10.4 Chemical composition of durum wheat (DW) semolina, split pea (SP) flour, faba bean (FB) flour and blends of 65% DW semolina and 35% SP or FB flour used to produce pasta

				Fibers			Vitamins						Minerals		
	Proteins	Starch	Lipids	Total	Soluble	Ash	B_1	B_2	B_3	B_5	B_6	B_9	Fe	Mg	P
Durum wheat semolina	13.3±0.2	77.6±0.3	1.7±0.04	2.4	0.7	1.1±0.02	0.28	<0.05	3.7	0.5	0.15	0.022	1.9	63	233
Split pea flour	21.4±0.4	47.9±0.5	2.5±0.09	13.4	1.1	2.9±0.02	0.83	0.14	3.1	1.5	0.15	0.025	5.4	115	495
Faba bean flour	29.0±0.8	44.4±0.4	2.2±0.09	7.3	1.0	3.4±0.02	0.66	0.20	2.5	0.7	0.23	0.145	6.0	124	630
65% DW semolina +35% SP flour[a]	16.1	67.0	1.9	6.2	0.8	1.7	0.47	0.05	3.5	0.8	0.15	0.023	3.1	81	324
65% DW semolina +35% SP flour[a]	18.8	66.0	1.8	4.1	0.8	1.9	0.41	0.07	3.3	0.5	0.18	0.065	3.2	84	372

Protein, starch, fiber, lipid and ash contents are expressed in b 100 g⁻¹ (d.b.); vitamin and mineral contents are expressed in mg 100 g⁻¹ (d.b.).
[a] Results obtained by calculation.
Source: Petitot et al. (2010).

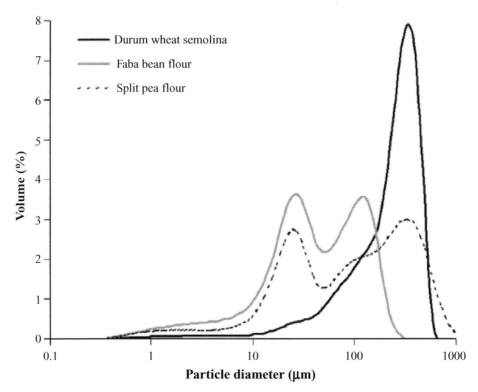

Figure 10.9 Particle size distribution of durum wheat semolina and legume flours. (Source: Petitot et al., 2010.)

is characterized by a mono modal particle size distribution (single peak at 340 μm) composed mostly of particles in the 160–1000 μm range as seen in Figure 10.9. In comparison, legume flours have a bimodal particle size distribution with two peaks: at 25 and 340 μm for split pea flour and at 25 and 120 μm for faba bean flour) characterized by a higher amount of fine (0–50 μm) and lower amount of larger (160–1000 μm) particles. As a consequence, fine particles of legume flours can absorb water faster than semolina particles which can generate heterogeneous dough. Petitot et al. (2010) concluded that pasta fortification with high levels of pulse flour presents a challenge and an opportunity to enhance the nutritional quality of a very popular food product and emphasize the optimization of the pasta-making process as the key to overcome sensory deficiencies when producing pulse-fortified pasta. Similar conclusions were reported by Chillo et al. (2008) upon the fortification of spaghetti in base amaranthus whole meal with quinoa, broad bean and chickpea flours.

Some examples of fortified pasta products with pulses include Barilla PLUS pasta (Barilla, Parma, Italy), and chickpeas flour to make non-wheat and non-gluten Del'Ugo Natrapasta (Ugo Foods, UK).

10.6.1 Pulse roasting

Roasting or puffing of pulses by subjecting them to high temperatures for a short time has been practiced in Asia, the Middle East and South America for many years. Roasted chickpea, commonly known as leblebi, is widely consumed as a traditional snack in the Mediterranean region and the Middle East. Leblebi can be classified as dehulled (Sari Leblebi and Girit Leblebi) or unhulled leblebi (e.g. Beyaz Leblebi and Sakiz Leblebi) (Coskuner and Karababa, 2004). However, the method of preparation may show considerable variation based on geographic location (Koeksel et al., 2004). The processing stages for all kinds of leblebi include cleaning and grading, soaking, tempering, boiling, roasting and dehulling. At the end of the production process, the chickpeas swell, soften and, after roasting, become crisp. Figure 10.10 shows the flowchart for the production of a dehulled roasted chickpea (Sari Leblebi). Microstructural observation indicates substantial structural changes occurring during processing as shown in Figure 10.11. Figure 10.11A shows that the raw chickpea posseses a tightly packed structure with no air spaces. However, during roasting (Fig. 10.11B), a large number of air spaces are formed in the cotyledon of the roasted chickpea. The compact structure of raw chickpeas subjected to roasting may cause an increase in the vapor pressure of water so that the steam that is generated causes the chickpeas to expand during roasting (Koeksel et al., 2004). High temperature during roasting (120–140 °C) may induce voids in the cellular matrix of grains. Leblebi can be characterized as a healthy snack with potential use as a natural "functional food" due to its chemical composition. It has high protein, cellulose and mineral content and is low in fat and calories. In addition, leblebi is characterized by its long shelf-life due to its low moisture content (Coskuner and Karababa, 2004). Product development is possible by coating leblebi with salt, sugar, spices, ginger and cloves.

10.6.2 Gluten-free products

The gluten-free market has grown by 27% since 2001, and this growth is projected to continue until at least 2010 (Heller, 2006). As gluten intolerance affects as many as one in 22 people all over the

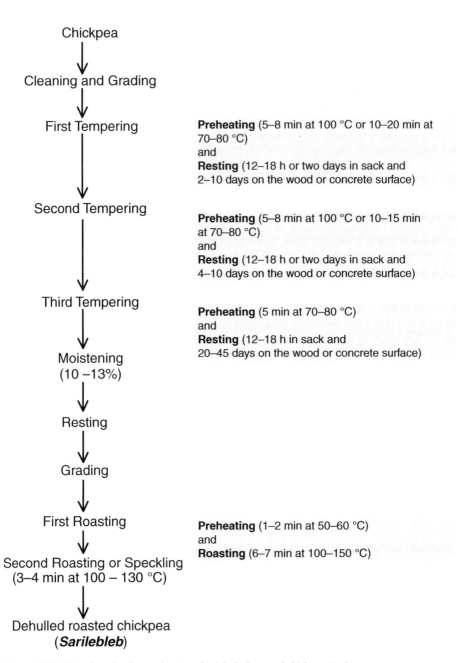

Figure 10.10 Flowchart for the production of a dehulled roasted chickpea (Sari Leblebi) (Coskuner and Karababa, 2004).

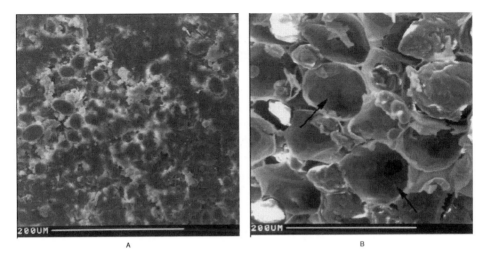

Figure 10.11 Scanning electron micrographs of (A) raw chickpea and (B) roasted chickpea (Petitot et al., 2010).

world, there are opportunities for developing specialty products for this market. Pulses are gluten-free and thus could provide some solutions for the gluten intolerance sector. Han et al. (2010) reported the development of a gluten-free, 100% pulse-based cracker snack using pulse fractions including chickpea, green and red lentil, yellow pea, pinto and navy bean flours and pea proteins. The developed products scored highly during consumer acceptance testing; however, the beany flavor was found to be the greatest challenge to the development of successful pulse-containing crackers. In addition, while good crispness of crackers containing chickpea flour, red lentil flour and pinto flour was reported, the gluten-free nature of these products could inhibit the development of a texture resembling the wheat-based crackers familiar to many general consumers. On the nutritional side, Han et al. (2010) reported that gluten-free crackers made with chickpeas showed three to six times higher levels of iron as compared to existing products on a per serving basis. Kinnikinnick Foods (Edmonton, Canada), a major producer of gluten-free foods, uses pea fiber in many of their products (Tosh and Yada, 2010).

10.6.3 Pulses and noodles

Pulse starches have occupied an important place in noodle preparation in several countries of the world, and mung bean (*Vigna radiate*) has been reported as the best raw material for starch noodle preparation (Chen, 1978). The ideal starch for starch noodle manu-facturing should have a high amylase content with a high iodine affinity value (6–7%), restricted swelling and a *C*-type Brabender viscosity pattern. Lii and Chang (1981) prepared noodles from azu-ki bean (*Vigna angularis*) starch and reported that noodles were of acceptable quality but not as good as mung bean starch noodles. Generally, pulse starch pastes are more viscous than those of cereal starches, indicating that pulse starches have a higher resistance to swelling and rupture than do cereal starches. Pulses provide a less expensive source for making noodles with physicochemical proper-ties comparable to those of cereal starches.

10.7 Developments in pulse-based fermented products

Biotechnological processes such as germination and fermentation are simple and economic methods that could be employed to im-prove the nutritive value of pulses by causing desirable changes in nutrient availability, texture and organoleptic properties. The popu-larity of these bioprocesses, which have been practiced in Eastern countries for centuries, is increasing nowadays in Western countries (Mwikya et al., 2000; Doblado et al., 2003; Granito et al., 2005).

Recently, pulses have also been studied for their antioxidant properties. The main antioxidant compounds in pulses are vitamins C and E, phenolic compounds and reduced glutathione (GSH). The effect of germination and fermentation on improving the antioxi-dant content of pulses has been examined. Fernandez-Orozco et al. (2009) reported an increase in total phenolic content and antioxi-dant capacity of chickpeas upon natural or induced fermentation (with 10% v/v *L. plantarum*) leading to the development of pro-cessed chickpea flours with added value that could be utilized by the food industry as functional ingredients.

Black beans are a nutritionally rich food that provides a plentiful supply of protein and calories and contain isoflavone, vitamin E, soaponin, carotenoid and anthocyanin. In China, black beans are used to prepare koji and further processed to manufacture tradi-tional fermented condiments such as *In-yu* black sauce and *In-si* or *Ttou-si*, the dried by-product of the mash of black bean sauce (Su, 1980). These fermented condiments are widely included in Chinese daily meals. Lee et al. (2008) reported that the DPPH free radical scavenging activity, Fe^{2+} iron-chelating ability and reducing power of the black bean can be enhanced using solid-state fermentation (SSF) with GRAS filamentous fungi commonly employed for the preparation of oriental fermented foods. The previous study also suggested that the incorporation of dried fermented black bean powder as an ingredient may enhance the health and nutrition of various food formulae.

Soybeans are traditionally fermented into tempeh by using fungi such as *Penicillium* spp., *Aspergillus* spp. and *Fusarium* spp. Recent developments in soybean fermentation investigated using *Rhizopus oligosporus* as a means of transforming the unpleasant beany taste due to lipoxygenase catalyzed oxidation of soybean oil into hexanal and hexanol. In contrast, soybean tempeh fermented by *R. oligosporus* has a pleasant mushroom odor (Feng et al., 2005). The co-inoculation with lactic acid bacteria during tempeh production was found to improve the hygienic safety of this process.

The SSF process represents a technological alternative for improving the nutritional value of legumes and to obtain edible products with palatable sensory characteristics.

SSF has been applied to produce tempeh flour from chickpea in order to enhance its nutritional properties, which include protease inhibitors, phytates and phenolic compounds. Angulo-Bejarano et al. (2008) showed that SSF of chickpeas using *Rhizopus oligosporus* strain enhanced net protein retention, protein efficiency ratio and protein digestibility corrected amino acid score. Chickpea fermented flour may be considered for the fortification of widely consumed legume-based food products.

Traditional fermented pulse-based products usually rely on uncontrolled fermentation using lactic acid bacteria which could require lengthy processing times. One such example is *"idli"*, which is a popular traditional fermented rice-black gram dhal-based snack food of India. The batter is prepared by using parboiled rice and black gram dhal in the proportion of 3:1. Preparation of *idli* is a time-consuming process as the fermentation process could reach up to 24 h. Reduction in fermentation time of the *idli* batter is of great commercial importance for large-scale *idli* production. Iyer and Ananthanarayan (2008) investigated the addition of α-amylase to speed up the fermentation process and reported that levels of α-amylase up to 15 U resulted in batters that fermented within 8 h at 30 °C and yielded sensory properties that were similar to the naturally fermented batter.

In developing countries, weaning foods are based on staples, usually cereals, that are processed into porridges. Cereal-based gruels are of poor nutritional value and have some implication in the incidence of protein-energy malnutrition, a major cause of high infant mortality rates in developing countries. Co-fermentation of cereal and grain legumes could improve the nutritional value of traditional weaning foods. Egounlety (2002) reported the natural fermentation of maize with soybeans, cowpea or groundnut in the development of high protein-energy pulse-fortified weaning foods referred to collectively as *Ogi* foods. The resulting products contained protein and energy levels that fluctuated between 15.55 and 19.30% and 17 505 and 18 726 kJ kg^{-1}, respectively, and were higher than recommended levels for weaning foods. Fermentation of a combination of cereals and pulses appears to be an appropriate technique for the production of high protein-energy, socially acceptable and nutrient-dense weaning foods at household levels in developing countries.

Bacillus spp. fermented pulses products include among others dawadawa and soumbala made from African locust bean, and fermented soy bean foods like Japanese natto, Thai Thua-nao and Indian kinema. Dawadawa is a culinary product that can be used to enhance or intensify meatiness in soups, sauces and other prepared dishes and is considered the most important food condiment in the entire West/Central African Savanna region. *Bacillus subtilis* fermentations are characterized by extensive hydrolysis of protein to amino acids, peptides and ammonia and a rise of the pH. Kiers et al. (2000) reported that fermentation of soy bean using several *Bacillus* spp. resulted in major biochemical changes including breakdown of protein and carbohydrate into water-soluble low molecular weight peptides, oligosaccharides and monosaccharides. The resulting modifications led to a product with high nutrient availability in which the need for degradation of nutrients by gastrointestinal enzymes is minimal.

Onyango et al. (2004) studied the effects of fermentation of maize-finger millet blend in the production of uji, the principal weaning food in Eastern Africa, followed by extrusion as a means of enhancing the functional and physical properties of ugi. Fermentation followed by extrusion results in the acidification of the feed material which causes depolymerization of starch which, in turn, affects reconstitutability, viscosity, expansion behavior and strength of the extruded product (Sriburi and Hill, 2000). The incorporation of extrusion with submerged culture fermentation offers a new possibility for utilizing pulse crops with further modification in their physical and nutritional values, thus widening the offering of pulse-based products.

10.8 Conclusions

The incorporation of pulses with cereals, through the application of different technologies as presented throughout this chapter, has far-reaching nutritional benefits to a wide range of the population worldwide. A number of traditional methods of pulse processing were also reviewed, including localized modifications to these methods to give better utilization of pulses produced. These developments have the potential to raise significantly the profile of pulses as a highly nutritious and globally available food product base.

References

Abd El-Hady, E.A., Habiba, R.A., 2003. Effect of soaking and extrusion conditions on antinutrients and protein digestibility of legume seeds. Lebensm. -Wiss. Technol. 36, 285–293.

Alonso, R., Orue, E., Marzo, F., 1998. Effects of extrusion and conventional processing methods on protein and antinutritional factor content in pea seeds. Food Chem. 63, 505–512.

Alonso, R., Grant, G., Dewey, P., Marzo, F., 2000. Nutritional assessment *in vitro* and *in vivo* of raw and extruded peas (*Pisum sativum* L.). J. Agric. Food Chem. 68, 2286–2290.

Angulo-Bejarano, P., Verdugo-Montoya, N., Cuevas-Rodriguez, E., 2008. Tempeh flour from chickpea (*Cicer arietinum* L.) nutritional and physiochemical properties. Food Chem. 106, 106–112.

Anton, A.A., Ross, K.A., Beta, T., Fulcher, R.G., Arntfield, S.D., 2008. Effect of pre-dehulling treatments on some nutritional and physical properties of navy and pinto beans (*Phaseolus vulgaris* L.). Lebensm. -Wiss. Technol. 41, 771–778.

Anton, A.A., Ross, Fulcher, R.G., Arntfield, S.D., 2009. Physical and nutritional impact of fortification of corn starch-based extruded snacks with common bean (*Phaseolus vulgaris* L.) flour: effects of bean addition and extrusion cooking. Food Chem. 113, 989–996.

Bazzano, L.A., Thompson, A.M., Tees, M.T., Nguyen, C.H., Winham, D.M., 2009. Non-soy legume consumption lowers cholesterol levels: a meta-analysis of randomized controlled trials. Nutr. Metab. Cardiovasc. Dis. doi: 10.1016/j.numecd.2009.08.012.

Berrios, J., De, J., Wood, D.F., Whitehand, L., Pan, J., 2004. Sodium bicarbonate and the microstructure, expansion and color of extruded black beans. J. Food Process. Preserv. 28, 321–335.

Boateng, J., Verghese, M., Walker, L.T., Ogutu, S., 2008. Effect of processing on antioxidant contents in selected dry beans (*Phaseolus* spp. L.). LWT – Food Sci. Technol. 41, 1541–1547.

Boye, J., Zare, F., Pletch, A., 2010. Pulse proteins: processing, characterization, functional properties and applications in food and feed. Food Res. Int. 43, 414–431.

Chen, C.Y., 1978. The noodle quality and physicochemical properties of various starches isolated by wet process. MS thesis, National Taiwan University.

Chillo, S., Laverse, J., Falcone, P.M., Protopapa, A., Del Nobile, M.A., 2008. Influence of the addition of buckwheat flour and durum wheat bran on spaghetti quality. J. Cereal Sci. 47, 144–152.

Coskuner, Y., Karababa, E., 2004. Leblebi: a roasted chickpea product as a traditional Turkish snack food. Food Rev. Int. 20, 257–274.

Da, S., Akingbala, J., Rooney, L., Miller, F., 1982. Evaluation of To quality in sorghum breeding program. In: Rooney, L., Murty, D., Mertin, J. (Eds.), Proceedings of International Symposium on Grain Quality. ICRISAT, Patancheru, India, pp. 11–23.

Dalgetty, D.D., Baik, B.K., 2006. Fortification of bread with hulls and cotyledon fibers isolated from peas, lentils and chickpeas. Cereal Chem. 83, 269–274.

Doblado, R., Frias, J., Munoz, R., Vidal-Valverde, C., 2003. Fermentation of *Vigna sinensis* var. carilla flours by natural microflora and Lactobacillus. J. Food Prot. 66, 2313–2320.

Egounlety, M., 2002. Production of legume-fortified weaning foods. Food Res. Int. 35, 233–237.

Eihusen, J., Albrecht, J.A., 2003. Dry bean intake of women ages 19–45. J. Am. Diet. Assoc. 103, 51–52.

Estevez, A., Luh, B., 1985. Chemical and physical characteristics of ready-to-eat dry beans. J. Food Sci. 50, 777–781.

FAO, 2005. Main findings of the study "Pulses: past trends and future prospects". http://www.fao.org

FAO, 2009. http://faostat.fao.org

Feng, X., Eriksson, R., Schnurer, J., 2005. Growth of lactic acid bacteria and *Rhizopus oligosporus* during barley tempeh fermentation. Int. J. Food Microbiol. 104, 249–256.

Fernandez-Orozco, R., Frias, J., Zielinski, H., 2009. Evaluation of bioprocesses to improve the antioxidant properties of chickpeas. Lebensm. -Wiss. Technol. 42, 885–892.

Gamez, E.J.C., Luyengi, L., Lee, S.K., Zhu, L., Zhou, B., Fong, H.H.S., 1998. Antioxidant flavonoid glycosides from *Daphniphyllum calycinum*. J. Nat. Prod. 61, 706–708.

Gowen, A., Abu-Ghannam, N., Frias, J., Oliveira, J., 2006a. Optimisation of dehydration and rehydration properties of cooked chickpeas (*Cicer arietinum* L.) undergoing microwave–hot air combination drying. Trends Food Sci. Technol. 17, 177–183.

Gowen, A., Abu-Ghannam, N., Frias, J., Oliveira, J., 2006b. Comparative study of quality changes occurring upon dehydration and rehydration of cooked chickpeas (*Cicer arietinum* L.) subjected to combined microwave–convective hot-air dehydration and convective hot-air dehydration. J. Food Sci. 71, E282–E289.

Gowen, A., Abu-Ghannam, N., Frias, J., Oliveira, J., 2008. Modeling dehydration and rehydration of cooked soybeans subjected to combined microwave–hot-air drying. Innovat. Food Sci. Emerg. Technol. 9, 129–137.

Granito, M., Torres, A., Frias, J., Guerre, M., Vidal-Valverde, C., 2005. Influence of fermentation on the nutritional value of two varieties of *Vigna sinensis*. Eur. Food Res. Technol. 220, 176–181.

Han, J., Janz, J., Gerlat, M., 2010. Development of gluten-free cracker snacks using pulse flours and fractions. Food Res. Int. 43, 627–633.

Havck, B., Huber, G., 1989. Single screw vs twin screw extrusion. Am. Assoc. Cereal Chem. 34, 930–939.

Heller, L., 2006. Gluten-free foods and beverages in the US Packaged Foods Inc. <marketresaerch.com>

Hnatowich, G., 2000. Pulse production manual 2000. Saskatchewan Pulse Growers, Saskatoon.

Iyer, B., Ananthanarayan, L., 2008. Effect of α-amylase addition on fermentation of idli – a popular south Indian cereal-legume-based snack food. Lebensm. -Wiss. Technol. 41, 1053–1059.

Johnson, J., Jimmerson, J., 2003. Chickpeas (Garbanzo beans). Agricultural Marketing Policy Center, Briefing No. 55.

Khalil, A., Zeb, A., Mahmood, F., Tariq, S., Khattak, A., Shah, H., 2007. Comparison of sprout quality characteristics of desi and kabuli type chickpea cultivars (*Cicer arietinum* L.). LWT – Food Sci. Technol. 40, 937–945.

Kiers, J.L., Van Laeken, A.E.A., Rombouts, F.M., Nout, M.J.R., 2000. *In vitro* digestibility of *Bacillus* fermented Soya bean. Int. J. Food Microbiol. 60, 163–169.

Koeksel, H., Sivri, D., Scanlon, M.G., Bushuk, W., 1998. Comparison of physical properties of raw and roasted chickpeas (leblebi). Food Res. Int. 31, 659–665.

Lee, I.H., Hung, Y.H., Chou, C.C., 2008. Solid-state fermentation with fungi to enhance the antioxidant activity, total phenolic and anthocyanin contents of black bean. Int. J. Food Microbiol. 121, 150–156.

Leterme, P., Carmenza Muñoz, L., 2002. Factors influencing pulse consumption in Latin America. Br. J. Nutr. 88 (S3), 251–254.

Lii, C., Chang, S., 1981. Characterisation of red bean (*Phaseolus radiatus* var. Aurea) starch and its noodle quality. J. Food Sci. 46, 79–85.

Madhujith, T., Shahidi, F., 2005. Antioxidant potential of pea beans (*Phaseolus vulgaris* L.). J. Food Sci. 70, S85–S89.

Malhotra, R.S., Pundir, R.P.S., Slinkard, A.E., 1987. Genetic resources of chickpea. In: Saxena, M.C., Singh, K.B. (Eds.), The chickpea. CAB International, Wallingford, pp. 67–81.

Marconi, E., Carcea, M., 2001. Pasta from non traditional materials. Cereal Food World 46, 522–530.

Messina, M., 1999. Legumes and soybeans: overview of their nutritional profiles and health effects. Am. J. Clin. Nutr. 70, 439S–450S.

Mintel, 2005. Frozen and canned fruit and vegetables – UK. Mintel Reports – UK series. Mintel Group.

Muehlbauer, F.J., Singh, K.B., 1987. Genetics of chickpea. In: Saxena, M.C., Singh, K.B. (Eds.), The chickpea. CAB International, Wallingford, pp. 99–125.

Mwikya, S., Camp, J., Yiru, Y., Hyghebaert, A., 2000. Nutrient and antinutrient changes in finger Mollet (*Eleusine coracana*) during sprouting. Lebensm. -Wiss. Technol. 33, 9–14.

NGCIA, 2009. http://www.entm.purdue.edu/ngica/default.html

Onyango, C., Henle, T., Ziems, A., Hofmann, T., Bley, T., 2004. Effect of extrusion variables on fermented maize-finger millet blend in the production of uji. Lebensm. -Wiss. Technol. 37, 409–415.

Pelembe, L., Erasmus, C., Taylor, R., 2002. Development of a protein-rich composite sorghum–cowpea instant porridge by extrusion cooking process. Lebensm. -Wiss. Technol. 35, 120–127.

Petitot, M., Boyer, L., Minier, C., Micard, V., 2010. Fortification of pasta with split pea and faba beans flours: pasta processing and quality evaluation. Food Res. Int. 43, 634–641.

Rampersad, R., Badrie, N., Comissiong, E., 2003. Physico-chemical and sensory characteristics of flavoured snacks from extruded cassava/pigeonpea flour. J. Food Sci. 68, 363–367.

Robertson, L., Johannessen, G., Gjerde, B., Loncarevic, S., 2002. Microbiological analysis of seed sprouts in Norway. Int. J. Food Microbiol. 75, 119–126.

Rodriguez, C., Frias, J., Vidal-Valverde, C., Hernandez, A., 2008. Correlations between some nitrogen fractions, lysine, histidine, tyrosine, and ornithine contents during the germination of peas, beans, and lentils. Food Chem. 108, 245–252.

Roy, A., Moktan, B., Sarkar, P., 2007. Microbiological quality of legume-based traditional fermented foods marketed in West Bengal, India. Food Control 18, 1405–1411.

Schneider, A., 2002. Overview of the market and consumption of pulses in Europe. Br. J. Nutr. 88, S243–S250.

Singh, K., 1997. Chickpea (*Cicer arietinum* L.). Field Crops Res. 53, 161–170.

Singh, S., Gamlath, G., Wakeling, L., 2007. Nutritional aspects of food extrusion: a review. Int. J. Food Sci. Technol. 42, 916–929.

Sosulski, K., Sosulski, F., 2005. Legume: horticulture, properties and processing. In: Hui, Y.H. (Ed.), Handbook of food science, technology and engineering, Vol. 3. CRC Press, pp. 18-1–18-14.

Sriburi, P., Hill, S.E., 2000. Extrusion of cassava starch with either variations in ascorbic acid concentration or pH. Int. J. Food Sci. Technol. 35, 141–154.

Su, Y., 1980. Traditional fermented food in Taiwan. In: Proceedings of the Oriental fermented foods, 15. Food Industry Research and Development Institute, Taipei.

Tharanathan, R., Mahadevamma, S., 2003. A review: grain legumes a boon to human nutrition. Trends Food Sci. Technol. 14, 507–518.

Tosh, S., Yada, S., 2010. Dietary fibres in pulse seeds and fractions: characterization, functional attributes, and applications. Food Res. Int. 43, 450–460.

Tupe, R., Chiplonkar, S., 2010. Diet patterns of lactovegetarian adolescent girls: need for devising recipes with high zinc bioavailability. Nutrition 26, 390–398.

US Patent 6090433. Leguminous snack food and process of making the same. US Patent Issued on July 18, 2000.

US Patent 7022369. Method of preparing nutritional legume product.

USDA, 2001. New technologies for decontaminating sprouting seeds and produce with easily damaged surfaces. http://www.reeis.usda.gov/web/crisprojectpages/404137.html

Van der Merwe, D., Osthoff, G., Pretorius, A., 2006. Evaluation and standardisation of small-scale canning methods for small white beans canned in tomato sauce. J. Sci. Food Agric. 86, 1115–1124.

Wang, H., Murphy, P., 1994. Isoflavone content in commercial soybean foods. J. Agric. Food Chem. 42, 1666–1673.

Williams, P., Singh, U., 1988. Quality screening and evaluation in pulse breeding. In: Summerfield, R.J. (Ed.), World crops: cool season food legumes. Kluwer Academic Publishers, London, pp. 445–456.

Winham, D., Hutchins, A., 2007. Baked bean consumption reduces serum cholesterol in hypercholesterolemic adults. Nutr. Res. 27, 380–386.

Young Kwon, D., Daily, J., Jin Kim, H., Park, S., 2010. Antidiabetic effects of fermented soybean products on type 2 diabetes. Nutr. Res. 30, 1–13.

Novel food and industrial applications of pulse flours and fractions

Zubair Farooq[1], Joyce I. Boye[2]
[1]McGill IR Group, Department of Food Science and Agricultural Chemistry, Macdonald Campus, McGill University, Sainte-Anne-de-Bellevue, QC, Canada
[2]Food Research and Development Centre, Agriculture and Agri-Food Canada, Saint-Hyacinthe, QC, Canada

11.1 Introduction

Pulses are among the important crops grown on cultivated land and are a rich source of carbohydrates, proteins, vitamins and minerals, providing readily available energy. Pulses are consumed as whole cooked grains, or they may be processed into flours or fractionated to obtain ingredients such as pulse starch flours, fibers, protein concentrates and isolates. The growing trend toward convenient,

Pulse Foods: Processing, Quality and Nutraceutical Applications. DOI: 10.1016/B978-0-1238-2018-1.00007-0

ready-to-eat, fast-cooking foods has fueled interest in the processing of pulses to obtain flours and fractions that can be easily used as ingredients in food product development.

In addition to being a rich source of energy, pulses contain bioactive compounds that may have health benefits as well as components that may impart unique functional properties to foods and food products. Moreover, pulses are rich in some essential amino acids lacking in cereals and are therefore an excellent complement to these foods. Thus, in both the developed and developing worlds, legume-fortified cereals may offer great potential for the creation of many novel nutritional foods. Examples of food products that can be prepared from pulses and pulse flours and fractions include bakery products, soups, pasta, noodles and canned products. The incorporation of pulse ingredients into meat products has also been a recent trend. This chapter summarizes the key properties of pulse flours and fractions and describes some of the trends in the application of pulse flours and fractions for the development of novel food and industrial products.

11.2 Brief description of the major types of pulse flours and fractions

11.2.1 Whole pulse flours

Pulses are normally consumed as whole seeds or dhals (splits) which are obtained by splitting the seeds during or after removal of the hulls. To improve the convenience of use and decrease cooking times, whole pulses or splits can be milled into flours. A variety of mills are available commercially for pulse flour milling. The type of mill used and the conditions used during milling as well as for pre–treatment determine the particle size distribution and product characteristics of the final product. Commercially available pulse flours include pea flour, chickpea flour, bean flour and lupin flour. The hulls of pulses are a rich source of fiber. Thus, whole pulse flours contain higher amounts of fiber than flours obtained from the dhal. When milled to obtain whole flour, pulse grains also contain other food ingredients that contribute to product functionality depending on the processing conditions used. In general, the composition of whole pulse flours does not vary very much from that of the pulse seed. The composition of flours obtained from the dhals will, however, vary depending on the processing equipment and milling conditions used.

11.2.2 Pulse fiber

Pulses, especially pea, chickpea and lentil, are rich sources of dietary fiber, which has various beneficial physiological effects for human health. Pulse fiber can be isolated by milling and fractionation of pulse seeds for incorporation into commercial food products to increase their fiber content and/or serve as functional ingredients. There are two main types of fiber, classified as soluble and insoluble fiber. The majority of the soluble fiber in pulses is present in the cotyledon (Meuser, 2001). Fiber is the main ingredient in the hull of pulses and is also an essential ingredient of flour composition. Pulse hull fiber is mostly insoluble fiber (Meuser, 2001). Commercial sources of pea fiber, some containing over 80% fiber, are available on the market today.

Fiber displays special functional properties because it absorbs water and oil, thereby introducing swelling properties that can change flour rheology and product behavior. Particle size is an important characteristic of pulse fibers as it can also influence its functionality. Furthermore, by developing complexes with other ingredients, especially proteins, fiber can be incorporated into gelling and foaming systems which are desirable attributes in special product formulations.

11.2.3 Pulse starch

Starch is another important component of pulses. Starch contains amylose (AM) and amylopectin (AMP) chains that create inter- and intramolecular complexes in variable physicochemical environments, causing molecular and physical changes that influence product functionality. During thermal treatment in the presence of suitable amounts of moisture, starch undergoes changes in its water and oil absorption capacities, resulting in multidynamic changes to the starch molecule. These changes result in the gelatinization, retrogradation and dextrinization of food products during processing.

Various processes have been developed to obtain high starch flours from pulses, particularly pea. Both wet processing techniques using solvents and dry processing techniques such as air classification are used. Pea starch products containing over 50–80% starch are available commercially and are widely used in a variety of food products.

In food applications, starch- and carbohydrate-rich fractions are often used as fat replacers in meat products and are good sources of slow-release carbohydrates. Legume starches are characterized by a high AM content (24–65%), which is responsible for a higher resistant starch (RS) content that may account for their lower digestibility. Resistant and slowly digestible starches contribute to a low glycemic index in starch-based food products, a feature that is beneficial in the management of diabetes and hyperlipidemia. These starches may also help prevent colon cancer, increase absorption of minerals, and are a substrate for the growth of probiotic organisms (Hoover and Zhou, 2003).

These important nutritional characteristics leading to the increased use of legume starches in dietetic foods increase their value in the industrial manufacture of low-fat, high-fiber foods that may help reduce the risks of obesity, cardiovascular disease and colon cancer (Pietrasik and Janz, 2010). Although the mechanisms are not completely understood, research carried out over the past several decades has demonstrated the beneficial effect of legume consumption in terms of lowering blood cholesterol levels and normalizing blood glucose and insulin levels. For example, serum cholesterol is reduced by consuming plant foods, including a daily serving of cooked beans, lentils or chickpeas (Tosh and Yada, 2010). The carbohydrate fraction of legumes (fiber and resistant starch) may contribute to these beneficial effects.

11.2.4 Pulse protein flours, concentrates, isolates

Pea, chickpea, bean and lentil contain 17–30% protein with varying concentrations of essential amino acids. Lupin contains 34–43% protein on a dry basis with very little starch and higher amounts of oil (5.4–10%). The major proteins found in pulses are globulins and albumins. Albumins are water soluble and comprise enzymatic proteins, protease inhibitors, amylase inhibitors and lectins, and have molecular masses (MM) ranging between 5000 and 80 000 Da. Globulins, on the other hand, are salt soluble. The major globulins found in pulse legumes are legumin (11S) and vicilin (7S). Other minor proteins found in pulses include prolamins and glutelins (Gupta and Dhillon, 1993; Saharan and Khetarpaul, 1994). Prolamins are alcohol soluble and are characterized by a high proportion of proline and glutamine. Glutelins, on the other hand, are soluble in dilute acid or alkali detergents, and in the presence of chaotropic or reducing agents (Osborne, 1924).

From a functionality perspective, proteins play a crucial role in the manufacture of food products. Proteins contribute to solubility, emulsification, foaming ability, gelling characteristics and oil absorption, influencing functional properties. In response to changing physicochemical processing environments, proteins individually and in combination with other ingredients can cause meaningful interactions that lead to diversified functional properties in the end-products.

Similar to pulse starch flours, pulse protein flours with protein contents ranging from >50 to >90% can be obtained from pulses using both dry (air classification) and aqueous extraction techniques. The most readily available commercial supplies of high protein-containing pulse flours are pea protein concentrates and isolates. Pulse proteins are rich in lysine, leucine, aspartic acid, glutamic acid and arginine, but lack methionine, cysteine and tryptophan. Thus, blending of pulse proteins with cereal flours significantly increases their nutritional value.

11.3 Functional properties of pulse flours and fractions

Functional properties are the intrinsic physicochemical characteristics of foods that affect product behavior during and after processing. Components such as fiber, starch and proteins significantly contribute to product functionality; as a result, their presence in an ingredient or food product (e.g. hulled vs. unhulled grains) could alter the functionality of the food. Addition of pulses can impart unique functional properties to foods and food products by modifying their texture. Moreover, molecular interactions between the major pulse components and other ingredients present in the formulation can result in the development of novel structures and food properties. The composition of some whole pulse flours and pulse fractions is presented in Table 11.1. Examples of key functional attributes of pulse flours are presented in Table 11.2.

Table 11.1 Proximate composition of flours from whole pulse seeds							
Component (%)	Yellow peas	Green lentils	Red lentils	Desi chickpea	Kabuli chickpea	Kidney bean	Red kidney bean	Green pea
Moisture	14.19±0.03	10.68±0.01	9.27±0.11	9.26±0.04	12.06±0.15			6.3g
Protein[a]	21.09±0.28 (24.57)	23.03±0.08 (25.78)	25.88±0.12 (28.52)	20.52±0.24 (22.62)	16.71±0.15 (19.00)	23.58	16.89	21.3g
Ash	2.42±0.01	2.39±0.03	2.34±0.02	3.04±0.01	2.76±0.01	3.83	1.14	2.7g
Fat	2.01±0.28	0.82±0.003	0.53±0.003	5.23±0.15	7.34±0.54	0.83	1.64[b]	0.6g (lipids)
Carbohydrate (calculated by difference)	60.29	63.08	63.10	61.94	61.14	60.01	–	69.2g

Boye et al., (2010a, b); Veenstra et al., (2010).
[a]Values in parentheses are dry basis; [b]ether extract.

Table 11.2 Functional properties of pulses

Properties	Beach pea	Black bean	Black gram	Chickpea	Cowpea	Grasspea	Jack bean	Kidney bean	Lentil	Mung bean	Navy bean	Northern bean	Pigeon pea
Gelatinization[a] (°C)	74.2	81.2–86.7	76.0–80.4	69.8–81.5	81.0	74.2–74.6	95.0	76.0–90.8	71.0–82.0	75.0	71.0–91.0	70.0–79.9	80.6–87.0
Pasting temperature (°C)		76[b]	50.3[c]	75[b]	76[b]	71.2–75.8[d]	84[b]	75.2[c]	72 [b]	76.5[c]	73.9[c]	79[b]	50.9[c]
Highest viscosity		1120[b] (Bu)	5147[c] (mPa s)	410–460[b] (Bu)	340[b] (Bu)	96–103[d] (Bu)	645[b] (Bu)	1980–2286[c] (CP)	540–560[b] (Bu)	591[c] (CP)	2746[c] (CP)	680[b] (Bu)	4025[c] (mPa s)
Swelling factor[e] (at 90 °C)	24.9			20.9–25.9	27.0	22.1–27.0		20.7–24.1	22.5–24.5	37.8	24.4	14.0	6.3
Swelling power[f] (at 90 °C)		10		9.2–10.4			33.9	13.5	10.9–11.9		13.5		11.8

Hoover et al. (2010).

[a]Measured by differential scanning colorimetry (DSC); [b]measured by Brabender viscoamylogram (BVA); [c] measured by rapid viscoanalyzer (RVA); [d]measured by microviscoamylogram (MVA); [e]granular swelling measured as swelling factor, which is reported as the volume of swollen starch granules to the volume of dry starch (measures only intragranular water); [f] granular swelling measured as swelling power (g/g) (measures both intergranular and intragranular water); CP, centipoises; Bu, Brabender units; mPa s, millipascals per second.

For food applications, important functional properties include solubility, water binding or hydration, foaming, emulsifying, thickening and gel-forming properties. These properties influence the overall quality and sensory perception of foods and are important in the manufacture of products such as baked goods, confectioneries, desserts, beverages, salad dressings and meat products. Understanding the factors that influence these functional properties and the mechanisms underlying the interactions that induce these properties is vital for the successful incorporation of whole pulses or pulse isolates, concentrates, hydrolysates or blends into various foods. Furthermore, this knowledge is required to identify ideal processing conditions (e.g. pH, temperature, and concentration) as well as ingredient interactions suitable for optimal product manufacturing. Factors which can influence the functionality of pulse flours and ingredients are discussed below.

11.3.1 Oil-absorption capacity

Fat- or oil-absorption capacity (FAC or OAC), also sometimes referred to as the fat- or oil-binding capacity (FBC or OBC) or fat- or oil-holding capacity (FHC or OHC), is calculated as the weight of oil absorbed per weight of sample. FAC varies for different pulse flours and fractions, usually ranging from ≈0.4 to $5\,\mathrm{g}\,\mathrm{g}^{-1}$.

Very few studies have systematically looked at the effect of processing on the FAC of pulse flours and fractions. The mechanism of fat absorption involves physical entrapment of oil. Thus, particle size, pulse flour composition, moisture content and microstructure are some of the factors that can impact FAC. Additionally, different protein compositional profiles and amounts of non-polar amino acid residues as well as differences in conformational features, and starch–protein–lipid binding could contribute to differences in oil retention characteristics in pulse flours (Lazou and Krokida, 2010).

FAC of pulse protein flours may also vary depending on the source and the conditions used for processing. Among pulse concentrates treated by ultrafiltration (UF), red lentil concentrate has the highest FAC, followed by yellow pea, green lentil, kabuli chickpea and desi chickpea; however, no significant differences were observed between the FAC of concentrates processed using isoelectric precipitation (IEP) (Boye et al., 2010a). FAC of chickpea protein isolate (PI) processed by micellization was $2 \, \mathrm{g} \, \mathrm{g}^{-1}$ but was only $1.7 \, \mathrm{g}$ g^{-1} for PI prepared by IEP (Paredes-López et al., 1991).

11.3.2 Water-holding capacity

Water-binding capacity (WBC), water-absorption capacity (WAC) and water-holding capacity (WHC) are the terms often used to describe the hydration properties of pulse flours and fractions and refer to the amount of water that can be absorbed per gram of sample material. These terms are frequently used interchangeably, although the measurement methods used may differ. WBC is an important measurement in food processing applications. Materials that have a low WBC may not be able to hold water effectively, whereas materials that have a high WBC or WAC may render food products brittle and dry, especially during storage (Boye et al., 2010a).

Water absorption results in swelling. The swelling power of flour depends on the concentration of protein, starch and fiber. Starch is known to exhibit high swelling power in water (Torruco-Uco and Betancur-Ancona, 2007), and when this power is combined with the swelling abilities of proteins and fiber, starch can contribute superior swelling ability. Thus, the composition of starch in pulse flours and fractions can influence their WBC and swelling properties.

The type, content and particle size of fiber can also influence the hydration properties of flours. In pea hulls, water retention and absorption, including the initial rate of water uptake, increased when mean particle size decreased from 950 to $300 \, \mu\mathrm{m}$, a phenomenon attributed to an increase in surface area and pore volume upon grinding. Conversely, water retention is slightly higher in coarsely ground navy bean hulls (particle size: $425–850 \, \mu\mathrm{m}$) than in finely ground hulls ($<150 \, \mu\mathrm{m}$) (Tosh and Yada, 2010).

The role of proteins in water absorption is equally important. WHC values for different pulse protein concentrates range between 0.6 and $2.7 \, \mathrm{g} \, \mathrm{g}^{-1}$ (Boye et al., 2010a). Interestingly, pea PI, which is the most readily available commercial pulse protein isolate, was found to have higher swelling ability than whole seed flour and fiber products, suggesting that proteins also contribute to swelling in seed flours, especially when the product is heated (Torruco-Uco and Betancur-Ancona, 2007; Agboola et al., 2010). Kaur and Singh (2007) found that protein isolates from different chickpea cultivars had higher WHC compared to the whole flours. WBC of whole pulse flour may, however, sometimes be higher than that of other derived flour products depending on processing and product composition. In another study, pea PI processed using dehulled seed had lower WHC values compared to whole seed flour perhaps owing to its lower levels of starch and fiber compared to fiber products or whole flour, where the flour is milled from whole seeds containing hulls (Agboola et al., 2010).

11.3.3 Protein (nitrogen) solubility index

Protein solubility is an important factor influencing product functionality and effective utilization. Solubility of whole pulse flours and fractions is greatly affected by environmental and processing conditions. The solubility of most pulse proteins is highest at low-acidic and high-alkaline pH values. Lowest solubility is observed at pH values close to the isoelectric point. This high solubility at acidic pH values could make pulse PIs very promising candidates for use in acidic and neutral beverages as well as in soup and salad dressing applications (Boye et al., 2010a,b). Opportunities may exist to enhance the solubility of pulse protein flours, concentrates and isolates through hydrolysis and physicochemical modification. For example, interactions of faba bean (*Vicia faba* L.) legumin protein and hydrolyzed legumin with polysaccharides (chitosan) increased its solubility at the isoelectric point and at higher pH values (Braudo et al., 2001). Thus, processing techniques that enhance pulse protein solubility could be explored to enhance application of pulse flours and fractions in foods requiring greater solubility.

11.3.4 Gelling properties

Gelation occurs when proteins and starches form a three-dimensional network that is resistant to flow under pressure. This ability to form gels is an important functional property in food processing and food formulation for products such as puddings and jellies and in many dessert and meat applications. Gelling capacity is frequently measured by the least gelling concentration (LGC), which may be defined as the lowest concentration required for a self-supporting gel to form. A lower LGC indicates that the sample has a better capacity to form gels.

Pulse starches and proteins have excellent gel-forming characteristics. The presence of moderate amounts of starch, fiber and protein in food systems results in molecular interactions that favor gel formation on heating and subsequent cooling. Moreover, intermolecular interactions can be substantially enhanced at temperatures colder than room temperature. This increase in gelling ability upon cooling results from a decrease in mobility of the molecules with decreasing temperature, which enhances bond formation within and between the biopolymer molecules.

Gelation is affected not only by protein and starch concentration but also by the type of protein and starch and the presence of nonprotein components. Changes in physicochemical conditions, such as pH and ionic strength, influence gelling properties. Furthermore, the manufacturing processes used to produce pulse flours and fractions can have an impact on gelling properties. For pulse protein extracts as an example, protein extracts produced using ultrafiltration had better gelling properties (i.e. lower LGC) than extracts produced using isoelectric precipitation. Differences between pulse flours and fractions as a result of pulse type, variety, conditions used for processing and drying, product purity and composition of the extracts, as well as storage conditions can influence gelling properties (Boye et al., 2010a).

11.3.5 Emulsifying properties

Emulsifying properties of pulse flours can be attributed mostly to the protein components of pulses. Proteins act as emulsifiers by forming a film or skin around oil droplets dispersed in an aqueous medium, thereby preventing structural changes such as coalescence, creaming, flocculation or sedimentation. The emulsifying properties of proteins are therefore affected by the hydrophobicity/hydrophilicity ratio of the proteins and by structural constraints that determine the ease with which the proteins can unfold to form a film or skin around dispersed oil droplets.

Emulsifying properties are generally evaluated by two indices, namely emulsifying activity (EA) and emulsifying stability (ES). In a simplified system, EA measures the amount of oil that can be emulsified per unit of protein, whereas ES measures the ability of the emulsion to resist changes to its structure over a defined time period (Boye et al., 2010a).

There is great variability in the emulsifying properties reported for different pulse flours and protein fractions. Pulse varieties and fractions having higher amounts of vicilin proteins are likely to have better emulsifying properties (Dagorn-Scaviner et al., 1987). Albumin proteins of great northern bean were found to be better emulsifiers than the globulins, and the total globulin and 7S fractions of pea varieties had better emulsifying properties and higher surface hydrophobicity (S_0) than the albumin and 11S fractions (Cserhalmi et al., 1998). Techniques such as hydrolysis as well as processing conditions can impact the emulsifying properties of foods. Hydrolysis improved the EA of pea PI by approximately twofold with a maximum occurrence at a degree of hydrolysis (DH) of 3.7. Similarly, hydrolysis plus interaction with chitosan improved the ES of faba bean (Braudo et al., 2001).

11.3.6 Foaming properties

Another characteristic of proteins is that they exhibit foaming properties. The most frequently used indices for measuring foaming properties are foam expansion (FE), foaming capacity (FC) and foam stability (FS). Foams are formed when proteins unfold to form an interfacial skin that keeps air bubbles in suspension and prevents their collapse. Foam formation is important in food applications such as beverages, mousses, meringue cakes and whipped toppings (Boye et al., 2010b).

There is great variation in the FE and FS values for different pulse proteins. Desi and kabuli chickpea concentrates processed by UF had higher FE values compared to yellow pea, green pea and red lentil protein concentrates processed by both IEP and UF (Boye et al., 2010a). Processing treatments such as boiling, fermenting, roasting and malting decrease FC. Hydrolysis, on the other hand, improved the FE of pea PI nearly threefold at a DH value of 2.5. In the acid and alkaline regions, FC and FS are improved, as evidenced by the higher solubility of pigeon pea (*Cajanus cajan* L.) proteins, and FC and FS also improved in both regions compared to the isoelectric region. The FS of pigeon pea concentrate also improved with an increase in protein concentration and ionic strength (Akintayo et al., 1999).

Foaming properties depend very much on the conditions under which biopolymer complexation occurs. Similar to the oil–water interface, increased surface hydrophobicity due to polysaccharide-induced conformational changes in the protein upon binding can result in greater absorption at the air–water interface. Furthermore, electrostatic interactions between biopolymers help strengthen the viscoelastic film at the interface, inducing conformational changes in the protein for better integration at the interface and increasing bulk phase viscosity (Liu et al., 2010). As an example, in admixtures of gum acacia, locust bean gum and xanthan gum with common bean (*P. vulgaris* L.) and scarlet runner bean (*P. coccineus* L.), improvements in the FS and bulk phase viscosity of the PIs were observed, creating a biopolymer network that prevented air bubble coalescence.

11.4 Physical properties of pulse flours in dough systems

Dough is a complex mixture of flour and other additives mixed in water to form a paste or sticky mass with high viscoelastic properties. Varying the proportions of ingredients such as flour, protein, fat and water can result in changes to the properties of the dough. Fat contributes to the plasticity of the dough, acting as a lubricant as well as preventing excessive development of flour protein during mixing. Similarly, the interactions of starch ingredients with each other and with the proteins also form a variety of complexes, resulting in changes in dough behavior. As with cereals, pulses influence dough systems owing to their high concentrations of proteins, starch and fiber. In order to understand the functionality of dough systems in various formulations, it is important to understand the physical and rheological properties of the components that comprise the flours.

11.4.1 Physical properties of flour that influence dough systems

Physical properties, such as dough elasticity, plasticity, stickiness, crumbliness and WHC, are the basic parameters for evaluating dough systems. These properties depend on the apparent physical characteristics of pulse seeds, such as appearance, color, shape, size, uniformity and hardness/softness, in addition to flour composition.

Characteristics contributing to dough rheology will vary from variety to variety and sometimes among cultivars. Two important characteristics that influence dough behavior are starch birefringence and particle size.

11.4.1.1 Pulse starch birefringence

The physical appearance of pulse grains reflects the status of internal ingredients, especially starch, which is an important component in terms of dough rheology. Varietal differences in terms of physical appearance exist among pulses. One parameter for measuring particle arrangement within grains is birefringence. Pulse starches exhibit birefringence, which is an indication of an orderly radial arrangement of starch crystallites. Native and modified pea, navy bean and lentil starches show decreased birefringence. In pea and navy bean starches, birefringence disappears in the center of the granules upon heat–moisture treatment (HMT) but is retained at the periphery. The extent of the above changes is more pronounced in pea and lentil starches than in navy bean starch (Chung et al., 2010).

There are many short branch chains derived from the branch linkages located inside the starch crystalline structure. A-type starch granules are larger in size and have a lenticular shape and are formed in the earlier stages of development as compared to B-type starch granules which are formed later on and are smaller and more spherical. A-type starches are also classified as being more amorphous whereas B-type starches are polymorphic (Jane et al., 1997). In starches containing the B-type unit cells, however, more branch points are clustered in the amorphous regions, and there are fewer short branch chains (Jane et al., 1997). These characteristics suggest that the crystalline areas in B-type starches are better ordered than those in A-type starches. The B-type polymorphic content, determined by X-ray diffraction pattern, is higher in navy bean starch (33.6%) than in pea (28.4%) and lentil (28.3%) starches (Chung et al., 2010); this difference suggests that the greater loss of birefringence and greater crystallite disruption seen with pea and lentil starches upon HMT is due to the higher content of A-type unit cells in those starches.

11.4.1.2 Pulse starch and flour particle size characteristics

Pulse starch granule particle size as well as the particle size distribution of the flours determine the solubility potential of flour and can influence dough rheology. Most pulse starches are composed of

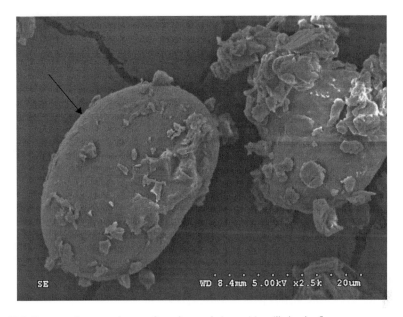

Figure 11.1 Electron microscopy image of starch granule (arrow) in milled pulse flour.

simple granules, although compound granules have been reported in smooth pea and wrinkled pea starches. Starch granules are generally oval in shape, although spherical, round, elliptical, and irregularly shaped granules are found. A scanning electron micrograph of milled pulse flour showing the intact starch granules is presented in Figure 11.1. The surfaces of starch granules are generally smooth with no evidence of fissures or pin holes. Wrinkled pea starch granules are generally extensively damaged, with splitting and exposure of the internal layering (Chung et al., 2008). The granule surface influences reactivity towards enzymes, chemical modifying reagents, and the rate and extent of hydration (Colonna et al., 1982; Bertoft et al., 1993a,b; Hoover and Manuel, 1995; Hoover and Ratnayake, 2002; Singh et al., 2004; Zhou et al., 2004; Chung et al., 2008; Hoover et al., 2010).

Unfortunately, very little work has been done on the effect of pulse flour particle size distribution and flour functionality. In general, the smaller the particle size the larger the surface area and the greater the WHC. The flour particle sizes of various pulses are given in Table 11.3. The type of mill used, as well as the conditions used during pretreatment and milling, will influence the particle size distribution of the flours.

Table 11.3 Particle size distribution (%) of pulse flours and their fractions

Particle size (μm)	Chickpea flour	Green lentil flour	Red lentil flour	Pinto bean flour	Navy bean flour	Yellow pea flour	Pea protein isolate	Pea fiber isolate	Pea starch isolate
>841	0.39	0.56	1.72	0.72	0.00	0.31	2.07	3.43	0.33
>400	0.64	21.59	26.13	9.50	2.31	0.14	0.05	0.09	0.00
>250	67.43	31.79	36.05	34.58	23.90	1.98	0.76	2.43	0.11
>177	27.55	15.22	14.95	26.04	39.49	16.66	1.26	8.53	2.86
>149	0.25	2.76	1.89	2.20	1.68	28.32	1.04	33.33	9.17
>125	2.06	29.80	14.07	21.59	27.60	33.00	8.54	42.53	27.92
<125	1.69	8.30	5.19	5.38	5.03	19.53	86.29	9.67	59.62

With permission from Han et al. (2010).

11.4.2 Effect of flour preprocessing (tempering) on dough systems

The moisture content of starch determines dough consistency. In wheat, moisture content varies in hard and soft varieties and is affected in many different ways during processing. Tempering is the addition of water to the seed grains before they are milled. Tempering causes softening of the cotyledon and ultimately of the starch, changing the characteristics of the starch and contributing to its viscoelastic properties.

11.4.3 Effect of proximate composition on dough systems

Pulses have variable proximate compositions, thus changes in ingredients will result in corresponding changes in dough behavior. The proximate composition of pulse flours varies significantly even among different cultivars of the same type. Thus, protein content and the composition of protein vary for different pulse types and cultivars. The soluble dietary fiber fraction of pulses, as another example, includes pectins, gums, mucilages, and some soluble hemicelluloses which contribute to dough systems. In both desi and kabuli chickpeas, this fraction constitutes a large proportion (about 55%) of the total dietary fiber; thus, subtle changes in their composition could affect dough rheology (Singh and Sokhansanj, 1984).

11.4.4 Effect of starch gelatinization, retrogradation and pasting on dough rheology

11.4.4.1 Starch gelatinization

When heated in water, starch undergoes a transition process, during which the granules break down into a mixture of polymers-in-solution, known as gelatinization (Ratnayake and Jackson, 2008). This phase transition is associated with the diffusion of water into the granule, water uptake by the amorphous background region, hydration and radial swelling of the starch granules, loss of birefringence, uptake of heat, loss of crystalline order, uncoiling and dissociation of double helices, and AM leaching (AML) (Waigh et al., 2000). The gelatinization transition temperatures and the enthalpy of gelatinization are influenced by the molecular architecture of the crystalline region, which corresponds to the distribution of AMP short chains, and not by the proportion of the crystalline region, which corresponds to the AM/AMP ratio (Noda et al., 1996). Amylopectin chains that are too short to form stable double helices require less energy to unravel and melt during gelatinization (Srichuwong and Jane, 2007; Chung et al., 2009). The gelatinization parameters also depend on the proportion of AMP long chains to AMP short chains and on leaching of the amylase. The application of these conditions causes amylase–AM and amylase–AMP interactions, the extent of which determines the level of gelatinization in various pulse starches. As a result of these interactions, new crystallites are formed, which also vary among pulse varieties depending on their starch content. In the case of navy bean starch, for example, this interaction is much lower.

Furthermore, the overall crystallinity (quality and amount of starch crystallites) of AMP is reflected by the ΔH (enthalpy of gelatinization) which causes the loss of double helical order but not crystallinity (Cooke and Gidley, 1992). The ΔH of pea, lentil and jack bean (*Canavalia ensiformis* L.) starches decreases during HMT, suggesting the probability of increased mobility of double helices (forming the crystalline structure) with the high temperature during HMT, leading to a disruption of some of the hydrogen bonds linking adjacent double helices (Chung et al., 2010).

11.4.4.2 Starch retrogradation

Retrogradation is a process in which the starch chains (AM and AMP) in the gelatinized paste interact on cooling after heating, leading to the formation of a more ordered structure of molecular interactions (Hoover et al., 2010). Retrogradation is accompanied by increases in the degree of crystallinity and gel firmness, the exudation of water (syneresis) and the appearance of a B-type X-ray pattern (Miles et al., 1985; Hoover, 1995). In the case of pulse starches, retrogradation is determined mainly by measuring the amount of water exuded (syneresis) when a frozen gelatinized starch gel is thawed at room temperature. A true picture of the retrogradation mechanism in pulse starches can be obtained with both macroscopic and molecular techniques. The extents of syneresis that have been reported for various pulse starches cannot be compared owing to differences in storage temperature, number of freeze–thaw cycles, thawing time, centrifugal speed, centrifugation time and method of measuring the exuded water (Hoover et al., 2010). Pulse starches retrograde to a greater extent than cereal or tuber starches because pulse starches have a higher AM content and/or a different molecular structure.

11.4.4.3 Starch pasting

Paste is defined as a viscous mass consisting of a continuous phase of solubilized AM and/or AMP and a discontinuous phase of granule ghosts and fragments (Hoover et al., 2010). Pasting refers specifically to changes in the starch upon further heating after gelatinization, including further swelling and leaching of polysaccharides from the starch granule, and increased viscosity, which occurs with the application of shear forces (Atwell et al., 1988; Tester and Morrison, 1990a,b). Most pulse starches exhibit a high pasting temperature, the absence of peak viscosity, increasing viscosity during the holding period, and a high setback when measured with a viscoamylograph or viscoanalyzer (Hoover et al., 2010). The pasting properties of pulse starches probably reflect their high AM content, the presence of only trace quantities of lipid-complexed amylase chains, a strong interaction between starch chains (AM–amylase and/or AM–AMP) within the native granule, and the orientation of AM chains relative to one another (Hoover and Sosulski, 1985). Pasting properties may also be influenced by AM and AMP chain length. The pasting temperature of pulse flours differs among type and cultivar, ranging from 61.7 to 75.2 °C, indicating the minimum temperature required to cook the flour sample to develop viscoelasticity (Kaur and Singh, 2005). All pulse flours show a gradual increase in viscosity with increasing temperature because of AM release from the swelling starch granules (Sanjeewa et al., 2010).

11.4.5 Effects of pulse starch composition on dough systems

The two major components of starch are AM and AMP. The ratio between AM and AMP varies depending on the starch source. Both AM and AMP contribute to dough rheology by forming inter- and intramolecular interactions.

11.4.5.1 Amylose activity in pulse flour dough
In starches, AM normally constitutes about 15–30% of total starch. However, AM accounts for approximately 0–5% in waxy starches and 35–70% in high-AM starches (Shannon and Garwood, 1984). In pea, lentil and navy bean starches, AM contents are 38.2%, 38.9% and 40.8%, respectively (Chung et al., 2010). The molecular weight of AM ranges from 1×10^5 to 1×10^6. The structure of pulse AM has not been well characterized.

Amylose plays a very important role in the gelatinization and pasting properties of flour (Sanjeewa et al., 2010). From a chain length perspective, navy bean starch has the greatest average chain length (20.9), while pea and lentil starch chain lengths are 20.3 and 20.5, respectively. The proportions of short branch chains follow the order pea (20.2%) > lentil (20.0%) > navy bean (19.6%). In contrast, the proportions of long branch chains follow the order navy bean (10.7%) > lentil (9.9%) > pea (9.5%). Amylose forms helical complexes with iodine, fatty acids and monoglycerides, and the degree of polymerization (DP) of AM varies from 690 to 6340 (Hoover and Ratnayake, 2002; Chung et al., 2010).

11.4.5.2 Amylopectin activity in pulse dough
Amylopectin is the major component of all starches, with an average molecular weight of 10^7–10^9, and it plays a significant role in pulse dough development during various processing phases. Amylopectin is composed of linear chains of α-(1,4)-D-glucose residues connected through α-(1,6) linkages (5–6%). The currently accepted

AMP structure involves short AMP chains forming double helices and associating into clusters that pack together to form a structure consisting of alternating crystalline and amorphous lamellae (Hizukuri, 1986). The AMP unit chains are classified as A, B and C. The A chains are the shortest and are linked by a single α-(1,6) linkage to the AMP molecule. The B chains are classified into B1, B2, B3 and B4 depending on their length and the number of clusters that they span. The most exterior chains (A and B1) have been shown to form double helices within the native granules, which are packed into lamellae crystallites, thereby affecting dough behavior. The B2–B4 chains act as connecting chains in the AMP molecule (Robin et al., 1974; Hizukuri, 1986). The C chains are classified as a mixture of A and B and have restricted swelling characteristics. Usually legume starches (desi type) have a viscosity pattern of C type (Sanjeewa et al., 2010).

11.4.5.3 Combined effects of amylose and amylopectin in dough systems

Both AM and AMP contribute to the rheological characteristics of pulse flour dough. The swelling factor (SF) and AML of native and modified pulse starches are variable at 80 °C, and the differences are influenced by AM content, AMP structure, and the extent of interaction between AM–AM and/or AM–AMP chains within the native granule (Chung et al., 2010). Larger proportions of long AMP branch chains in navy bean starch may contribute to a stronger crystalline structure, since these long chains can form long double helices. A strong crystalline structure hinders granule swelling. Furthermore, the lower extent of AML in native navy bean starch in spite of its higher AM content reflects a stronger interaction between AM–AM and/or AM–AMP chains. Strong interactions not only reduce the amount of free hydroxyl groups available for hydration but also decrease the ingress of water into the granule interior. Strong interactions also help explain the order of granular swelling among native pulse starches (Chung et al., 2010).

Changing the physicochemical processing environment, as in HMT, also affects AM and AMP functionality, thereby decreasing the SF and AML of pulses and increasing crystallinity, AM–lipid interactions, and interactions involving AM–AM and/or AM–AMP chains. This phenomenon seems plausible because of increased AM chain mobility resulting from the higher thermal energy during

HMT. Dual modification can also aggravate the decrease in SF owing to interplay between increased AM–AM and/or AM–AMP interactions and crystallite change. The SF and AML further decrease on annealing. In native and modified starches, HMT–annealing treatment results in the lowest SF and AML in all pulse starches (Hoover and Vasanthan, 1994; Chung et al., 2010).

11.5 Flour and semolina fortification with pulse ingredients

Pulses are rich in essential amino acids, including lysine, threonine, isoleucine, leucine, phenylalanine and valine, and have a good mineral profile, containing K, Fe, Cu, Mg, Zn and Mn. In contrast, cereals are deficient in certain essential amino acids and also lose some of their mineral content when milled, owing to the removal of the aleurone layer, which is a rich source of minerals. To improve their amino acid profile and mineral complementarity, cereals may be fortified with pulses. Ensuring an optimal level of fortification requires studying the characteristics of flour and semolina, especially their particle size and composition, for acceptable product development. Legume-fortified cereals therefore offer a broader spectrum for those who wish to improve the nutritional quality of their diet. Vitamin and mineral profiles of some pulses and pulse flours are given in Table 11.4.

The ultimate goal for the industry is to obtain desired product characteristics at industrial processing levels. Processing conditions for milling, mixing and processing, including hydration to limit particle agglomeration during mixing, need to be optimized along with the selection of flour/semolina type and pulse varieties. The final impact on cooking/baking and textural quality, and the impact on the ability of gluten to form a protein network, also need to be studied.

Flour and semolina are characterized on the basis of particle size. To achieve good mixing of the pulse flour for an enriched nutritive profile, the particle size of both flours should match. Similarly, the compositions of the flour to be fortified and the fortifying flour should also be known to obtain good complementarity.

Particle size is the determinant of product end-use quality. Particle size distribution during and after mixing determines viscoelastic properties owing to the likely formation of a

Table 11.4 Vitamin and mineral composition of some pulses (mg $100\,g^{-1}$)

Component	Split pea flour	Faba bean flour	Pearl millet	Starchy legumes	Soybeans	Peanuts
Vitamins						
B$_1$ (Thiamine)	0.83[a]	0.66[a]	–	0.81–1.32	0.90	0.99
B$_2$ (Riboflavin)	0.14[a]	0.20[a]	–	0.11–0.41	0.23	0.13
B$_3$ (Niacin)	3.1[a]	2.5[a]	–	0.85–3.21	2.0	12.8–16.7
B$_5$ (Pantothenic acid)	1.5[a]	0.7[a]	–	–	1.2	2.72
B$_6$ (Pyridoxine)	0.15[a]	0.23[a]	–	0.3–0.66	0.64	0.30
B$_9$ (Folic acid)	0.025[a]	0.145[a]	–	–	–	–
Minerals						
Fe (Iron)	5.4[a]	6.0[a]	2.1–11.7	3.3–8.0	10	1.4–1.8
Mg (Magnesium)	115[a]	124[a]	–	–	–	–
P (Phosphorus)	495[a]	630[a]	631–1353	380–570	660	415–470
K (Potassium)	–	–	366–543	1320–1780	1670	618–634
Ca (Calcium)	–	–	7–117	70–210	280	74–88
Cu (Copper)	–	–	0.42–0.58	0.5–1.4	1.2	1.1–1.3
Zn (Zinc)	–	–	0.10–3.8	1.9–6.5	–	6.1–6.2

Swanson (2007); Petitot et al. (2010).
[a]Dry basis (d.b.).

heterogeneous dough. The dough of durum wheat (*Triticum turgidum* L. var. *durum*) is characterized by high amounts of small particles (<1 mm) and almost no large lumps (>4 mm). When durum wheat dough is supplemented with 35% legume flour, however, there appear to be fewer fine particles (<1 mm) and more large dough lumps (>4 mm) (Petitot et al., 2010). Fortification makes the dough sticky and leads to the formation of large lumps. Differences in the particle size distribution and chemical composition of semolina and legume flours impact their hydration properties (Petitot et al., 2010), resulting in heterogeneous particle hydration and large dough lump formation. This occurs because the fine particles of legume flours can absorb water more quickly than semolina particles, resulting in the generation of heterogeneous dough (Petitot et al., 2010).

During processing, pressure application is also affected in a very complex way by different intrinsic dough parameters, including the nature of the flour (such as particle size or protein content), and the humidity, temperature or viscosity of the dough (Mohamed, 1990; Dalbon et al., 1996). A reduction in the specific mechanical energy (SME) at a given hydration level in the presence of non-traditional ingredients such as flaxseed (*Linum usitatissimum* L.) flour and buckwheat (*Fagopyrum esculentum* Moench) bran flour was found to reduce dough strength, thereby inducing higher cooking losses as a result of the lower density and a higher amount of water-soluble components (Manthey et al., 2004).

The chemical composition of semolina and legume flours also determines the nature and level of fortification. Pulses are important sources of plant proteins and carbohydrates in the human diet. Lentil, dry pea, chickpea, field pea and dry bean are considered foods of great nutritional value and are also important sources of dietary fiber, an essential part of a healthy diet. Dietary fiber not only naturally promotes bowel health but also helps lower cholesterol, balances blood glucose, and promotes healthy physiology and well-being. Unlike other food components, such as protein, carbohydrate or fat, which the body breaks down and absorbs, fiber is not digested by the body. Instead, fiber passes almost unchanged through the stomach and small intestine into the colon. The carbohydrate–oligosaccharide fraction of pulses includes starch, soluble sugars and dietary fiber. Many health benefits are attributed to these components of pulse seeds, and pulses are also considered to be a rich source of minerals.

The optimal levels of fortification based on quality and descriptive sensory analysis differ from pulse to pulse. For example, the optimal levels for green pea or lentil flour and chickpea or yellow pea flour are 15% and 20%, respectively, in pulse-containing wheat flour spaghetti.

11.6 Pulses and pulse fractions in food applications

Pulses and pulse fractions can be used in a variety of food products, the most popular of which are bakery products, meat products, pasta, noodles, soups and canned products. Among food grains, pulses provide good sources of protein, fiber, vitamins and minerals. Pulse proteins are relatively low in sulfur-containing amino acids such as methionine, cysteine and tryptophan but high in lysine and threonine, two essential amino acids that are limited in durum wheat. The introduction of legume flour into cereal products can therefore increase the protein, insoluble fiber, vitamin (B_1, B_5, B_6 and B_9) and mineral (Fe, Mg, and P) contents, thereby affecting the nutritional quality of food products. Legumes and cereals are thus nutritionally complementary, and the level and type of supplementation depend on the nature of the product.

11.6.1 Bakery product applications

Popular worldwide, a variety of bakery products are manufactured by modifying heating and product formulation depending on the type of end-product desired. Physicochemical, rheological and sensory characteristics of baked goods are affected by the ingredients used in their production. Modifications to product formulation and processing conditions can, therefore, be considered to maximize the potential of bakery products in terms of both satiety and health benefits. Thanks to the rich nutrient profile of pulses and the wide variety of ingredients that could be derived from the fractionation of pulses, significant opportunities exist to develop and introduce healthier bakery items onto the market.

The particle size distribution of pulse flours and fractions has the potential to affect substantially the palatability of bakery products. Therefore, when pulse seeds are milled into flours, the hulls can be separated from the cotyledons and ground through screens to the desirable particle size for use as a dietary fiber source in food applications. Breads, baked goods and cereal-based products have been the traditional vehicles for fiber enrichment, but pulse fibers can also be incorporated into various other processed foods for added nutritional benefits.

For bakery products containing pulses and pulse fractions, sufficient moisture is required to gelatinize starch, a phenomenon that decreases cooking time and allows the inclusion of pulses as a food ingredient without the need for a precooking step. Moisture is the main player contributing to the sensory appeal of bakery products. The incorporation of pea starch and other pulse fractions into a dough formula is simple because additional moisture is not required, compared to the case when pea protein and fiber are used. The resulting dough has good mechanical handling properties (i.e. mixing and sheeting), and the final product has excellent sensory characteristics, including a brighter color, good flavor and good texture. Similarly, sponge and layer cakes were prepared with chickpea–wheat flour blends at different substitution levels to increase fiber and lysine content (Gómez et al., 2008). This supplementation caused a decrease in cake volume and symmetry with increased substitution of chickpea flour, along with a firmer, gummier and less cohesive cake texture.

Cake doughnuts formulated using high-protein flours from black bean and navy bean have desirable flavor and exterior/interior resistance to biting and chewing, although they are less tender and darker in color. Similarly, the addition of chickpea flour (24.83% protein), lupin flour (26.30% protein) and soybean flour (56.36% protein) was used to replace wheat flour in a cookie formulation (Faheid and Hegazi, 1991). The substitution of lupin flour for wheat flour increased the protein and fiber contents and amino acid scores, as well as the levels of certain minerals. Bread can be fortified with pulse ingredients such as chickpea and faba bean flours and powders at the 15% and 20% levels (Abdel-Aal et al., 1987). Chickpea powder produces more acceptable fortified bread than wheat flour. Likewise, bread fortified with faba bean protein concentrate as partial replacement for wheat flour was preferred over wheat flour-fortified products. The incorporation of protein concentrate into products such as breads and cookies generally resulted in a less beany flavor and yielded acceptable products at substitution levels up to 10%.

11.6.2 Meat product applications

Meat products to which pulses and pulse flours are added are a new trend in food science. Protein and starch can imbibe water and form gel matrices upon heating. Pulse macromolecules can also interact with other components in the food matrix to create networks. In the presence of meat protein, for example, pulse macromolecules can form a complex three-dimensional gel network involving various forces such as van der Waals', electrostatic and hydrogen bonding forces, trapping fine particles of emulsified meat or the meat matrix, with the starch and non-meat proteins as fillers. Additionally, flour components help retain water and fat during the cooking process, thereby increasing the cooking yield.

Sanjeewa et al. (2010) found highest cooking yields for bologna and sausages containing kabuli chickpea, desi chickpea, konjac (elephant yam; *Amorphophallus konjac* C. Koch) flour blends, common bean flours and wheat flour. The lowest cooking yield was observed for low-fat sausages manufactured without any binders. In contrast, increasing the level of substitution of chickpea flour in pork, beef and mutton sausages enhanced cooking losses, a phenomenon that can be counteracted by the addition of insoluble fiber, which favors water-binding properties and FAC. Water is bound to insoluble polysaccharides by hydrogen, ionic and/or hydrophobic interactions and by surface tension in the pores of the matrix.

Generally, reducing the fat content decreases the hydration properties of bologna, resulting in significantly higher expressible moisture and purge losses in low-fat formulations compared to high-fat products (Pietrasik and Janz, 2010). Increased weight losses occur when fat reduction is accompanied by an increase in moisture content. Adding pea ingredients to low-fat formulations decreases the percentage of water loss, indicating that the binders improve water retention in low-fat bologna sausages (Pietrasik and Janz, 2010).

The amylose (AM) content of starch plays a very important role in the gelatinization and pasting properties of flour in low-fat pork batter containing kabuli chickpea, desi chickpea and pea flour. In addition to AM, the type and level of flour also affect the meat batter, ultimately influencing product quality. More water and less fat can lead to decreased batter viscosity for the comminuted meat products in the absence of any plant-based extenders (Claus et al., 1990; Claus and Hunt, 1991).

Bologna sausages with low fat content supplemented with pea flour were found to be less hard and chewy, whereas the addition of 4% pea starch produced a firmer, chewier texture (Pietrasik and Janz, 2010). Fat content can be reduced by increasing the water content while keeping the protein level virtually constant (Bloukas and Paneras, 1993; Cavestany et al., 1994; Carballo et al., 1995, 1996; Colmenero et al., 1995). Usually, low-fat comminuted products tend to be tougher than higher-fat formulations. Low-fat bologna exhibits a similar trend in terms of chewiness, with lower chewiness values as compared to high-fat products. Similarly, fat reduction in frankfurters and bologna results in products that are firmer, chewier and more rubbery; when a decrease in fat content is accompanied by an increase in water, however, products with the opposite characteristics are obtained, a result that may cause handling problems in the pre-stuffing and stuffing processes. Furthermore, fat level has no significant effect on springiness, and low-fat formulations have poorer binding properties than full-fat products, a problem that can be overcome by the addition of various binders such as chickpea flour. The addition of pea ingredients other than pea flour considerably improves the textural properties of the product, restoring the values of a low-fat product to levels equivalent to a regular-fat bologna (Pietrasik and Janz, 2010). Pietrasik and Janz (2010) observed that the binder ingredients like flour, starch and fiber ingredients favorably affected hydration properties and thermal stability, yielding lower cooking and purge losses and higher WHC.

The addition of pulse flours and fractions modifies the sensory appeal of meat products. Addition of 5% desi chickpea or pea flour produced bologna with a harder texture. Generally, low-fat pork bologna containing chickpea and pea flours were found to have significantly higher hardness and cohesiveness values and firmer texture. Adding common bean flour to beef sausages results in lower shear force and hardness if a binder is not used (Dzudie et al., 2002). Likewise, bologna containing kabuli chickpea, desi chickpea, pea and wheat flours scored lower for sustained juiciness. The addition of desi chickpea flour also promoted graininess in the product because of the contribution of finely ground seed coat materials. It is, however, not necessary to dehull kabuli chickpea during flour preparation for use in meat processing, especially at lower levels of flour addition (Sanjeewa et al., 2010).

The overall flavor intensity of meat products containing pulse flour is not significantly affected, although there is an effect on product color. The reddish-pink color of thermally processed meat products is due to the combined effects of thermally denatured myoglobin and hemoglobin blended with the other non-meat additives. Bologna formulations with pea, wheat and kabuli chickpea flours are significantly lighter in color than formulations with desi chickpea flour (Sanjeewa et al., 2010). Similarly, adding common bean flour to beef sausages affects the color of the final product (Dzudie et al., 2002).

Color attributes of cooked meat products are derived from meat pigments and their conversion during cooking. When the myoglobin content of comminuted products is held constant by controlling the level of meat protein in the formulation, color is then influenced by fat content and the addition of water and non-meat ingredients. When the fat content of cured meat products, such as frankfurters and bologna, is reduced, the products become darker and redder. The color of lean-meat bologna is minimally affected by the addition of pea ingredients. Supplementing beef sausages with common bean flour significantly increased lightness and yellowness while reducing redness (Pietrasik and Janz, 2010).

Ingredient type is more important than fat level for the remaining consumer-rated characteristics. Pulse fractions have different effects on consumer acceptability of emulsified meat products. Adding 10% cowpea (*V. unguiculata* L.) flour, for example, decreased flavor scores for chicken nuggets, whereas meatballs extended with other legume flours, i.e. pea, had relatively high acceptability scores (Prinyawiwatkul et al., 1997). Similarly, none of the other legume flours had detrimental effects on the sensory properties of buffalo burgers. The acceptability of mutton sausages was not affected by the inclusion of chickpea flour, whereas pork and beef sausages with added chickpea flour were significantly less acceptable (Shand, 2000).

For calorie-conscious people, a significant reduction in fat and total calories derived from fat can be achieved without affecting consumer acceptability in low-fat bologna, where pea and wheat flour can be combined for product development, because the fat is replaced with water and because the moisture content is directly proportional to the amount of added water. The levels of moisture, fat, protein and net calories vary depending on the variety and the level of fortification (Pietrasik and Janz, 2010).

11.6.3 Pasta and noodle products

Pasta and noodles are popular foods consumed all over the world. They are routinely made from wheat flour, but their dietary value can be increased through the addition of pulse flours at optimal supplementation levels. Added levels of pulse flours can result in some challenges in product processing and extrusion due to differences in the particle sizes of the biomolecules. Major product problems are, however, related to sensory characteristics. Despite this, pulse flours can be successfully incorporated when suitable supplementation levels are adopted.

Pasta products are traditionally manufactured from durum wheat semolina, which is known to be the best raw material for pasta production. Pasta and noodles are good sources of carbohydrates (74–77%, dry basis [d.b.]) and proteins (11–15%, d.b.), and interest in their consumption is increasing owing to their nutritional properties, particularly their low glycemic index (Monge et al., 1990). The incorporation of pulses into cereals provides an opportunity to use non-traditional raw materials to increase the nutritional quality of pasta. Introduction of pulse flour into pasta ultimately increases the protein, insoluble fiber, vitamin (B_1, B_5, B_6 and B_9) and mineral (Fe, Mg and P) contents, although it impacts the ease of processing and quality attributes of pasta.

Pulse flour supplementation causes some difficulties in product processing which can be overcome by optimizing the substitution level. For example, substituting split pea or faba bean flour for durum wheat semolina at a high level (35%, d.b.) requires adapting the pasta-making process at the pilot scale (Petitot et al., 2010). Higher hydration levels and mixing speeds are required to limit particle agglomeration during mixing of durum wheat semolina with faba bean flour and facilitate the extrusion of the dough. Moreover, fortifying pasta with split pea or faba bean flour has a noticeable impact on the cooking quality of pasta, decreasing the optimal cooking time for low-temperature dried pasta and resulting in lower water uptake and higher cooking losses. Cooking losses can also be due to both AML and the solubilization of some salt-soluble proteins. To control cooking losses, a drying temperature strategy can be applied which reduces water uptake without changing the optimal cooking time. A high drying temperature promotes protein aggregation and leads to a strong protein network, entrapping starch granules, preventing AML in the cooking water, and decreasing stickiness, thereby positively affecting pasta quality.

Cooking significantly decreases K, tannin, sucrose, oligosaccharide and ash contents, through the diffusion of these compounds into the cooking water. In the case of beans, field pea and chickpea, cooking has no effect on the Fe, Zn and phytic acid contents; an increase in protein and starch contents was attributed to the loss of soluble solids and the increase in insoluble dietary fiber during cooking (Wang et al., 2008). The increase in soluble dietary fiber increases the RS content in chickpea and decreases it in cooked beans. The presence of fiber causes a physical disruption of the gluten matrix, which may facilitate the penetration of water to the core of the pasta (Wang et al., 2010).

The cooking quality and textural properties of durum wheat pasta are also affected by the ability of gluten to form a protein network. Durum wheat proteins are composed mainly of glutenins and gliadins that form intra- and intermolecular disulfide bonds during processing. This phenomenon leads to the formation of a three-dimensional gluten network responsible for the unique textural properties of pasta. In contrast, pulse proteins are composed mainly of salt-soluble globulins and water-soluble albumins. The addition of non-gluten material dilutes the gluten strength and likely weakens the overall structure of the pasta. As a consequence, more solids leach from the pasta into the cooking water (Rayas-Duarte et al., 1996).

The texture of pasta and noodle products is determined by the hardness, resilience and cohesiveness of the product. Hardness corresponds to the force required to compress a pasta strand between molar teeth. Resilience assesses the ability of the pasta strand to regain its original shape after the first compression (elastic recovery). Cohesiveness is the ability of the material to resist to two successive compressions (Epstein et al., 2002). Fortifying durum wheat pasta with split pea, faba bean or buckwheat bran flour significantly increases pasta hardness, a result that can be attributed to the higher protein content and lower water uptake. Such fortification has no impact on the flour's cohesiveness but has a moderate impact on resilience. Moreover, increasing the drying temperature affects cohesiveness and resilience, given that the denaturation and aggregation of proteins at high temperatures promote the formation of a strengthened structure. Breaking energy, another determinant of pasta and noodle quality which varies among pulses, is lower with the inclusion of pulse fiber fractions that promote the formation of discontinuities or cracks inside the pasta/noodle strand, thus resulting in a weakened structure. Fortification with legume flours such as navy bean, pinto bean, lentil or green pea protein concentrates causes an increase in firmness, which represents the cutting force required to penetrate product strands (Bahnassey and Khan, 1986; Zhao et al., 2005).

During pasta and noodle processing, flour fortification makes the dough sticky and leads to the formation of large dough lumps that render the extrusion step difficult. This problem can be counteracted by reducing the amount of water added compared to durum wheat dough preparation, to improve mixing and extrudability. In addition, reducing the specific mechanical energy (SME) at certain hydration levels causes a decrease in dough strength with the ultimate low density of extruded spaghetti, thereby inducing higher cooking losses (Manthey et al., 2004).

From a sensory evaluation perspective, pasta and noodles fortified with lupin or chickpea flour up to a 10% substitution level are generally well accepted, and there appears to be no impact on appearance, taste or texture at higher substitution levels. Product color is another essential sensory quality attribute. Generally, consumers prefer a bright yellow color, but fortification with pulse flour significantly decreases brightness because of the higher ash content of legume flours, especially green pea, chickpea, lentil and faba bean flours (Wood, 2009). The practices used to improve the culinary and sensory properties of fortified pasta include high drying temperatures, from 50 °C to 90 °C, but that technique increases pasta redness owing to the Maillard reaction, which yields a brownish color that could be disagreeable to consumers and which could also impact the nutritional properties of the pasta, especially by reducing lysine availability (Petitot et al., 2010). Green pea, yellow pea, lentil and chickpea flours, which contain about 20% protein, can be used as a nutritional additive in the production of fortified spaghetti noodles made with wheat flour semolina. Further optimization studies to improve processing, quality and sensory attributes of pasta and noodle products supplemented with pulse flours and ingredients are, however, still needed.

11.6.4 Extruded snacks

Pulse flours can be used as ingredients in expanded snack foods and extruded breakfast cereals. Fabrication of value-added, pulse-based snacks with high dietary fiber contents is an excellent way to provide the population with the recommended amounts of beneficial food components (Berrios et al., 2010). A major concern with extrusion is the potential loss of available lysine. Most grain legumes are relatively high in protein and lysine and low in cysteine and methionine. Combining legume proteins with low-lysine cereal proteins results in a more nutritionally complete protein. Lysine is, however, one of the most sensitive and most limiting amino acids in the human diet and harsh processes such as extrusion can significantly decrease its concentration. During processing, lysine and other amino acids with a free amino group may react with reducing sugars via the Maillard reaction. Lysine losses from the Maillard reaction may also occur in the absence of simple sugars owing to starch fragmentation during extrusion. Puffing and dark roasting result in the greatest destruction of available lysine when ready-to-eat breakfast cereals are processed from wheat products (McAuley et al., 1987). Available lysine was significantly reduced when pinto bean flour was extruded at higher moisture content (25.0% vs. 20.5%) and a lower temperature (133 vs. 156 °C) due to lysine retention (Gujska and Khan, 2002).

Extrusion conditions can also impact on final product quality. Pulse extrudate quality declined with increasing protein and moisture contents at a barrel set temperature of 120 °C during the extrusion–expansion of pea flour. In comparison, extrudates prepared from other pulses, including whole pinto bean meal and black bean flour, had substantially lower expansion indices (EIs) due to their lower starch concentrations and higher protein and fiber concentrations. EIs of pea flour extrudates were increased by incorporating a leavening agent into the extruder feed (Hood-Niefer and Tyler, 2010).

During extrusion cooking, losses in protein solubility can occur due to denaturation and the resulting structural changes that enable hydrophilic groups such as –OH, –NH$_2$, –COOH and –SH to form cross-links with starch. Moreover, all the macromolecular modifications that occur during extrusion cooking are affected by the shearing forces developed in the extruder barrel. The shear rate depends on the specific mechanical energy (SME), a system parameter that is affected by extrusion conditions (feed rate, extrusion temperature and screw speed) and material characteristics (feed moisture content, ingredients and viscosity) (Colonna et al., 1989; Liang et al., 2002; Fernández-Gutiérrez et al., 2004). During the preparation of corn (*Zea mays* L.)–lentil (*Lens culinaris* Medik.) snacks on a twin-screw extruder, the incorporation of proteins into extrudates decreased protein solubility, but the addition of 50% lentil increased this index in extrudates (Lazou and Krokida, 2010). Extrudate quality declines as protein and moisture contents increase (Hood-Niefer and Tyler, 2010). The optimal conditions identified for the extrusion–expansion of pea flour are 6% protein and 15% moisture at a barrel set temperature of 120 °C, with the actual product temperature at 155 °C (Hood-Niefer and Tyler, 2010).

11.6.5 Pastes, soups, beverages and other food applications

Pulse flours and fractions can be used in a variety of other applications such as in pastes, soups, beverages and glazes. Although limited studies have been conducted in these areas, the functional properties of pulse ingredients lend themselves as potential material for innovative value-added product development. The low viscosity of soluble pulse fibers makes them suitable as dietary fiber supplements for non-viscous products such as juices or functional beverages (Tosh and Yada, 2010). Addition of flour, starch and fiber ingredients to low-fat batters, soups and beverages increases cooking yield, because these additives have water-binding properties.

Whole beans are commonly used as a filling in baked goods, including pastries and buns. The beans are normally cooked and used to prepare a sweetened paste. Pastes from chickpea flour, smooth pea flour (var. Columbian) and wrinkled pea flour were found to be as acceptable as the commercial staple made from adzuki bean (*Vigna angularis* Willd) (Klamczynska et al., 2001).

Whole pulses can also be used for preparation of baby foods (e.g. pulse PIs or concentrates). Boiled mung bean (*V. radiata* L.) with rice soup can be used as a protein supplement, and porridge-type weaning food could be prepared from boiled, mashed and sieved mung bean with the addition of selected cereals and sugar (Singh and Singh, 1992). Lupin fractions have also been used as egg glazes (Sironi et al., 2005).

11.7 Industrial applications of pulses and pulse fractions

Pulse fractions such as bran, cotyledon and germ have different proportions of biomolecules as compared to whole grains or milled flour. Depending on the process used, the final product of the milling process may be whole flour, straight grade flour, bran, or shorts of high starch and protein fractions. Pulse flours can be further down-stream processed to obtain highly purified starch and protein fractions. In theory, pulse flours and fractions can be used by a variety of industries, including the food processing, pharmaceutical, nutraceutical, biomedical and animal feed sectors.

In the food processing industry, a wide variety of food products can be made from pulses as listed above. However, pulse fractions could also potentially be used as additives, binders, emulsifiers, and thickening or gelling agents, further expanding their use. Further research is, however, needed to identify the appropriate conditions for use and technologies for processing.

Some interesting potential areas of application in the pharmaceutical and nutraceutical sectors are the production of amylase inhibitors, phytates and polyphenolics, and dietary fiber components (Bravo et al., 1998; Hoover and Zhou, 2003; Tharanathan and Mahadevamma, 2003).

Pulses and pulse ingredients are already extensively used as animal feed (Boye et al., 2010b). Animal feeds in the form of concentrates, feed cakes and flakes are processed mainly from the by-products of the milling and food industries. The major portion of animal feeds comes from the hulls or husks and bran layers of pulse grains. The by-products of the food industry are also further processed to produce animal feeds. Research is ongoing to identify ways to remove antinutritional components and enhance the digestibility of pulse ingredients.

Components with pesticide properties have also been identified in pulses and pulse fractions. Proteinaceous cysteine proteinase inhibitor (CPI), an insecticidal protein found in pulses, can be used to control the proteolytic activity of endogenous digestive cysteine proteinase in the mid-gut of some insects (Hines et al., 1992). A 36 000 Da lectin-like alpha amylase inhibitor in hyacinth bean has been identified which inhibited conidial germination and hyphal growth of *Aspergillus flavus*, a fungal pathogen (Fakhoury and Woloshuk, 2001). Other insecticidal proteins toxic to pea pests have been reported (Bodnaryk et al., 1999). Thanks to their enormous wealth of useful components that can be exploited through research and the development of appropriate processing techniques, pulses may offer an interesting source of biomass for the development of a variety of industrial products.

11.8 Conclusions

Pulses are a rich source of energy, carbohydrates, protein, fats, vitamins, minerals and many bioactive compounds, which can be used to fortify food formulations and functional foods or be used in nutraceutical and/or pharmaceutical applications. A wide variety of food products can be envisaged in this regard which could provide many health benefits to consumers. Moreover, incorporation of pulses into various products builds on the growing trend toward more environmentally friendly food production and a better balance between the provisions of foods derived from animal and plant sources. However, there continues to be a need for further research to improve the knowledge base of the properties of pulses and pulse fractions in order to identify the appropriate conditions to optimize their functionality when used in foods. Additionally, cheaper technologies for processing pulses into flours and for fractionating them into value added ingredients such as starches, fibers and protein powders are needed and will go a long way to stimulate the utilization of these ingredients in novel food and industrial food applications.

References

Abdel-Aal, E.M., Youssef, M.M., Shehata, A.A., El-Mahdy, A.R., 1987. Some legume proteins as bread fortifier and meat extender. Alexandria J. Agric. Res. 32, 179–189.

Agboola, S.O., Mofolasayo, O.A., Watts, B.M., Aluko, R.E., 2010. Functional properties of yellow field pea (*Pisum sativum* L.) seed flours and the *in vitro* bioactive properties of their polyphenols. Food Res. Int. 43, 582–588.

Akintayo, E.T., Oshodi, A.A., Esuoso, K.O., 1999. Effects of NaCl, ionic strength and pH on the foaming and gelation of pigeon pea (*Cajanus cajan*) protein concentrates. Food Chem. 66, 51–56.

Atwell, W.A., Hood, L.F., Lineback, D.R., Varriano-Marston, E., Zobel, H.F., 1988. The terminology and methodology associated with basic starch phenomena. Cereal Foods World 33, 306–311.

Bahnassey, Y., Khan, K., 1986. Fortification of spaghetti with edible legumes. II. Rheological, processing, and quality evaluation studies. Cereal Chem. 63, 216–219.

Berrios, J.D.J., Morales, P., Cámara, M., Sánchez-Mata, M.C., 2010. Carbohydrate composition of raw and extruded pulse flours. Food Res. Int. 43, 531–536.

Bertoft, E., Manelius, R., Qin, Z., 1993a. Studies on the structure of pea starches – Part I. Initial stages in α-amylolysis of granular smooth pea starch. Starch 45, 215–220.

Bertoft, E., Manelius, R., Qin, Z., 1993b. Studies on the structure of pea starches. Part 2. α-Amylolysis of granular wrinkled pea starch. Starch 45, 258–263.

Bodnaryk, R., Fields, P., Xie, Y., Fulcher, K., 1999. Insecticidal factors from field pea. United States Patent 5,955,082.

Boye, J.I., Aksay, S., Roufik, S., 2010a. Comparison of the functional properties of pea, chickpea and lentil protein concentrates processed using ultrafiltration and isoelectric precipitation techniques. Food Res. Int. 43, 537–546.

Boye, J., Zare, F., Pletch, A., 2010b. Pulse proteins: processing, characterization, functional properties and applications in food and feed. Food Res. Int. 43, 414–431.

Bloukas, J.G., Paneras, E.D., 1993. Substituting olive oil for pork backfat affects quality of low-fat frankfurters. J. Food Sci. 58, 705–709.

Braudo, E.E., Plashchina, I.G., Schwenke, K.D., 2001. Plant protein interactions with polysaccharides and their influence on legume protein functionality – a review. Nahrung 45, 382–384.

Bravo, L., Siddhuraju, P., Saura-Calixto, F., 1998. Effect of various processing methods on the *in vitro* starch digestibility and resistant starch content of Indian pulses. J. Agric. Food Chem. 46, 4667–4674.

Carballo, J., Baretto, G., Colmenero, F.J., 1995. Starch and egg white influence on properties of bologna sausage as related to fat content. J. Food Sci. 60, 673–677.

Carballo, J., Fernandez, P., Baretto, G., Solas, M.T., Colmenero, F.J., 1996. Morphology and texture of bologna sausage as related to content of fat, starch and egg white. J. Food Sci. 61, 652–655.

Cavestany, M., Colmenero, F.J., Solas, M.T., Carballo, J., 1994. Incorporation of sardine surimi in bologna sausage containing different fat levels. Meat Sci. 38, 27–37.

Chung, H.J., Liu, Q., Donner, E., Hoover, R., Warkentin, T.D., Vandenberg, B., 2008. Composition, molecular structure, properties, and *in vitro* digestibility of starches from newly released Canadian pulse cultivars. Cereal Chem. 85, 471–479.

Chung, H.J., Liu, Q., Hoover, R., 2009. Impact of annealing and heat-moisture treatment on rapidly digestible, slowly digestible and resistant starch levels in native and gelatinized corn, pea and lentil starches. Carbohydr. Polym. 75, 436–447.

Chung, H.J., Liu, Q., Hoover, R., 2010. Effect of single and dual hydrothermal treatments on the crystalline structure, thermal properties, and nutritional fractions of pea, lentil, and navy bean starches. Food Res. Int. 43, 501–508.

Claus, J.R., Hunt, M.C., 1991. Low-fat, high added-water bologna formulated with texture-modifying ingredients. J. Food Sci. 56, 643–648.

Claus, J.R., Hunt, M.C., Kastner, C.L., Kropf, D.H., 1990. Low-fat, high-added water bologna: effects of massaging, preblending, and time of addition of water and fat on physical and sensory characteristics. J. Food Sci. 55, 338–341.

Colmenero, F.J., Carballo, J., Solas, M.T., 1995. The effect of use of freeze–thawed pork on the properties of bologna sausages with two fat levels. Int. J. Food Sci. Technol. 30, 335–345.

Colonna, P., Buléon, A., LeMaguer, M., Mercier, C., 1982. *Pisum sativum* and *Vicia faba* carbohydrates. Part IV. Granular starches of wrinkled pea starch. Carbohydr. Polym. 2, 43–59.

Colonna, P., Tayeb, J., Mercier, C., 1989. Extrusion cooking of starch and starchy products. In: Mercier, C., Linko, P., Harper, J.M. (Eds.), Extrusion cooking. American Association of Cereal Chemists, pp. 247–320.

Cooke, D., Gidley, M.J., 1992. Loss of crystalline and molecular order during starch gelatinization. Origin of the enthalpic transition. Carbohydr. Res. 227, 103–112.

Cserhalmi, Z.S., Czukor, B., Gajzago-Schuster, I., 1998. Emulsifying properties, surface hydrophobicity and thermal denaturation of pea protein fractions. Acta Aliment. 27, 357–363.

Dagorn-Scaviner, C., Gueguen, J., Lefebvre, J., 1987. Emulsifying properties of pea globulins as related to their adsorption behaviours. J. Food Sci. 52, 335–341.

Dalbon, G., Grivon, D., Pagani, M.A., 1996. Continuous manufacturing process. In: Kruger, J.E., Matsuo, R.B., Dick, J.W. (Eds.), Pasta and noodle technology. American Association of Cereal Chemists, St Paul, pp. 13–58.

Dzudie, T., Scher, J., Hardy, J., 2002. Common bean flour as an extender in beef sausages. J. Food Eng. 52, 143–147.

Epstein, J., Morris, C.F., Huber, K.C., 2002. Instrumental texture of white salted noodles prepared from recombinant inbred lines of wheat differing in the three granule bound starch synthase (Waxy) genes. J. Cereal Sci. 35, 51–63.

Faheid, S.M.M., Hegazi, N.A., 1991. Effect of adding some legume flours on the nutritive values of cookies. Egypt J. Food Sci. 19, 147–159.

Fakhoury, A.M., Woloshuk, C.P., 2001. Inhibition of growth of *Aspergillus flavus* and fungal alpha-amylases by a lectin-like protein from *Lablab purpureus*. Molec. Plant Microbe Interact. 14, 955–961.

Fernández-Gutiérrez, J.A., San Martín-Martínez, E., Martínez-Bustos, F., Cruz-Orea, A., 2004. Physicochemical properties of casein–starch interaction obtained by extrusion process. Starch – Stärke. 56, 190–198.

Gómez, M., Oliete, B., Rosell, C.M., Pando, V., Fernández, E., 2008. Studies on cake quality made of wheat–chickpea flour blends. LWT – Food Sci. Technol. 41, 1701–1709.

Gujska, E., Khan, K., 2002. Effect of extrusion variables on amino acids, available lysine and *in vitro* protein digestibility of the extrudates from pinto bean (*Phaseolus vulgaris*). Polish J. Food Nutr. Sci. 11, 39–43.

Gupta, R., Dhillon, S., 1993. Characterization of seed storage proteins of lentil (*Lens culinaris* M.). Ann. Biol. 9, 71–78.

Han, J., Janz, J.A.M., Gerlat, M., 2010. Development of gluten-free cracker snacks using pulse flours and fractions. Food Res. Int. 43, 627–633.

Hines, M.E.C., Osuala, I., Nielsen, S.S., 1992. Screening for cysteine proteinase inhibitor activity in legume seeds. J. Sci. Food Agric. 59, 555–557.

Hizukuri, S., 1986. Polymodal distribution of the chain lengths of amylopectins and its significance. Carbohydr. Res. 147, 342–346.

Hood-Niefer, S.D., Tyler, R.T., 2010. Effect of protein, moisture content and barrel temperature on the physicochemical characteristics of pea flour extrudates. Food Res. Int. 43, 659–663.

Hoover, R., 1995. Starch retrogradation. Food Rev. Int. 11, 331–346.

Hoover, R., Manuel, H., 1995. A comparative study of the physicochemical properties of starches from two lentil cultivars. Food Chem. 53, 275–284.

Hoover, R., Ratnayake, W.S., 2002. Starch characteristics of black bean, chick pea, lentil, navy bean and pinto bean cultivars grown in Canada. Food Chem. 78, 489–498.

Hoover, R., Sosulski, F.W., 1985. Studies on the functional characteristics and digestibility of starches from *Phaseolus vulgaris* biotypes. Starch – Stärke 37, 181–191.

Hoover, R., Vasanthan, T., 1994. Effect of heat-moisture treatment on the structure and physicochemical properties of cereal, legume, and tuber starches. Carbohydr. Res. 252, 33–53.

Hoover, R., Zhou, Y., 2003. *In vitro* and *in vivo* hydrolysis of legume starches by α-amylase and resistant starch formation in legumes – a review. Carbohydr. Polym. 54, 401–417.

Hoover, R., Hughes, T., Chung, H.J., Liu, Q., 2010. Composition, molecular structure, properties, and modification of pulse starches: a review. Food Res. Int. 43, 399–413.

Jane, J., Wong, K., McPherson, A.E., 1997. Branch-structure difference in starches of A- and B-type X-ray patterns revealed by their Naegeli dextrins. Carbohydr. Res. 300, 219–227.

Kaur, M., Singh, N., 2005. Studies on functional. Thermal and pasting properties of flours from different chickpea (*Cicer arietinum* L.) cultivars. Food Chem. 91, 403–411.

Kaur, M., Singh, N., 2007. Characterization of protein isolates from different Indian chickpea (*Cicer arietinum* L.) cultivars. Food Chem. 102, 366–374.

Klamczynska, B., Czuchajowska, Z., Baik, B.K., 2001. Composition, soaking, cooking properties and thermal characteristics of starch of chickpeas, wrinkled peas and smooth peas. Int. J. Food Sci. Technol. 36, 563–572.

Lazou, A., Krokida, M., 2010. Functional properties of corn and corn–lentil extrudates. Food Res. Int. 43, 609–616.

Liang, M., Huff, H.E., Hsieh, F.H., 2002. Evaluating energy consumption and efficiency of a twin-screw extruder. J. Food Sci. 67, 1803–1807.

Liu, S., Elmer, C., Low, N.H., Nickerson, M.T., 2010. Effect of pH on the functional behaviour of pea protein isolate–gum Arabic complexes. Food Res. Int. 43, 489–495.

Manthey, F.A., Yalla, S.R., Dick, T.J., Badaruddin, M., 2004. Extrusion properties and cooking quality of spaghetti containing buckwheat bran flour. Cereal Chem. 81, 232–236.

McAuley, J.A., Kunkel, M.E., Acton, J.C., 1987. Relationships of available lysine to lignin, color and protein digestibility of selected wheat-based breakfast cereals. J. Food Sci. 52, 1580–1582, 1610.

Meuser, F., 2001. Technological aspects of dietary fibre. In: McCleary, B.V., Prosky, L. (Eds.), Advanced dietary fibre technology. Blackwell Science, Oxford, pp. 248–269.

Miles, M.J., Morris, V.J., Orford, P.D., Ring, S.G., 1985. The roles of amylose and amylopectin in the gelation and retrogradation of starch. Carbohydr. Res. 135, 271–281.

Mohamed, S., 1990. Factors affecting extrusion characteristics of expanded starch-based products. J. Food Process. Preserv. 14, 437–452.

Monge, L., Cortassa, G., Fiocchi, F., Mussino, G., Carta, Q., 1990. Glycoinsulinaemic response, digestion and intestinal absorption of the starch contained in two types of spaghetti. Diabetes Nutr. Metab. 3, 239–246.

Noda, T., Takahata, Y., Sato, T., Ikoma, H., Mochida, H., 1996. Physicochemical properties of starches from purple and orange fleshed sweet potato roots at two levels of fertilizer. Starch – Stärke 48, 395–399.

Osborne, T.B., 1924. The vegetable proteins, 2nd edn. Longmans Green and Co, London.

Paredes-López, O., Ordorica-Falomir, C., Olivares-Vázquez, M.R., 1991. Chickpea protein isolates: physicochemical, functional and nutritional characterization. J. Food Sci. 56, 726–729.

Petitot, M., Boyer, L., Minier, C., Micard, V., 2010. Fortification of pasta with split pea and faba bean flours: pasta processing and quality evaluation. Food Res. Int. 43, 634–641.

Pietrasik, Z., Janz, J.A.M., 2010. Utilization of pea flour, starch-rich and fiber-rich fractions in low fat bologna. Food Res. Int. 43, 602–608.

Prinyawiwatkul, W., McWatters, K.H., Beuchat, L.R., Phillips, R.D., 1997. Physicochemical and sensory properties of chicken nuggets extended with fermented cowpea and peanut flours. J. Agric. Food Chem. 45, 1891–1899.

Ratnayake, W.S., Jackson, D.S., 2008. Starch gelatinization. Adv. Food Nutr. Res. 55, 221–268.

Rayas-Duarte, P., Mock, C.M., Satterlee, L.D., 1996. Quality of spaghetti containing buckwheat, amaranth, and lupin flours. Cereal Chem. 73, 381–387.

Robin, J.P., Mercier, C., Charbonniere, R., Guilbot, A., 1974. Lintnerized starches. Gel filtration and enzymatic studies of insoluble residues from prolonged acid treatment of potato starch. Cereal Chem. 51, 389–406.

Saharan, K., Khetarpaul, N., 1994. Protein quality traits of vegetable and field peas: varietal differences. Plant Foods Hum. Nutr. 45, 11–22.

Sanjeewa, W.G.T., Wanasundara, J.P.D., Pietrasik, Z., Shand, P.J., 2010. Characterization of chickpea (*Cicer arietinum* L.) flours and application in low-fat pork bologna as a model system. Food Res. Int. 43, 617–626.

Shand, P.J., 2000. Textural, water holding, and sensory properties of low-fat pork bologna with normal or waxy starch hull-less barley. J. Food Sci. 65, 101–107.

Shannon, J.C., Garwood, D.L., 1984. Genetics and physiology of starch development. In: Whistler, R.L., BeMiller, J.N., Paschal, E.F. (Eds.), Starch chemistry and technology. Academic Press, New York, pp. 26–79.

Singh, D., Sokhansanj, S., 1984. Cylindrical concave mechanism and chemical treatment for dehulling of pigeon pea. Agric. Mechanizat. Asia Africa Latin Am. 15, 53–58.

Singh, N., Sandhu, K.S., Kaur, M., 2004. Characterization of starches separated from Indian chickpea (*Cicer arietinum* L.) cultivars. J. Food Eng. 63, 441–449.

Singh, U., Singh, B., 1992. Tropical grain legumes as important human foods. Econ. Bot. 46, 310–321.

Sironi, E., Sessa, F., Duranti, M., 2005. A simple procedure of lupin seed protein fractionation for selective food applications. Eur. Food Res. Technol. 221, 145–150.

Srichuwong, S., Jane, J., 2007. Physicochemical properties of starch affected by molecular composition and structure: a review. Food Sci. Biotechnol. 16, 663–674.

Swanson, R.B., 2007. Grains and legumes. In: Hui, Y.H. (Ed.), Food chemistry (2nd ed.). Science TEchnology Systems, West Sacramento, CA, pp. 20-1–20-24.

Tester, R.F., Morrison, W.R., 1990a. Swelling and gelatinization of cereal starches. I. Effects of amylopectin, amylose, and lipids. Cereal Chem. 67, 551–557.

Tester, R.F., Morrison, W.R., 1990b. Swelling and gelatinization of cereal starches. II. Waxy rice starches. Cereal Chem. 67, 558–563.

Tharanathan, R.N., Mahadevamma, S., 2003. Grain legumes – a boon to human nutrition. Trends Food Sci. Technol. 14, 507–518.

Torruco-Uco, J., Betancur-Ancona, D., 2007. Physicochemical and functional properties of makal (*Xanthosoma yucatanensis*) starch. Food Chem. 101, 1319–1326.

Tosh, S.M., Yada, S., 2010. Dietary fibres in pulse seeds and fractions: characterization, functional attributes, and applications. Food Res. Int. 43, 450–460.

Veenstra, J.M., Duncan, A.M., Cryne, C.N., Deschambault, B.R., Boye, J.I., Benali, M., Marcotte, M., Tosh, S.M., Farnworth, E.R., Wright, A.J., 2010. Effect of pulse consumption on perceived flatulence and gastrointestinal function in healthy males. Food Res. Int. 43, 553–559.

Waigh, T.A., Gidley, M.J., Komanshek, B.U., Donald, A.M., 2000. The phase transformations in starch during gelatinisation: a liquid crystalline approach. Carbohydr. Res. 328, 165–176.

Wang, N., Hatcher, D.W., Gawalko, E.J., 2008. Effect of variety and processing on nutrients and certain anti-nutrients in field peas (*Pisum sativum*). Food Chem. 111, 132–138.

Wang, N., Hatcher, D.W., Tyler, R.T., Toews, R., Gawalko, E.J., 2010. Effect of cooking on the composition of beans (*Phaseolus vulgaris* L.) and chickpeas (*Cicer arietinum* L.). Food Res. Int. 43, 589–594.

Wood, J.A., 2009. Texture, processing and organoleptic properties of chickpea-fortified spaghetti with insights to the underlying mechanisms of traditional durum pasta quality. J. Cereal Sci. 49, 128–133.

Zhao, Y.H., Manthey, F.A., Chang, S.K.C., Hou, H.J., Yuan, S.H., 2005. Quality characteristics of spaghetti as affected by green and yellow pea, lentil, and chickpea flours. J. Food Sci. 70 (6), S371–S376.

Zhou, Y., Hoover, R., Liu, Q., 2004. Relationship between α-amylase degradation and the structure and physicochemical properties of legume starches. Carbohydr. Polym. 57, 299–317.

By-product utilization

Ankit Patras[1], B. Dave Oomah[2], Eimear Gallagher[1]
[1]University College Dublin, Dublin, Ireland
[2]National Bioproducts and Bioprocesses Program, Pacific Agri-Food Research Centre, Agriculture and Agri-Food Canada, Summerland, BC, Canada

12.1 Introduction

Many food-processing methods produce waste streams. In the past, it has been relatively easy to dispose of the waste components of food processing operations by dumping them. Today, however, this is becoming much more expensive, and various businesses are examining the options available for them to add value to these materials. Waste utilization is both a necessity and a challenge. In the food industry, the recovery and modification of wastes is becoming increasingly important. From the current knowledge worldwide, it is quite evident that many of these contain low levels of suspended solids and low concentrations of dissolved materials. Apart from the environmental challenges posed, such streams represent considerable amounts of potentially reusable materials and energy. Much of the material generated as waste by food-processing industries throughout the world – and about to be generated within biofuels programs – contains components that could be utilized as substrates and nutrients in a variety of microbial/enzymic processes, to give

Pulse Foods: Processing, Quality and Nutraceutical Applications. DOI: 10.1016/B978-0-1238-2018-1.00007-0

rise to added-value products. These by-products are promising sources of bioactive compounds which may be used because of their favorable technological or nutritional properties. Due to increasing production, disposal represents a growing problem since the plant material is usually prone to microbial spoilage, thus limiting further exploitation. On the other hand, costs of drying, storage and shipment of by-products are economically limiting factors (Schieber et al., 2001). Therefore, agro-industrial waste is often utilized as feed or as fertilizer on farms. The problem of disposing of by-products is further aggravated by legal restrictions within the EU. Thus, efficient, inexpensive and environmentally sound utilization of these materials is becoming more important (Lowe and Buckmaster, 1995).

Today, by-products are considered to be a promising source of functional compounds (Carle et al., 2001). Searching for possibilities for better utilization of valuable components from food by-products, a number of research studies have been carried out on improving physical, sensory and nutritional characteristics of extruded food prepared from food by-products including cereal and legume grains. For example, Stojceska et al. (2008) studied the incorporation of cauliflower trimmings into ready-to-eat expanded products (snacks) and their effect on the textural and functional properties of extrudates. Addition of cauliflower significantly increased the dietary fiber and levels of proteins. Extrusion cooking significantly ($P < 0.0001$) increased the level of phenolic compounds and antioxidants but significantly ($P < 0.001$) decreased *in vitro* protein digestibility and fiber content in the extruded products. The expansion indices, total cell area and wall thickness were inversely correlated to the level of cauliflower (5–20%). Sensory test panel scores indicated that cauliflower could be incorporated into ready-to-eat expanded products up to the 10% level. Vegetable waste material contains significant amounts of dietary fiber. For example, the high crude fiber content of vegetable pomace (in total 20–65% DM) suggests its utilization as a crude fiber "bread improver". In bread and bakery goods, as well as in pastry, cereals and dairy products, carrot pomace can be incorporated as a stabilizer. Besides crude fiber, carrot pomace is rich in provitamins, color and natural acids with several beneficial functional properties. In addition to substituting sourdough in bread, carrot pomace acts as an acidifying agent, preservative or antioxidant in several food products (Filipini and Hogg, 1997; Masoodi and Chauhan, 1998).

Several research groups throughout the world are actively focusing their efforts on exploiting pulse by-products, which have remained unexplored, underutilized or only localized in a particular region. A detailed section dedicated to this topic can be found in this chapter (12.2.1). For alleviating hunger and to overcome malnutrition, particularly in children and pregnant women, there is an increased demand in developing countries to explore these by-products for proteins and phytochemicals. There is a rapidly growing body of literature covering the role of plant secondary metabolites in food and their potential effects on human health. Furthermore, consumers nowadays are increasingly aware of diet-related health problems, demanding natural functional ingredients which are expected to be safe and to promote health and wellness. Pulses are an important source of macronutrients, such as proteins, bioactive carbohydrates and dietary fiber (Tharanathan and Mahadevamma, 2003). In addition to being a source of macronutrients and minerals, pulses contain plant secondary metabolites, i.e. polyphenols (flavanoids, phenolic acids, tannins, saponins) that are increasingly being recognized for their potential benefits for human health.

The marketability and market value of pulse by-products can be improved by identifying suitable technologies for safe handling, classification and by establishing their functional and phytochemical qualities for further processing. This chapter focuses on nutraceutical composition of the by-products of pulse processing. In addition, novel and traditional ways to extract functional bioactive components are discussed. This includes a description and application of the extraction techniques and an updated overview of the principal applications of these techniques for obtaining functional ingredients from pulse processing by-products.

12.2 Overview of pulse processing by-products

Pulses are generally consumed after seed coat removal by industrial or small-scale milling. Soaking is a preliminary step common to almost all methods of preparing legumes, prior to cooking, fermentation and germination. Soaking aids in the removal of the seed coat to shorten the cooking time. Various operations involved in primary processing of pulses, such as cleaning, grading, drying/conditioning, storage and milling, have been discussed in previous chapters. Dhal milling is the third largest food grain processing industry in India (Ramakrishnaiah et al., 2004). Commercial dhal milling efficiency is approximately 75%, producing 25% by-product yield consisting of husk, powder, broken, shriveled and unprocessed grains. In India, dhal mills generate an estimated 2.5 million tonnes of by-product annually, containing about 40% of valuable cotyledon material, and a technology platform has been developed to recover and increase the value to these by-products (Fig. 12.1) (Ramakrishnaiah et al., 2004). The lentil milling industry is well established in Turkey, and has expanded to Sri Lanka, Egypt, Syria, the United Arab Emirates and, most recently, to Australia and Canada. In western Canada, peas are commercially processed or fractionated into component products, such as hulls, flour, starch and protein concentrate. Pea hulls are used in high-fiber products; the starch fraction has applications in adhesives and carbonless paper, whereas the pea protein concentrate has limited application in human food and in protein isolate production. Pea products are manufactured by both the dry and wet milling processes, with higher potential human food applications for the more expensive products from the wet process, due to elimination of bitter pea flavor components. However, the wet process also produces wet pea pulp and pulp bran, mostly destined for animal nutrition products.

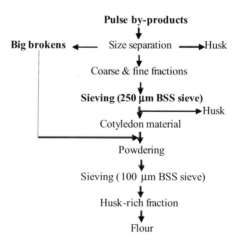

Figure 12.1 Recovery of husks, brokens and husk-rich fraction from pulse by-products.

12.2.1 Pulse milling by-products

A typical flowchart of pulse processing is shown in Figure 12.2. Milling of pulses involves removal of the outer husk and splitting the grain into two. In traditional pulse milling, the loosening of husk by conditioning (adding small amounts of water) is insufficient. Therefore, a large amount of abrasive force is applied for the complete dehusking of the grains which results in high losses in the form of broken grains and powder. It is therefore necessary to improve the traditional milling processes in order to increase the total yield of dehusked and split pulses and reduce losses. In general, the outer layers of legume cotyledons are rich sources of protein; removal during dehulling results in considerable losses in protein (Singh and Jambunathan, 1980). Processing methods are known to affect greatly the composition of cereal products, and considerable amounts of nutrients, namely protein, amino acids, minerals and vitamins, may be lost if refined cereal products are consumed instead of whole grain products (Pedersen and Eggum, 1983). Singh et al. (1989) investigated nutrient (principal chemical constituents, amino acids, minerals and trypsin inhibitors) losses due to scarification of pigeon pea cotyledons and examined the distribution of such nutrients in dhal (scarified cotyledons) and powder fractions of pigeon pea. Protein, soluble sugars and ash of the dhal fraction decreased with increasing scarification time, while starch content increased. Considerable amounts of calcium (20%) and iron (30%) were removed by scarification for 4 min, but the process did not adversely affect protein quality in terms of amino acids. Trypsin inhibitors were not removed substantially by scarification.

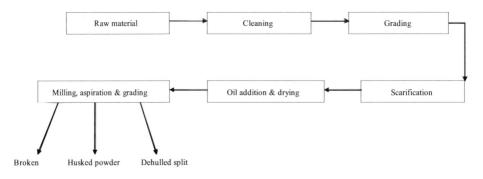

Figure 12.2 Flowchart of dry milling process of pulses.

Table 12.1 Proportion of seed coat, cotyledons and embryo in relation to whole in different legumes

Legume	Proportion of whole grain (%)		
	Seed coat	Cotyledon	Embryo
Pigeon pea	15.50	83.00	1.50
Peas	10.00	89.28	1.72
Lentils	8.05	89.97	1.98
Mung bean	12.09	85.61	2.30
Cowpea	10.64	87.23	2.13

Adapted from Salunkhe et al. (1985).

The proportion of seed coat, cotyledon and embryo in different legumes is shown in Table 12.1 (Salunkhe et al., 1985). The by-products of pulse milling include broken grains (6–13%), germ, powder (7–12%) and husk (4–14%). After isolation of the main constituent, there are abundant residues which represent an inexpensive material that has been undervalued until now (Mateos-Aparicio et al., 2010a,b). These include hull, powder, large and small broken, shrivelled and under-processed grains, accounting for approximately 20–30% of total grains processed depending upon the machinery and other processing factors (Narasimha et al., 2004). In a study conducted by these authors, a process to retrieve the cotyledon material from commercial dhal mill by-products, which contain approximately 50% cotyledon material amounting to 1 million t, was investigated. The process consisted of destoning, size separation and air classification followed by refining and thermal stabilization of edible material. In their study, approximately 30–35% of the cotyledon material was recovered from the by-products and used to prepare traditional pulse-based products. They suggested that cotyledon material could be substituted in traditional pulse-based products such as *vada*, *rasam/sambar* and *papads* (Indian snacks) by up to approximately 50%.

Dehulling of legume seeds (into dhal) and splitting of the cotyledons are often carried out for better product profile and acceptability. Dehulling also removes tannins that lower protein digestibility (Bressani and Elfas, 1980). Hulls contribute a significant portion of the insoluble dietary fiber (DF) in whole pulses. Pulse hulls are rich in DF, ranging from dry weight contents of 75% (chickpeas) to 87% (lentils) and 89% (peas) (Dalgetty and Baik, 2006). Characterization of the cell wall polysaccharides in pulse hulls and cotyledons is important to understand better their function as dietary fibers; however, only limited research has been carried out in this area. Some studies have demonstrated the presence of important functional components in pulse by-products. Sosulski and Dabrowski (1984) quantified free and hydrolyzable phenolic acids in flours and hulls of ten legume samples. The phenolic constituents in the defatted flours and hulls of legume species were fractionated into free acid, soluble ester and residue components, followed by alkaline hydrolysis and quantification by capillary gas–liquid chromatography. The flours contained only soluble esters, and hydrolysis revealed the presence of trans-ferulic, trans-p-coumaric and syringic acids in nearly all species. Mung bean, field pea, lentil, faba bean and pigeon pea contained only 2–3 mg of phenolic acids $100 \, \mathrm{g}^{-1}$ of flour, but higher levels were obtained in navy bean, lupine, lima bean, chickpea and, especially, cowpea. Their results demonstrated that hulls contained p-hydroxybenzoic, protocatechuic, syringic, gallic, trans-p-coumaric and trans-ferulic acids in the soluble ester fraction and, to a lesser extent, in the insoluble residue. Polyphenols discussed above have several health benefits. It has been suggested that the possible health benefits of polyphenol derive from their antioxidant and anti-inflammatory properties (Joseph et al., 2006). Evidence for their role in the prevention of degenerative diseases is still emerging. Experimental studies on animals and human cell lines have demonstrated that polyphenols can play a role in the prevention of cancer and cardiovascular diseases (Scalbert et al., 2005).

12.2.2 Food applications of pulse processing by-products

Interest in the use of pulses and their constituents in food formulation is growing in many developed countries (Boye et al., 2010). Factors contributing to this growth include their reported nutritional and health benefits, changes in consumer preferences, increasing demand for variety and balance, change in demographics (age, racial diversity), rise in the incidence of food allergies and ongoing research on production and processing technologies (Boye et al., 2010). Some studies have been carried out on by-product utilization of pulses; various authors have attempted to incorporate pulse by-products in a range of food products. For example, Tiwari et al. (2011) substituted wheat flour with pigeon pea (*Cajanus*

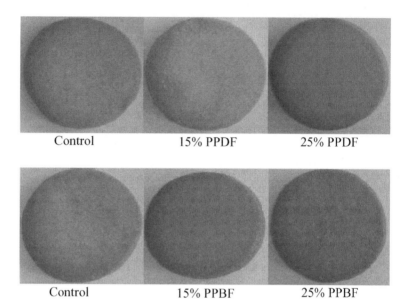

Figure 12.3 Biscuits made from wheat flour (control) and composite flour (Tiwari et al., 2011).

cajan L.) dehulled flour (PPDF) and pigeon pea by-product flour (PPBF) and formulated biscuits. Considerable increases in protein content of the biscuits and significant influence on physical and sensory quality were observed (Fig. 12.3). The study demonstrated the potential use of pigeon pea milling by-products in preparation of biscuits. It is envisaged that supplementing pulse fractions in bakery products should enhance the nutritive value of the products while, at the same time, making appropriate use of by-products. Table 12.2 displays the proximate composition of hulls and cotyledon fibers used to fortify wheat flour for bread production. The fiber content of hulls ranged from 74.8 to 88.9%, with chickpea hulls containing the lowest, and pea containing the highest amounts of fiber. The purity of insoluble cotyledon fibers ranged from 81.5 to 83.7%. Various levels of starch, ash and protein comprised the other 16.3–18.5% of the insoluble fiber isolates. The purity of soluble cotyledon fibers ranged from 64.5 to 70.6%.

Tiwari et al. (2009) incorporated by-products of legumes and cereal processing in the development of low-cost nutritious deep-fried snacks. Their study demonstrated that the incorporation of legume flour (red gram, green gram and black gram) with cereal flour (rice) improves the nutritional value of the products. Sensory results and

Table 12.2 Proximate composition of isolated legume fiber

Fraction	Starch[a]	Protein	Fiber (%)
Hull			
Pea	2.6[b]	5.2[c]	88.9[a]
Lentil	1.3[b]	9.7[b]	86.7
Chickpea	7.4[a]	12.1[a]	74.8[c]
Insoluble fiber			
Pea	4.3[b]	8.4[b]	82.9[ab]
Lentil	5.1[a]	9.4[a]	81.5[b]
Chickpea	2.9[c]	10.1[a]	83.7[a]
Soluble fiber			
Pea	nd	23.6[a]	64.7[b]
Lentil	nd	24.0[a]	64.4[b]
Chickpea	nd	16.1[b]	70.6[a]

[a] Means in a column with different letters are significantly different ($P < 0.05$).
Adapted from Dalgetty and Baik (2006).

shelf-life studies showed complementary results. The study indicated the feasibility of developing such products.

Jackson (2009) studied protein nutritional quality of cowpea and navy bean residue fractions. The protein quality of cowpea residue-wheat and navy bean residue-wheat diets was determined using *in vivo* and *in vitro* protein digestibility assays with an American Institute of Nutrition growth diet (AIN-93G diet) as control. The diets were fed to laboratory rats over 4 weeks. All cowpea residue diets, the 30% and 70% navy bean residue diets, and the control diet supported growth, while the 100% navy bean residue diet induced weight loss. *In vitro* digestibility ranged from 77.8 to 84.5% and 66.5 to 79.6% for the cowpea residue and the navy bean residue diets, respectively. Since the cowpea residue diets had both higher protein digestibility and higher amino acid composition than the navy bean residue diets, it is feasible to conclude that the cowpea residue diets had higher protein quality than the navy bean residue diets. The author also suggested that legume residues after protein extraction could be recommended for human food if complemented with a cereal, particularly as it meets the amino acid pattern.

Pea hull fibers have been extensively used in baked products. The incorporation of commercial pea hull fibers (containing approximately 80–88% total dietary fiber and 2–10% soluble dietary fiber) has been the most widely studied among the pulse fibers. Sosulski and Wu (1988) carried out one of the earlier and more extensive studies comparing dough and loaf quality characteristics of high-fiber wheat breads enriched with various levels (5–20%) and sources of fiber, including pea hulls. With increasing levels of pea hulls, the water-retention capacity of the composite flours increased (although mixing properties were not markedly affected), crust color became lighter and more yellow, and loaf and specific volumes decreased. A satisfactory high-fiber formulation containing about 15% total dietary fiber (TDF) (compared to 3% TDF in the control) was obtained with 15% (by weight) of pea hulls. Hence, it is quite evident that these by-products could be a useful source of ingredients and have the potential to fortify and increase the functionality of various food products.

In a separate study (Wang et al., 2002), it was found that adding 3% pea hulls to wheat flour slightly increased water absorption and dough resistance and decreased dough extensibility. Although loaf volume was reduced, crumbs were softer and other bread quality characteristics were acceptable. Addition of the hulls resulted in bread with 5% TDF (compared to 3% TDF in the control bread without added fiber). Similar results for dough rheology and bread quality were obtained by Gómez et al. (2003) with the addition of 2% pea hulls. After 2 days, storage, crumb firmness did not increase to the same degree as the control bread, which was attributed to the water-retention capacity of the fiber and possible interaction between the fiber and starch that delayed starch retrogradation.

Insoluble and soluble pea lentil and chickpea cotyledon fiber fractions were isolated by Dalgetty and Baik (2006) and compared with pulse hull fibers for wheat bread enrichment. This comprehensive study examined mixing and baking characteristics of the doughs and bread staling during storage. Breads enriched with pulse hulls and fibers were higher in moisture content than control bread regardless of the type, source or enrichment level. Incorporation of 5% hulls or insoluble fibers, or 3% soluble fibers, into wheat flour was successful in terms of increased total fiber content and improved moistness of bread, and did not significantly increase crumb firmness during storage.

Pulse fiber can also be incorporated into meat-based products. Pea cotyledon fiber (inner fiber) was identified from among 10 commercial starch-, fiber- and protein-based products as having the most potential as an ingredient for retaining fat in ground beef during high temperature heating (Anderson and Berry, 2001a). This unique application was designed for the US Army which required calorie-dense beef-based rations to address the calorie under-consumption problem by some personnel. Formulations were further optimized in a subsequent study (Anderson and Berry, 2001b) and fat retention in high-fat ground beef increased from 33% to between 85% and 98% when pea fiber was added at 10–16%; cooking yield was also significantly greater. The pea fiber product (Swelite®, Consucra, Belgium) contained approximately 48% dietary fiber, 44% starch and 7% protein; fiber components were approximately 66% insoluble cellulose and 33% soluble pectic material.

Texture modification can be observed when pulse fiber (pea cotyledon) is incorporated. For example, pea cotyledon fiber was investigated by Anderson and Berry (2000) as a texture-modifying ingredient in lower-fat (10% and 14% fat) beef patties to enhance eating quality. Pea fiber did not influence fat retention in beef patties at this lower fat level, but improved tenderness and cooking yield, and had no negative effect on juiciness or beef flavor intensity. Although the commercial production and applications of fiber fractions from peas have been reasonably well studied, little research on chickpea, lentil or dry bean fibers has been published. A clear understanding of the physicochemical properties of these fiber fractions is also needed to optimize the use of pulse fibers in processed foods.

12.2.3 Non-food applications of pulse processing by-products

There is little information in the scientific literature on non-food applications of pulses or pulse fractions. The use of agri-food waste, including pulse residues, as potential feedstock for bioethanol production was explored by Del Campo et al. (2006). Figure 12.4 illustrates a typical experimental procedure carried out with the agri-food wastes for ethanol production. A dilute acid hydrolysis process was investigated to pretreat the waste materials; sugar recovery from pulse materials required more intensive pretreatment due to high starch contents.

12.3 Nutritional value of pulse milling by-products

Processing of pulses and legumes in the early 1980s was designed for the reduction/elimination of antinutritional factors (ANF), including condensed tannins, protease inhibitors, alkaloids, lectins, pyrimidine glycosides and saponins. The presence of ANF has partially hampered the use of grain legumes in animal nutrition. Possible negative effects

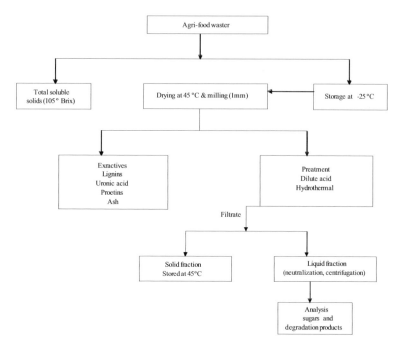

Figure 12.4 Experimental procedure used for characterization and pretreatment of the selected wastes.

related to ANF include feed refusals (tannins, alkaloids), reduced nutrient digestibility (tannins, protease inhibitors, lectins) or even toxic effects (alkaloids). Several technologies including breeding and genetic manipulation, feed formulation, primary and secondary processing were advocated at the time to reduce or eliminate ANF. Major developments in technology have taken place since then, and the new era of safe functional foods and nutraceuticals and heightened sensibility for environmental concern has made the use of industrial by-products a major priority.

Agricultural and industrial residues are attractive sources of natural antioxidants (Moure et al., 2001). By-products which remain after pulse processing still contain a huge amount of phenolic compounds (Sosulski and Dabrowski, 1984). Despite the recognized benefits of pulse and pulse products in the development of functional foods and nutraceuticals (Bassett et al., 2010), and animal nutrition (Jezierny et al., 2010), commercial pulse by-products, particularly related to phytochemicals, have not been adequately investigated. Pea hull fiber, depending on processing method, is 75–90% fiber of which 95% is insoluble. Centara III® pea hull fiber, produced by Parrheim Foods, Portage la Prairie, MB, is an approved novel fiber source in Canada with proven efficacy in fecal bulking. Adding a moderate amount of ground pea hull fiber (2–3 g per serving) to three to four foods each day provided a beneficial laxative effect and increased bowel frequency in institutionalized elderly individuals over a 6-week intervention period, with the added benefit of reducing dysphagia (difficulty chewing and swallowing) (Dahl et al., 2001). Exlite® and Centara III® are finely ground, off-white fiber sources with TDF contents of greater than 78% and 90%, respectively, but differ in their processing. Exlite® is produced by abrasion dehulling and grinding, whereas Centara III® undergoes an additional washing and drying step prior to grinding (Fitzpatrick, 2007). These fiber products have recently been characterized for their phenolic content and antioxidant activities (Table 12.3). The phenolics were mostly concentrated in the protein isolate and starch-fiber flour, whereas the flavonoids were sequestered in the major fiber fraction (Agboola et al., 2010).

Tannins (1.2%) and trypsin inhibitors (4.5 mg g^{-1}) are also concentrated in the protein fraction of air classified zero-tannin faba bean processed at the pilot plant scale (Gunawerdena et al., 2010). However, the presence of these components does not affect the nutrient digestibility of the faba bean protein fraction in growing pigs (Gunawerdena et al., 2010). Nutrient digestibility in growing pigs is not reduced by the faba bean protein fraction that had significantly higher tannin content than commercial soybean and field pea protein concentrates (0.45 and 0.75%, as is basis), respectively (Gunawerdena et al., 2010). Therefore, these concentrates are promising increased quality proteins and starches for weaned and grower pigs because of the greater nutrient digestibility and similar digestible nutrient

Table 12.3 Phenolics and antioxidant activities of pea products

Pea product	Yield (mg g^{-1})	Concentration[a]		Antioxidant activity[b]	
		Total phenolics	Flavonoids	TEAC	DPPH
Whole seed	546.20	5.67[a]	1.35[a]	15.62[a]	10.89[a]
Protein isolate	194.35	2.84[b]	0.44[c]	4.13[b]	1.89[c]
Fiber-Centara III	49.96	0.56[c]	0.81[b]	1.56[d]	0.63[c]
Fiber-Centara IV	34.72	0.47[c]	0.16[e]	0.54[e]	0.24[d]
Starch-fiber flour	33.27	2.36[b]	0.24[d]	3.54[c]	2.79[b]

[a] Means in a column with different letters are significantly different ($P < 0.05$). Concentration of phenolic compounds is expressed as mmol gallic acid or (+) catechin equivalents 100 g^{-1} of sample dry weight for total phenolics and flavonoids, respectively.
[b] Antioxidant activity expressed as mmol Trolox 100 g^{-1} sample dry weight.
Adapted from Agboola et al. (2010).

profile as international feedstuff standards. High-quality feedstuffs can thus be produced as value-added products from pulse seeds.

Pulses contain a wide range of polyphenolic compounds, including flavonols, flavone glycosides, flavanols and oligomeric and polymeric proanthocyanidins, located essentially in their seed coat which can contribute to their antioxidant capacity. Among pulses, broad bean presents a very high content of free flavanols (average concentration of 154.5 mg total flavanols $100 g^{-1}$ fresh weight), reflected in high Trolox equivalent antioxidant capacity (TEAC) (Table 12.4). Most of the antioxidants (40–80%) of pulses (with the exception of broad beans) are in bound form releasable only through alkaline hydrolysis and detected particularly through their ferric reducing antioxidant power (FRAP).

Chickpea and horse gram have been fractionated by mechanical dehulling and their phytochemical composition evaluated for potential applications (Table 12.5). The hulls were rich in phenolics and trypsin and alpha-amylase inhibitory activities (Sreerama et al., 2010). Chickpea dehulling by-products from ten chickpea dehulling and packaging units in East Azerbaijan, Iran, contained 0.01, 0.14, 3.1 and 5.1% (dry matter) of total tannins, total phenolics, ether extracts and soluble sugars, respectively (Maheri-Sis et al., 2007).

A highly thermostable peroxidase, an important enzyme for various industrial (commercial catalyst for phenolic resin synthesis), analytical and biomedical (component of medical diagnosis kits) applications, has been resourced from black gram husk, a by-product of the milling industry (9% yield of total) (Ajila and Prasada Rao, 2009).

Lentil seed coat tannins have been found to consist mainly of catechin and gallocatechin units, with the polymer fraction being the most abundant proanthocyanidin (65–75%) compared to monomers and oligomer fractions (Dueñas et al., 2003). Our recent investigation

Table 12.4 Ferric reducing antioxidant power (FRAP) and trolox equivalent antioxidant capacity (TEAC) of pulse extracts

Pulse	FRAP (mmol Fe^2 kg^{-1})			TEAC (mmol Trolox kg^{-1})		
	Total	Bound (%)	Free (%)	Total	Bound (%)	Free (%)
Bean	9.59	62	38	3.30	54	46
Broad bean	17.32	40	60	13.26	42	58
Chickpea	5.50	62	38	2.90	75	25
Lentil	41.65	79	21	9.30	60	40
Pea	8.59	59	41	3.73	65	35

Adapted from Pellegrini et al. (2006).

Table 12.5 Concentration of phytochemicals in milled fractions of chickpea and horse gram

Phytochemicals	Chickpea			Horse gram		
	Cotyledons	Embryonic axis	Hulls	Cotyledons	Embryonic axis	Hulls
Phenolics	15.2[c]	46.2[b]	75.9[a]	13.8[z]	159.4[y]	484.6[x]
Phytic acid (mg g^{-1})	9.8[a]	3.4[b]	0.8[c]	8.4[x]	3.8[y]	1.0[z]
TIA-buffer extract	6090[a]	938[b]	861[c]	9856[x]	2018[y]	1134[z]
TIA-methanol extract	1542[c]	1964[b]	2170[a]	2474[z]	3663[y]	10434[x]
αAIA-buffer extract	461.5[a]	75.2[b]	26.0[c]	56.9[x]	12.3[y]	4.1[z]
αAIA-methanol extract	210[c]	223[b]	263[a]	632[y]	640[y]	937[x]

Means in a column with different letters are significantly different ($P < 0.05$). Concentration of phenolic compounds is expressed as mg of gallic acid g^{-1} dry matter; TIA and αAIA are trypsin and α-amylase inhibitory activities, respectively, expressed in units g^{-1}.
Adapted from Sreerama et al. (2010).

(Oomah, B.D., and Quettier, N., 2010, unpublished report) on by-product obtained from industrial lentil milling showed that the phenolic and antioxidant activities of the hulls were dependent mainly on processing and extraction procedures (Tables 12.6 and 12.7). Phenolics and their antioxidant activity are the hallmark of the current functional foods and nutraceutical markets.

Mateos-Aparicio et al. (2010c) highlighted the potential use of two legume by-products, pea pod (*Pisum sativum* L.) and broad bean

Table 12.6 Phenolic content and antioxidant of commercial lentil hulls extracted with various solvents

Solvents	Concentration[a]				ORAC	
	Total phenolics	Tartaric esters	Flavonols	Anthocyanins	mg g^{-1}	μmol g^{-1}
	Acetone					
Raw	62.45[b]	1.70[a]	1.56[a]	0.076[b]	171.15[b]	683.82[b]
Cryo	69.70[a]	1.66[b]	1.52[b]	0.083[a]	202.41[a]	808.70[a]
Mean	66.06[x]	1.68[x]	1.54[x]	0.079[x]	186.78[x]	746.26[x]
	Water					
Raw	21.52[p]	1.19[q]	1.11[q]	0.060[p]	36.33[b]	145.13[b]
Cryo	20.72[q]	1.36[p]	1.34[p]	0.060[p]	39.65[a]	158.41[a]
Mean	21.12[y]	1.27[y]	1.23[y]	0.060[y]	37.99[y]	151.77[y]

Means in a column with different letters are significantly different ($P < 0.05$). Concentration of phenolic compounds is expressed as mg (+) catechin, caffeic acid, quercetin or cyanidin-3-glucoside equivalents g^{-1} of sample for total phenolics, tartaric esters, flavonols and anthocyanins, respectively.

Table 12.7 Effect of processing on phenolic content of commercial lentil hulls

Processing	Concentration[a]			
	Total phenolics	Tartaric esters	Flavonols	Anthocyanins
Microwave treatment				
Ethanol 50%	97.20[a]	6.62[a]	4.39[a]	0.81[a]
Water	40.20[b]	2.72[b]	2.46[b]	0.52[b]
Enzyme treatment – 24 h				
Untreated	11.22[y]	1.04[x]	1.12[x]	0.024[x]
Protease	15.20[x]	1.16[x]	1.20[x]	0.032[x]

Means in a column with different letters are significantly different ($P < 0.05$). Concentration of phenolic compounds is expressed as mg (+) catechin, caffeic acid, quercetin or cyanidin-3-glucoside equivalents g^{-1} of sample for total phenolics, tartaric esters, flavonols and anthocyanins, respectively.

pod (*Vicia faba* L.). These by-products have in common that their major fraction is dietary fiber (pea pod: 58.6 g $100 g^{-1}$; broad bean pod: 40.1 g $100 g^{-1}$). Sucrose, glucose and fructose are the most important soluble sugars in both pods. Protein is also a considerable component. Both pods showed low levels of fat. Linoleic acid is the most important fatty acid; oleic acid is remarkable in pea pod and linolenic acid in broad bean pod. Potassium, calcium and iron are the main minerals present in these by-products. Therefore, the presence of dietary fiber and other nutritive compounds provides an important added value to these by-products and the possibility to balance food according to nutritional requirements and consequently an advantage against the use of isolate functional ingredients.

From the ongoing research worldwide, and with current databases of information, there is clear evidence that some of the pulse by-products do possess high levels of phytochemicals. Application of modern processing methods along with incorporation of traditional knowledge will provide a substantial base for the commercial exploitation of these by-products for developing new novel foods (or for biofortification), as well as for use in the pharmaceutical industry.

12.4 Strategies for recovery of bioactive compounds

Extraction is one of the most widely used unit operations in the food industry. Its main use is for releasing components initially retained in a food matrix such as cereals, pulses, fruits, vegetables and their by-products (Barzana et al., 2002). Molecules obtained by extraction may be used as food additives or for exerting peculiar beneficial effects on human health (Osada et al., 2001). Various traditional and novel techniques, including ultrasound-assisted extraction, microwave-assisted extraction, supercritical fluid extraction and accelerated solvent extraction, have been developed for the extraction of nutraceuticals/bioactive compounds from plant matrix and its by-products in order to shorten the extraction time, decrease the solvent consumption, increase the extraction yield and enhance the quality of extracts. Extraction efficiency is known to be a function of process conditions. Several factors affect the concentration of desired components in the extract, e.g. temperature, liquid–solid ratio, flow rate and particle size (Pinelo et al., 2005). The practical issues of numerous extraction methods will be discussed in this section. Potential uses of these methods for the extraction of bioactive compounds (polyphenols), proteins and bioactive carbohydrates from pulse by-products will be finally summarized.

12.4.1 Conventional extraction techniques

12.4.1.1 Aqueous alkaline extraction (AAE)
Aqueous alkaline extraction followed by isoelectric precipitation (IEP) is another frequently used technique for the extraction of legume proteins and other bioactive components (Chakraborty et al., 1979; Bahnassey et al., 1986; Szymkiewicz and Jedrychowski, 1998).

12.4.1.1.1. Mechanism and application of AAE This unique extraction technique takes advantage of the solubility of legume proteins which is high at alkaline pH and low at pH values close to their isoelectric point (pH 4–5) (Boye et al., 2010). In AAE, ground pulse flour (with or without hulls) is dispersed in water in ratios ranging from 1:5 to 1:20. The pH of the mixture is adjusted to alkaline (pH 8–11) and the mixture is allowed to stand for 30–180 min to maximize protein solubilization. During this time the pH is maintained at the desired value and the temperature may be elevated (up to 55–65 °C) further to enhance protein solubilization and extraction. The mixture is subsequently filtered to remove any insoluble material and the pH of the extract is adjusted to the isoelectric point to induce protein precipitation, followed by centrifugation to recover the protein, washing to remove salts, neutralization and drying (Boye et al., 2010).

Fernandez-Quintela et al. (1997) investigated alkaline assisted extraction to quantify proteins in faba bean and pea. In their study, dehulled faba bean flour (var. Muchamiel) and pea flour (var. Verde Holanda) were dispersed in water (1:5 w/v ratio, pH 9, 20 min, room temperature) followed by centrifugation and protein precipitation from the supernatant (pH 4, room temperature, 20 min). The extracts obtained contained 81.2% and 84.9% protein, respectively. Flink and Christiansen (1973) produced protein isolates from seeds of *Vicia faba* (var Kleine Thuringer) using a 1:5 (w/v) bean to solvent ratio, pH 8–10, 23 °C, 10 min, followed by centrifugation and precipitation of protein from the supernatant at pH 3.5. The protein contents of the isolates ranged between 80% and 90%. Using slightly modified methods (extraction at pH 7–10, 30 min, 10 and 20 °C, 1:5 bean to solvent ratio, with and without 0.3 M NaCl, followed by precipitation at pH 4–5.3) isolates with protein contents ranging from 76.4% to 94.0% protein were obtained (McCurdy and Knipfel, 1990).

Recently, pea hulls have received interest as a fiber-rich product of commercial value which is richer in complex carbohydrates than wheat bran, is virtually tasteless, and has useful physicochemical properties when incorporated as a food ingredient (Weightman et al., 1994). The pea testa contains a parenchymatous layer, approximately ten cells deep, but quantitatively the majority of the total cell wall material may originate in the palisade layer, only one cell deep but with apparently thickened cell walls (Gassner, 1973). Pectins and hemicelluloses have been sequentially extracted from the alcohol insoluble residue (AIR) of pea hulls using chemical agents (Weightman et al., 1994). AIR was rich in carbohydrates (942 mg g^{-1}), notably cellulosic glucose, and was poorly lignified (6 mg g^{-1}) suggesting low (approximately 4 and 6%) AIR sequential extraction using CDTA (cyclohexane-trans-1,2-diamine-$NNN'N'$-tetra acetate) and HCl. After such depectination treatments, 4, 6 and 4% of the AIR were extracted by increasing concentrations of KOH (0.05 M (OHSP-I), 1 M (1OH-I) and 4 M (4OH-I), respectively). By contrast, alkali treatments alone (KOH 0.05 M (OHSP-II), 1 M (1OH-II) and 4 M (4OH-II)) extracted 2, 4 and 5% of the AIR, respectively. In their study, hemicelluloses extracted by 1 M and 4 M KOH in both series were further fractionated through precipitation, upon neutralization of the extracts. Precipitates accounted for 82, 60, 74 and 60% of total hemicelluloses in the IOH-I, 4OH-I, IOH-II and 4OH-11 extracts, respectively, and appeared to be essentially pure, acidic xylans. Interestingly, the final residue accounted for around 61% of the AIR after depectination and alkali treatment, and contained 91% cellulose. But after alkali treatment alone, it accounted for about 84% of the AIR and was predominantly rich in glucose ($\approx$70%) and uronic acids ($\approx$16%).

Mateos-Aparicio et al. (2010c) isolated and characterized cell wall polysaccharides from legume by-products (pea pod and broad bean pod) using sequential extraction with different extractive solutions (alkaline extraction). Isolation of the cell wall polysaccharides of pea pod and broad bean pod yielded three pectin-rich extracts, two containing pectin–hemicellulosic complexes and a cellulose-rich residue. Cellulose was the main polymer and some pectic polymers were released in the hemicellulosic fractions of the studied by-products, in which a large proportion of xyloglucans exists. Okara and broad bean pod comprised slightly methylated rhamnogalacturonan regions with arabinans, galactans and/or arabinoga-lactans in the pectin-rich fractions. Pectic acids from primary cell wall appeared in the rest of the fractions because of the high ramified non-solubilized pectic material is in association with other polymers, mostly cellulose. Pea pod, mainly enclosed insoluble fiber, contained cellulose, xyloglucans and xylans, overall. A great amount of glucuronic and galacturonic acids as major components was comprised by broad bean pod in accordance with its most soluble fiber-rich by-products.

12.4.1.2 Acid extraction (ACE)

The principle of acid extraction is similar to that of alkaline extraction except that initial extraction is conducted under acidic conditions (Boye et al., 2010). As previously indicated for alkaline extraction, the processing conditions used for acid extraction can influence the yield and purity of the finished product.

12.4.1.2.1. Applications of ACE Solubility of pulse proteins is high under very acidic conditions (i.e. pH < 4). This low pH range can, therefore, be used to solubilize proteins prior to their precipitation by IEP, cryo-precipitation (refrigeration) or membrane separation (Boye et al., 2010). Gof et al. (2001) extracted, purified and chemically characterized xylogalacturonans from pea hulls under acidic conditions. Their study revealed that pea hulls contained 925 mg g^{-1} sugar including 659 mg g^{-1} cellulosic glucose and 90 mg g^{-1} uronic acid. They were de-esterified by NaOH (pH > 13 at 4 °C, 2 h) and treated with HCl (0.1 mol l^{-1}, 80 C, 24 h). The HCl-soluble fraction represented 95 mg g^{-1} initial pea hulls. The authors found that it was rich in galacturonic acid (259 mg g^{-1}), xylose (93 mg g^{-1}) and rhamnose (91 mg g^{-1}), which co-eluted in anion-exchange chromatography, Large amounts of xylose were present in pea hulls as discussed above, originating mainly from xylans (Ralet et al., 1993). However, xylose-rich pectins were also identified in the copper precipitate of the hot acid extract, suggesting that xylogalacturonans could be present.

Mualikrishnaa and Tharanathana (1994) characterized and isolated pectic polysaccharides from the husks of field bean (*Dolichos lablab*), cowpea (*Vigna sinensis*) and pea (*Pisum sativum*), using HCl (pH 2.0) and 0.5% EDTA as extractants at 70 °C, in yields varying from 1.43 to 5.37%. Pectic polysaccharide obtained by 0.5% EDTA extraction was more viscous and gave a higher yield than acid-extracted ones. The acid extraction technique is generally used less frequently compared to the alkaline extraction technique (Boye et al., 2010).

12.4.1.3 Protein and starch recovery from pulse processing by-products

12.4.1.3.1. Air classification techniques (AC) Air classification is a technological process of separating particulates according to their size, as represented by particle equivalent diameter. The latter may be either geometric or hydrodynamic/aerodynamic equivalent diameter, depending strictly on the size-measurement method used. We will discuss AC method for separating pulse fractions.

12.4.1.3.2. Mechanism and application of AC In simplistic terms, air classification is a means of using air to effect a dry separation of objects having certain characteristics. These characteristics include physical properties such as size, shape, density and physicochemical nature. Air classification is an operation that applies the technology to fractionation of non-homogeneous particles, suspended in an air stream, into classes of fairly uniform size based on a common criterion of density or mass (Emami and Tabil, 2001). An interior space of a classifying device, where solid particles interact with the air stream, is the separation zone. Four basic separation zones in AC systems are known, which include gravitational–counterflow, gravitational–crossflow, centrifugal–counterflow and centrifugal–crossflow zone. A schematic diagram of gravitational–counterflow

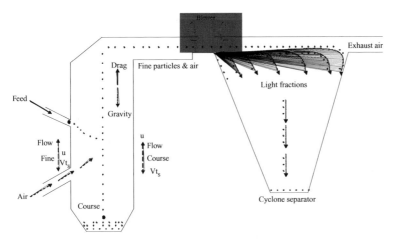

Figure 12.5 Schematic diagram of gravitational–counterflow zone AC systems. (Adapted from Shapiro and Galperin, 2005.)

zone is illustrated in Figure 12.5. A gravitational–counterflow zone exists in the air stream, rising up with velocity (u) inside a vertical chamber with parallel walls. In this system, particles experience gravity (G) and drag forces (Fd) acting in opposite directions. Coarse particles, having terminal settling velocity (VT_S) larger than the air flow velocity, move downwards, against the air stream, and fine ones rise with the stream (Shapiro and Galperin, 2005). AC is commonly used to produce protein concentrates in cereals and pulses (Emami and Tabil, 2001). It is a milling technique that allows for the fractionation of grains/seeds into high starch and high protein flours. Pulses are milled into flours having particles of two discrete sizes and densities. Air classification separates the light fine fraction (protein) from the heavy coarse fraction (starch) (Boye et al., 2010). Opposing centrifugal force and density are employed to separate protein and starch granules.

Quality of particle separation is characterized by mass contents of each fraction in the suitable product (Shapiro and Galperin, 2005). Denote the mass contents of the coarse fraction (in per cent) in the feed, coarse and fine products by C_0, C_c and C_f, respectively. Similarly, F_0, F_c and F_f are the mass contents of the fine fraction in the corresponding materials. Fractional cleanness is the mass content of any useful fraction in the corresponding product, namely, fine fraction content in the fine product F_f, and coarse fraction content in the coarse product C_c. Conversely, the content of fine in the coarse product F_c, or coarse in the fine product C_f is fractional dirtiness. The mass balance may be expressed as:

$$m_o = m_c + m_f \qquad (12.1)$$

where m_o, m_c and m_f are masses of initial, coarse and fine products, respectively. Product yield γ is the mass of a suitable product relative to the feed, i.e. coarse product yield is:

$$\gamma_c = m_c / m_o \qquad (12.2)$$

Fine product yield is expressed as shown below:

$$\gamma_f = m_f / m_o \qquad (12.3)$$

It is clear that $\gamma_c + \gamma_f = 1$ or 100%.

The coarse and fine fraction recoveries are expressed by:

$$\gamma_c = m_c C_c / m_o C_o \qquad (12.4)$$

and

$$\gamma_f = m_f F_f / m_o F_o \qquad (12.5)$$

It should be noted that using the fraction contents, one can estimate the recoveries of coarse and fine fractions by the following respective expressions (Rumpf, 1990):

$$\eta_c = C_c (C_o - C_f) / [C_o (C_c - C_f)], \ \eta_f = F_f (F_o - F_c) / [F_o (F_f - F_c)] \qquad (12.6)$$

During air classification process, whole or dehulled seed is ground into very fine flour, and the flour is subsequently classified in a spiral air stream to separate the starch from the protein (Boye et al., 2010). The process can be repeated several times to improve the separation efficiency, as protein bodies can still adhere to the surface of starch granules after the initial run. Agglomerates present in this initial starch fraction consist of starch granules embedded in a protein matrix, but by repeated pin-milling and air classification, further purification can be achieved easily (Gueguen et al., 1984). The type of milling used for air classification must, therefore, be capable of producing a very fine grind, yet selective enough to break up cells and cell fragments without severely damaging the starch granules (Jones et al., 1959). In spite of these efforts, various studies have shown that a portion of the protein cannot be milled free of the starch granules despite repeated milling.

Tyler et al. (1981) fractionated cowpea, great northern bean, lentil, mung bean, navy bean, faba bean and pea using air classification and measured the protein and starch contents of the fractions. The results obtained varied for the different pulses, with the highest and lowest protein purities observed for faba bean and lentil, respectively. In a separate study, Wu and Nichols (2005) showed that high-intensity grinding and air classification of whole or dehulled pea resulted in good separation of protein and starch into a high protein fraction and a high starch fraction. Their study documented that both whole pea and dehulled pea responded well to fine grinding and air classification, and dehulled pea gave higher protein content and higher starch content than the corresponding fraction from whole pea. The authors suggested that the high protein fraction can be used in food, and the high starch fraction can be fermented into ethanol. It is important to note that AC systems can be employed to fractionate different milling by-products and these by-products can be further used to obtain enriched flour for food product application. Further studies are warranted in this area. For further information on the processing of pulses by air classification, the reader is referred to Patel et al. (1980), Tyler and Panchuk (1982), Tyler (1984) and Aguilera et al. (1984).

12.4.2 Novel extraction technologies

12.4.2.1 Ultrasound extraction

The application of ultrasound as a laboratory-based technique for assisting extraction from plant material is widely reported. Earlier studies have focused on the extraction of plant metabolites (Knorr, 2003), flavonoids from foods using a range of solvents (Zhang et al., 2003) and bioactives from herbs (Vinatoru, 2001). The range of published extraction applications includes herbal, oil, protein and bioactives from plant materials (e.g. flavones, polyphenolics). Ultrasonic assisted extraction (UAE) in food-processing technology is of interest for enhancing extraction of components from plant and animal materials. This section shows that UAE technology can potentially enhance the extraction of health beneficial components when used as a pretreatment step in a unit process.

12.4.2.1.1. Mechanism and application of UAE Ultrasound is a form of vibrational energy in the frequency range of 20–100 kHz with a sound intensity of 10–1000 W cm^{-2}. Generally, power ultrasound employed in food processing uses lower frequencies (20–100 kHz) and causes cavitation with sound intensities of 10–1000 W cm^{-2} (Feng and Yang, 2005). The ultrasonic transducers convert electrical or mechanical energy to sound energy. There are three types of ultrasonic transducers in common usage: liquid-driven transducers, magnetostrictive transducers and piezoelectric transducers (Mason, 1998), with piezoelectric being the most common. For ultrasonic baths, power is often low in order to avoid cavitational damage to the tank walls and the power density is low due to large volume or processing liquid. When high-power ultrasound propagates in a liquid, cavitation bubbles will be generated due to pressure changes. These microbubbles will collapse violently in the succeeding compression cycles of a propagated sonic wave. This result in regions of high localized temperatures up to 5000 K and pressure of up to 50 000 kPa, resulting in high shearing effects (Mason, 1991) and a localized sterilizsation effect. The ultrasound power level or energy transmitted to a food medium can be expressed as ultrasound power (W), ultrasound intensity (W cm^{-2}), acoustic energy density (W ml^{-1}) or cavitational intensity. Ultrasonic intensity or acoustic energy density can be determined calorimetrically (Mason, 1991) using Equations 12.1–12.3. The absolute ultrasonic power P is given as follows:

$$P = mc_p \, (\mathrm{d}T/\mathrm{d}t)_{t=0} \qquad (12.7)$$

where m is the mass, c_p is the specific heat capacity and $(\mathrm{d}T/\mathrm{d}t)$ is the rate change of temperature during sonication time.

The intensity of ultrasonic power dissipated from a probe tip with diameter D is given by:

$$UI = 4P/\pi D^2 \qquad (12.8)$$

Most applications of ultrasound-assisted leaching comprise systems using a bath or an ultrasonic probe. This kind of equipment has been used for leaching organic and inorganic compounds in plant and animal matrices. Among the common ultrasonic system types is the ultrasonic bath, which appeared first for cleaning purposes and is equipped with a transducer at the bottom or submersed in a conventional tank. Because this type of system is inexpensive and easily available, it is commonly used. Probe systems are generally used in the laboratory, with the capacity to act directly within the solid–liquid mixture medium and deliver large amounts of power, which varies according to the variation of amplitude. The characteristic intensity distribution of an ultrasonic standing wave is in the axial direction, with higher intensity near the probe, which increasingly dissipates in the radial direction (Thompson and Doraiswamy, 1999). The advantage of ultrasonic probes over baths is the localized energy that provides more efficient liquid cavitation (Luque-Garcia and Luque de Castro, 2003).

In tissues where the desired components are located within cells, pre-ultrasound treatment by size reduction to maximize surface area is critical for achieving rapid and complete extraction (Vinatoru, 2001). Ultrasound-induced cavitation bubbles present hydrophobic surfaces within the extraction liquid (Feng and Yang, 2005), thereby increasing the net hydrophobic character of the extraction medium. The disruption of tissue surface structures is revealed with microscopic examination (Vinatoru, 2001; Chemat et al., 2004). Recently, the ultrasonic method has been widely employed to extract polysaccharide from plant material due to its high extraction efficiency (Yang et al., 2008).

Lai et al. (2010) investigated ultrasonic assisted extraction and antioxidant activity of water-soluble polysaccharide (MP1 (mannose) and MP2 (rhamnose and galactose)) from mung bean (*Vigna radiata* L.) hull. The study revealed that MP1 showed strong antioxidant potential according to the *in vitro* evaluation of its reduction power, free radical-scavenging activity and self-oxidation of 1,2,3-phentriol inhibitory activities in comparison with BHA (butylated hydroxyanisole) (Fig. 12.6). Moreover, both MP1 and MP2 exhibited a strong reduction power, which was equivalent to that of BHA at the dosage of $20\,mg\,ml^{-1}$. Although MP2 exhibited stronger hydroxyl radical-scavenging activity, it was less effective in scavenging superoxide radical, DPPH radical and self-oxidation of 1,2,3-phentriol than MP1 (Fig. 12.7). MP1 possessed excellent antioxidant properties as extracted by the ultrasonic method. From the above study, it is quite evident that ultrasound can enhance existing extraction processes and enable new commercial extraction opportunities, processes and novel antioxidants. It should be noted that the potential of UAE for extracting biomolecules warrants further investigation.

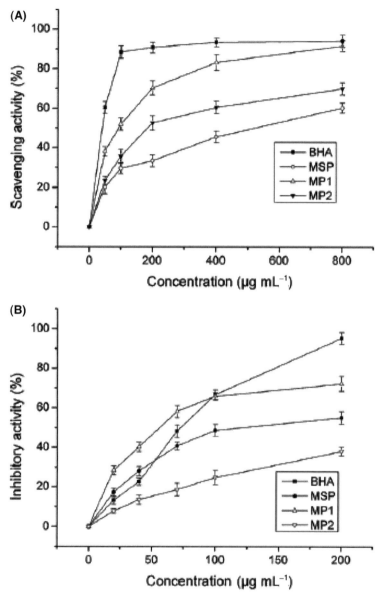

Figure 12.6 (A) DPPH radical-scavenging capacity of MSP, MP1 and MP2 from mung bean hull. (B) Inhibitory effects of MSP, MP1 and MP2 on self-oxidation of 1,2,3-phentriol (Lai et al., 2010).

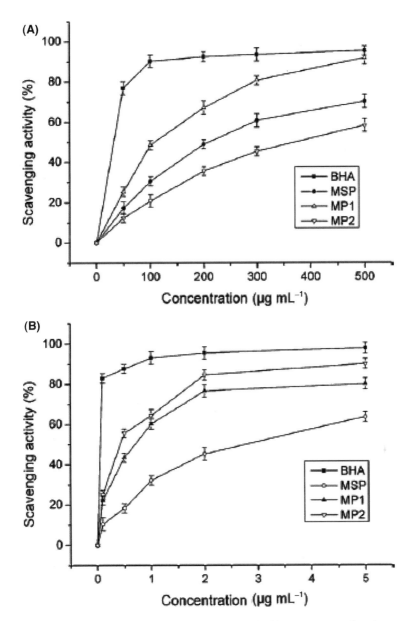

Figure 12.7 (A) Scavenging activity of superoxide radical by MSP, MP1 and MP2 from mung bean hull. (B) Hydroxyl radical-scavenging activity of MSP, MP1 and MP2 from mung bean hull (Lai et al., 2010).

12.4.2.2 Microwave assisted extraction (MAE)

Microwaves are non-ionizing electromagnetic energy with frequency ranging from 0.3 to 300 GHz. The energy is transmitted as waves, which can penetrate into materials and interact with polar molecules inside the materials, such as water, to generate thermal energy. MAE is a process that uses the effect of microwaves to extract biological materials. It has been considered an important alternative to low-pressure extraction because of its advantages: lower extraction time, lower solvent usage, selectivity, and volumetric heating and controllable heating process (Eskilsson and Bjorklund, 2000).

12.4.2.2.1. Mechanism and application of MAE Materials may be classified according to their ability to absorb the microwave energy: materials like metals are conductors, and their surfaces reflect the microwaves; transparent materials, such as plastics, are insulators and are used to support the material to be heated; and materials that absorb the microwave energy, which, therefore, are easily heated, such as polar liquids, are named dielectrics (Haque, 1999).

The physical principle of this novel extraction technique is based on the ability of polar chemical molecules to absorb microwave energy according to its nature or behavior, mainly the dielectric constant. This absorbed energy is proportional to the medium dielectric constant, resulting in dipole rotation in an electric field and migration of ionic species. The ionic migration generates heat as a result of the resistance of the medium to the ion flow, causing collisions between molecules because the direction of ions changes as many times as the field changes the sign. Rotational movements of the polar molecules occur while these molecules are trying to line up with the electric field, with consequent multiple collisions that generate energy and increase the medium temperature (Luque de Castro et al., 1999; Romanik et al., 2007). The electrical component of the waves changes 4.9×10^9 times per second and the frequency of 2.45 GHz corresponds to a wavelength of 12.2 cm and energy of 0.94 J mol^{-1} (Letellier and Budzinski, 1999). Apparently, a higher dielectric constant leads to a higher absorbed energy by the molecules, promoting a faster solvent heating and extraction at higher temperatures, from 423 to 463 Kelvin. However, other solvents with low dielectric constants are also utilized and, in these cases, the matrix is heated and the microwave heating leads to the rupture of cell walls by expansion, promoting the delivery of the target compounds into a cooler solvent; this technique is used for the extraction of thermally labile compounds of low polarity (Letellier and Budzinski, 1999). The effect of microwaves in the material is strongly dependent on the dielectric susceptibility of both the solvent and the solid matrix. The dielectric constant (ε') and dielectric loss factor (ε'') are values that express the dielectric response of materials in an applied microwave field. The thermal heat generation in the material not only depends on the dielectric constant, but also is in part dependent on the dissipation factor ($\ln \delta$), which is the ratio of the material dielectric loss to its dielectric constant:

$$\ln \delta = \varepsilon'' / \varepsilon' \qquad (12.9)$$

Sutivisedsak et al. (2010) investigated microwave-assisted extraction of phenolics from bean (*Phaseolus vulgaris* L.) hulls and cotyledons. Total phenolic content in samples of eight common beans (navy, pinto, great northern, dark red kidney, light red kidney, black, small red and pink) demonstrated that colored beans contained higher concentrations of extractable phenolics, and the concentration of phenolics was generally much higher in the hull (testa) than in the cotyledon. Extraction efficiency was superior at higher extraction temperatures. Their results indicated that the most effective extraction was achieved at a temperature of 150 °C using 50% ethanol (Table 12.8). Total phenolic content determined by microwave-assisted extraction with water at 100 °C was two to three times that determined by conventional extraction with water at the same temperature (Table 12.9). The development of a microwave-assisted extraction method for application in analyzing the concentration of total phenolics may aid in increasing the market potential for edible beans and bean constituents.

12.4.3 Other extraction techniques

Some other extraction techniques of interest are salt extraction (micellization), ultrafiltration and water extraction. These extraction techniques have been widely used to process pulse protein flours, concentrates and isolates. Madhukumar and Muralikrishna (2010) structurally characterized and determined prebiotic activity of purified xylo-oligosaccharides obtained from Bengal gram husk. Water-extractable polysaccharides were well isolated from Bengal gram husk in their study. Water extraction was useful in extracting xylo-oligosaccharides from husk. Some research is still needed to

Table 12.8 Total phenolic contents of bean samples as determined by microwave assisted extraction with 50% ethanol in water

Bean	Temperature (°C)	Total phenols (mgGAE 100 g^{-1})	
		Cotyledon[b]	Hull[c]
Navy	25	2.00 ± 0.01	1.20 ± 0.02
	50	2.92 ± 0.05	2.97 ± 0.01
	100	4.51 ± 0.01	3.55 ± 0.03
	150	8.67 ± 0.08	8.34 ± 0.31
Pinto	25	2.23 ± 0.01	1.28 ± 0.00
	50	3.19 ± 0.02	22.68 ± 0.25
	100	4.89 ± 0.26	38.64 ± 1.06
	150	10.42 ± 0.60	52.94 ± 1.47
Great northern	25	1.86 ± 0.04	1.16 ± 0.01
	50	2.73 ± 0.01	2.83 ± 0.06
	100	3.39 ± 0.08	3.11 ± 0.13
	150	9.03 ± 0.04	10.25 ± 0.04
Dark red kidney	25	2.11 ± 0.06	1.44 ± 0.00
	50	2.87 ± 0.03	22.55 ± 0.28
	100	3.32 ± 0.02	42.67 ± 0.14
	150	10.92 ± 0.11	63.77 ± 0.64
Light red kidney	25	2.15 ± 0.04	5.53 ± 0.87
	50	2.85 ± 0.04	26.94 ± 0.01
	100	4.71 ± 0.03	49.63 ± 0.04
	150	9.53 ± 0.28	56.52 ± 0.25
Small red	25	2.02 ± 0.08	3.79 ± 0.04
	50	3.16 ± 0.02	34.10 ± 0.38
	100	3.37 ± 0.03	53.93 ± 0.11
	150	3.34 ± 0.37	70.57 ± 1.36
Pink	25	1.97 ± 0.01	1.88 ± 0.00
	50	2.92 ± 0.01	22.97 ± 0.23
	100	3.13 ± 0.03	38.55 ± 0.33
	150	8.87 ± 0.05	56.21 ± 1.01
Black	25	2.10 ± 0.01	1.21 ± 0.08
	50	2.81 ± 0.01	17.17 ± 0.35
	100	3.22 ± 0.02	36.04 ± 0.54
	150	8.46 ± 0.90	60.77 ± 0.16

GAE = gallic acid equivalents; values expressed as mean ± standard deviation.
[b] Least significance difference = 0.44.
[c] Least significance difference = 1.05.
Sutivisedsak et al. (2010).

assess the efficacy of these extraction techniques to fractionate and isolate proteins and other bioactive compounds with nutraceutical properties. It should be noted that the traditional extraction methods (AAE, AEC) used to obtain these type of products have several drawbacks: they are time consuming and laborious, and have low

Table 12.9 Comparison of microwave-assisted and conventional extraction of total phenolics from bean samples with water at 100 °C

Beans	Total phenolic (mg g^{-1} bean GAE)		Rate of microwave to conventional			
			Microwave-assisted		Conventional	
	Cotyledon	Hull	Cotyledon	Hull	Cotyledon	Hull
Navy	5.48	3.45	1.71	1.76	3.20	1.96
Pinto	5.42	29.17	1.53	15.41	3.54	1.89
Small	5.77	39.98	1.74	20.35	3.31	1.97
Black	4.69	29.37	1.86	14.69	2.52	2.00
Great	3.77	3.50	1.55	1.61	2.44	2.17
Northern pink	5.94	31.54	2.02	18.04	2.94	1.75
Light red kidney	5.44	24.54	2.00	12.48	2.72	1.97
Dark red kidney	4.66	25.62	2.03	12.41	2.30	2.06

[a] GAE = Gallic acid equivalents.
Sutivisedsak et al. (2010).

selectivity and/or low extraction yields. Moreover, these traditional techniques employ large amounts of toxic solvents. At present, extraction methods able to overcome the above-mentioned drawbacks are being studied; among them, supercritical fluid extraction (SFE) and subcritical water extraction (SWE) are the more promising processes (King, 2000). These extraction techniques provide higher selectivities and shorter extraction times (3–40 min, depending on the solid matrix and compounds extracted) and do not use toxic organic solvents. In general, the art of separation is still improving and developing, with new methods and new procedures rapidly following each other worldwide. But not only are the procedures changing, the ongoing quest for new materials, new natural sources, is also ever increasing. Promising techniques including supercritical, subcritical and sonic extraction are still progressing rapidly. Extensive validation and verification, accuracy and cost effectiveness, controls and monitoring capabilities, are some of the key elements that will justify the adoption of the developing novel extraction techniques.

12.5 Challenges and opportunities

In recent years, many food companies have devoted effort to find value-added applications for these food by-products. As a result of much research, some biologically active compounds have been identified which can be incorporated into food material to increase its functionality. The exploitation of by-products of pulse processing as a source of functional compounds and their application in food is a promising field which requires interdisciplinary research by food technologists, food chemists, nutritionists and toxicologists. In the near future, we are challenged to respond to the following research needs: first, food processing technology should be optimized in order to minimize the amounts of waste at the outset; secondly, methods for the complete utilization of by-products resulting from pulse processing on a large scale and at affordable levels need to be developed.

Potential applications of proteins, polyphenols, bioactive carbohydrates, etc., in pulse bioprocessing leftovers as bioactive materials can increase the value of processing by-products. Apart from previously published research, it could be envisaged that most of the by-products generated from the pulses might also contain ample amounts of undiscovered phytochemicals that not only might have therapeutic usage, but also could be used for the development of new low-cost functional/medicinal foods. So far, a limited number of bioactivities have been identified from isolated compounds and further research is needed to develop methods to apply them for human health promotion. The need to extract biological functional components from pulse by-products prompts a continued search for economically and ecologically feasible extraction technologies (e.g. supercritical, subcritical, sonic, ultrasound, microwave and pulsed electric field).

This chapter has discussed the potential of the most important by-products of pulse processing as a source of valuable compounds. Studies on the stability and interactions of phytochemicals extracted from by-products with other food ingredients need to be developed. Further work is warranted wherein these underutilized pulse processing by-products could also be explored for their effective utilization as food ingredients. Identification of nutraceutical potential of natural compounds is a growing field and use of pulse processing by-products makes a new approach to be able to develop more commercial applications. There is a good potential to determine the performance of the developed products by nutrition intervention studies with the target group for certification and popularization.

References

Agboola, S.O., Mofolasayo, O.A., Watts, B.M., Aluko, R.E., 2010. Functional properties of yellow field pea (*Pisum sativum* L.) seed flours and the *in vitro* bioactive properties of their polyphenols. Food Res. Int. 43, 582–588.

Aguilera, J.M., Crisafulli, E.B., Lusas, E.W., Uebersax, M.A., Zabik, M.E., 1984. Air classification and extrusion of navy bean fractions. J. Food Sci. 49, 543–546.

Ajila, C.M., Prasada Rao, U.J.S., 2009. Purification and characterization of black gram (*Vigna mungo*) husk peroxidase. J. Mol. Catal. B: Enzym. 60, 36–44.

Anderson, E.T., Berry, B.W., 2000. Sensory, shear, and cooking properties of lower-fat beef patties made with inner pea fiber. J. Food Sci. 65, 805–810.

Anderson, E.T., Berry, B.W., 2001a. Identification of nonmeat ingredients for increasing fat holding capacity during heating of ground beef. J. Food Qual. 24, 291–299.

Anderson, E.T., Berry, B.W., 2001b. Effects of inner pea fiber on fat retention and cooking yield in high fat ground beef. Food Res. Int. 34, 689–694.

Bahnassey, Y., Khan, K., Harrold, R., 1986. Fortification of spaghetti with edible legumes. I. Physicochemical, antinutritional, amino acid, and mineral composition. Cereal Chem. 63, 210–215.

Barzana, E., Rubio, D., Santamar, R.I., Garcia-Correa, O., Garca, F., Ridaur-Sanz, V.E., 2002. Enzyme-mediated solvent extraction of carotenoids from marigold flower (*Tagetes erecta*). J. Agric. Food Chem. 50, 4491–4496.

Bassett, C., Boye, J., Tyler, R., Oomah, B.D., 2010. Molecular, functional and processing characteristics of whole pulses and pulse fractions and their emerging food and nutraceutical applications. Food Res. Int. 43, 397–659.

Boye, J., Zare, F., Pletch, A., 2010. Pulse proteins: processing, characterization, functional properties and applications in food and feed. Food Res. Int. 143, 414–431.

Bressani, R., Elfas, L.G., 1980. Nutritional values of legume crops for humans and animals. In: Summerfield, R.J., Bunting, A.H. (Eds.), Advances in legume science. FAO, London, pp. 57–66.

Carle, R., Keller, P., Schieber, A., et al. 2001. Method for obtaining useful materials from the byproducts of fruit and vegetable processing. Patent application, WO 01/78859 A1.

Chakraborty, P., Sosulski, F., Bose, A., 1979. Ultracentrifugation of salt soluble proteins in ten legume species. J. Sci. Food Agric. 30, 766–771.

Chemat, S., Lagha, A., AitAmar, H., Bartels, V., Chemat, F., 2004. Comparison of conventional and ultrasound-assisted extraction of carvone and limonene from caraway seeds. Flavour Fragrance J. 19, 188–195.

Dahl, W.J., Furkalo, R., Kish Greer, A., McGladdery, C., 2001. Addition of pea hull fiber to pureed foods for use with institutionalized elderly. Can. J. Diet. Pract. 62S (2), 108.

Dalgetty, D.D., Baik, B.-K., 2006. Fortification of bread with hulls and cotyledon fibers isolated from peas, lentils, and chickpeas. Cereal Chem. 83, 269–274.

Del Campo, I., Alegría, I., Zazpe, M., Echeverría, M., Echeverría, I., 2006. Diluted acid hydrolysis pretreatment of agri-food wastes for bioethanol production. Indust. Crops Prod. 24, 214–221.

Dueñas, M., Sun, B., Hernández, T., Estrella, I., Spranger, M.I., 2003. Proanthocyanidin composition in the seed coat of lentils (*Lens culinaris* L.). J. Agric. Food Chem. 51, 7999–8004.

Emami, S., Tabil, L.G., 2001. Processing of starch-rich and protein-rich fractions from chickpeas – a review. Presented September 27–28, 2002, ASAE/CSAE North-Central Intersectional Meeting, Paper No. MBSK 02-212, ASAE, St Joseph, MI, USA.

Eskilsson, C.S., Bjorklund, E., 2000. Analytical-scale microwave-assisted extraction. J. Chromatogr. A 902, 227–250.

Feng, H., Yang, W., 2005. Power ultrasound. In: Hui, Y.H. (Ed.), Handbook of food science, technology and engineering. CRC Press, New York, p. 3632.

Fernandez-Quintela, A., Macarulla, M.T., Del Barrio, A.S., Martinez, J.A., 1997. Composition and functional properties of protein isolates obtained from commercial legumes grown in northern Spain. Plant Foods Hum. Nutr. 51, 331–342.

Filipini, M., Hogg, T., 1997. Upgrading of vegetable wastes and applications in the food industry. 11 Forum for Applied Biotechnology. Gent (Belgium). 25–26 September 1997. Mededelingen-Faculteit-Landbouwkundige-en-Toegepaste-Biologische-Wetenschappen-Universiteit-Gent (Belgium), pp. 1329–1331.

Fitzpatrick, K., 2007. Novel fibers in Canada: regulatory developments. Cereal Foods World 52, 22–23.

Flink, J., Christiansen, I., 1973. The production of a protein isolate from Vicia faba. Lebensm. Wiss. Technol. 6, 102–106.

Gassner, G., 1973. Mikroskopische Untersuchung Pflanzlicher Lebensmittel. Gustav Fischer Verlag, Stuttgart, pp. 66–69.

Goff, A.Le., Renard, C.M.G.C., Bonnin, E., Thibault, J.F., 2001. Extraction, purification and chemical characterisation of xylogalacturonans from pea hulls. Food Chem. 45, 325–334.

Gómez, M., Ronda, F., Blanco, C.A., Caballero, P.A., Apesteguía, A., 2003. Effect of dietary fiber on dough rheology and bread quality. Eur. Food Res. Technol. 216, 51–56.

Gueguen, J., Vu, A.T., Schaeffer, F., 1984. Large-scale purification and characterisation of pea globulins. J. Food Sci. Agric. 35, 1024–1033.

Gunawardena, C.K., Zijlstra, R.T., Beltranena, E., 2010. Characterization of the nutritional value of air-classified protein and starch fractions of field pea and zero-tannin faba bean in grower pigs. J. Anim. Sci. 88, 660–670.

Haque, K.E., 1999. Microwave energy for mineral treatment processes – a brief review. Int. J. Miner. Process. 57, 1–24.

Jackson, J.C.A., 2009. Protein nutritional quality of cowpea and navy bean residue fractions. Afr. J. Food Agric. Nutr. Dev. 9, 764–778.

Jezierny, D., Mosenthin, R., Bauer, E., 2010. The use of grain legumes as a protein source in pig nutrition: a review. Anim. Feed Sci. Technol. 157, 111–128.

Jones, R.W., Taylor, N.W., Senti, F.R., 1959. Electrophoresis and fractionation of wheat gluten. Arch. Biochem. Biophys. 84, 363–376.

Joseph, J.A., Shukitt-Hale, B., Casadesus, G., 2005. Reversing the deleterious effects of aging on neuronal communication and behavior: beneficial properties of fruit polyphenolic compounds. Am. J. Clin. Nutr. 81, 313S–316S.

King, J.W., 2000. Advances in critical fluid technology for food processing. Food Sci. Technol. Today 14, 186–191.

Knorr, D., 2003. Impact of non-thermal processing on plant metabolites. J. Food Eng. 56, 131–134.

Lai, F., Wen, Q., Li, L., Wu, H., Li, X., 2010. Antioxidant activities of water-soluble polysaccharide extracted from mung bean (*Vigna radiata* L.) hull with ultrasonic assisted treatment. Carbohydr. Polym. 81, 323–329.

Letellier, M., Budzinski, H., 1999. Microwave assisted extraction of organic compounds. Analysis 27, 259–271.

Lowe, E.D., Buckmaster, D.R., 1995. Dewatering makes big difference in compost strategies. Biocycle 36, 78–82.

Luque de Castro, M.D., Jiménez-Carmona, M.M., Fernández-Pérez, V., 1999. Towards more rational techniques for the isolation of valuable essential oils from plants. Trends Anal. Chem. 18, 708–716.

Luque-Garcia, J.L., Luque de Castro, M.D., 2003. Ultrasound: a powerful tool for leaching. Trends Anal. Chem. 22, 41–47.

Madhukumar, M.S., Muralikrishna, G., 2010. Structural characterisation and determination of prebiotic activity of purified xylo-oligosaccharides obtained from Bengal gram husk (*Cicer arietinum* L.) and wheat bran (*Triticum aestivum*). Food Chem. 118, 215–223.

Maheri-Sis, N., Chamani, M., Sadeghi, A.A., Mirza-Aghazadeh, A., Safaei, A.A., 2007. Nutritional evaluation of chickpea wastes for ruminants using *in vitro* gas production technique. J. Anim. Vet. Adv. 6, 1453–1457.

Mason, T.J., 1991. Practical sonochemistry user's guide to application in chemistry and chemical engineering. Ellis Horwood, New York.

Mason, T.J., 1998. Power ultrasound in food processing, the way forward. In: Povey, M.J.W., Mason, T.J. (Eds.), Ultrasound in food processing. Thomson Science, London, pp. 105–126.

Masoodi, F.A., Chauhan, G.S., 1998. Use of apple pomace as a source of dietary fiber in wheat bread. J. Food Process. Preserv. 22, 255–263.

Mateos-Aparicio, I., Redondo-Cuenca, A., Villanueva-Suárez, M.J., 2010b. Isolation and characterisation of cell wall polysaccharides from legume by-products: okara (soymilk residuc), pca pod and broad bean pod. Food Chem. 122, 339–345.

Mateos-Aparicio, I., Redondo-Cuenca, A., Villanueva-Suárez, M.J., Zapata-Revilla, M.-A., Tenorio-Sanz, M.-D., 2010c. Pea pod, broad bean pod and okara, potential sources of functional compounds. LWT – Food Sci. Technol. 43, 1467–1470.

Mateos-Aparicio, I., Mateos-Peinado, C., Jiménez-Escrig, A., Rupérez, P., 2010a. Multifunctional antioxidant activity of polysaccharide fractions from the soybean by-product okara. Carbohydr. Polym. 82, 245–250.

McCurdy, S.M., Kripfel, J.E., 1990. Investigation of faba bean protein recovery and application to pilot scale processing. J. Food Sci. 55, 1101.

Moure, A., Cruz, J.M., Franco, D., 2001. Natural antioxidants from residual sources. Food Chem. 72, 145–171.

Mualikrishnaa, G., Tharanathana, R.N., 1994. Characterization of pectic polysaccharides from pulse husks. Food Chem. 50, 87–89.

Narasimha, H.V., Ramakrishnaiah, N., Pratape, V.M., Sasikala, V.B., 2004. Value addition to by product from dhal milling industry in India. J. Food Sci. Technol. 41, 492–496.

Osada, K., Hoshina, S., Nakamura, S., Sugano, M., 2001. Cholesterol oxidation in meat products and its regulation by supplementation of sodium nitrite and apple polyphenol before processing. J. Agric. Food Chem. 48, 3823–3829.

Patel, K.M., Bedford, C.L., Youngs, C.W., 1980. Amino acid and mineral profile of air-classified navy bean flour fractions. Cereal Chem. 57, 123–125.

Pedersen, B., Eggum, B.O., 1983. The influence of milling on the nutritive value of flour from cereal grains. Plant Foods Hum. Nutr. 33, 299–311.

Pellegrini, N., Serafini, M., Salvatore, S., Del Rio, D., Bianchi, M., Brighenti, F., 2006. Total antioxidant capacity of spices, dried fruits, nuts, pulses, cereals and sweets consumed in Italy assessed by three different in vitro assays. Mol. Nutr. Food Res. 50, 1030–1038.

Pinelo, M., Fabbro, P.D., Manzocco, L., Nuñez, M.J., Nicoli, M.C., 2005. Optimization of continuous phenol extraction from *Vitis vinifera* byproducts. Food Chem. 92, 109–117.

Ralet, M.- C., Della Valle, G., Thibault, J.F., 1993. Raw and extruded fibre from pea hulls. I. Composition and physico-chemical properties. Carbohydr. Polym. 20, 17–23.

Ramakrishnaiah, N., Pratape, V.M., Sashikala, V.B., Narasimha, H.V., 2004. Value addition to by-products from dhal milling industry in India. J. Food Sci. Technol. 41, 492–496.

Romanik, G., Gilgenast, E., Przyjazny, A., 2007. Techniques of preparing plant material for chromatographic separation and analysis. J. Biochem. Biophys. Meth. 70, 253–261.

Rumpf, H., 1990. Particle technology. Chapman and Hall, London.

Salunkhe, D.K., Kadam, S.S., Chawan, J.K., 1985. Post-harvest biotechnology of food legumes. CRC Press, Boca Raton, pp. 29–52.

Scalbert, A., Manach, C., Morand, C., Rémésy, C., Jiménez, L., 2005. Dietary polyphenols and the prevention of diseases. Crit. Rev. Food Sci. Nutr. 45, 287–306.

Schieber, A., Stintzing, F.C., Carle, R., 2001. By-products of plant food processing as a source of functional compounds – recent developments. Trends Food Sci. Technol. 12, 401–413.

Shapiro, M., Galperin, V., 2005. Air classification of solid particles: a review. Chem. Eng. Process. 44, 279–285.

Singh, U., Jambunathan, R., 1980. A survey of the methods of milling and the consumer acceptance of pigeonpea in India. Proceedings, International Workshop on Pigeon Peas, Vol. 2, 15–19 December, Patancheru, Andhra Pradesh, pp. 419–425.

Singh, U., Rao, P.V., Seetha, R., Jambunathan, R., 1989. Nutrient losses due to scarification of pigeonpea (*Cajanus cajan* L.) cotyledons. J. Food Sci. 54, 974.

Sosulski, F.W., Dabrowski, K.J., 1984. Composition of free and hydrolyzable phenolic acids in the flours and hulls of ten legume species J. Agric. Food Chem. 32, 131–133.

Sosulski, F.W., Wu, K.K., 1988. High-fiber breads containing field pea hulls, wheat, corn, and wild oat brans. Cereal Chem. 65, 186–191.

Sreerama, Y.N., Neelam, D.A., Sashikala, V.B., Pratape, V.M., 2010. Distribution of nutrients and antinutrients in milled fractions of chickpea and horse gram: seed coat phenolics and their distinct modes of enzyme inhibition. J. Agric. Food Chem. 58, 4322–4330.

Stojceska, V., Ainsworth, P., Plunkett, P., Ibanoğlu, E., Ibanoğlu, S., 2008. Cauliflower by-products as a new source of dietary fiber, antioxidants and proteins in cereal based ready-to-eat expanded snacks. J. Food Eng. 87, 554–563.

Sutivisedsak, N., Cheng, H.N., Willett, J.L., Lesch, W.C., Tangsrud, R.R., Biswas, A., 2010. Microwave-assisted extraction of phenolics from bean (*Phaseolus vulgaris* L.). Food Res. Int. 43, 516–519.

Szymkiewicz, A., Jedrychowski, L., 1998. Evaluation of immunoreactivity of selected legume seed proteins. Pol. J. Food Nutr. Sci. 7, 539–544.

Tharanathan, R.N., Mahadevamma, S., 2003. Legumes – a boon to human nutrition. Trends Food Sci. Technol. 14, 507–518.

Thompson, L.H., Doraiswamy, L.K., 1999. Sonochemistry: science and engineering. Ind. Eng. Chem. Res. 38, 1215–1249.

Tiwari, U., Gunasekaran, M., Jaganmohan, R., alagusundaram, K., Tiwari, B.K., 2009. Quality characteristic and shelf life studies of deep-fried snack prepared from rice brokens and legumes byproduct. Food Bioprocess. Technol. doi: 10. 1007/s11947-009-0219-6.

Tiwari, B. K., Tiwari U., Brennan, C.S., Jagan Mohon, R., Surabi, A., Alagusundaram, K., 2011. Utilisation of pigeon pea (*Cajanus cajan* L) byproducts in preparation of biscuits. LWT – Food Sci. Technol. doi:10.1016/j.lwt.2011.01.018.

Tyler, R.T., 1984. Impact milling quality of grain legumes. J. Food Sci. 49, 925–930.

Tyler, R.T., Panchuk, B.D., 1982. Effect of seed moisture content on the air classification of field peas and faba beans. Cereal Chem. 59, 31–33.

Tyler, R.T., Youngs, C.G., Sosulski, F.W., 1981. Air classification of legumes I – Separation efficiency, yield, and composition of the starch and protein fractions. Cereal Chem. 58, 144–148.

Vinatoru, M., 2001. An overview of the ultrasonically assisted extraction of bioactive principles from herbs. Ultrason. Sonochem. 8, 303–313.

Wang, J., Rosell, C.M., Benedito de Barber, C., 2002. Effect of the addition of different fibers on wheat dough performance and bread quality. Food Chem. 79, 221–226.

Weightman, R.M., Renard, C.M.G.C., Thibault, J.F., 1994. Structure and properties of the polysaccharides from pea hulls. Part 1: Chemical extraction and fractionation of the polysaccharides. Carbohydr. Polym. 24, 139–148.

Wu, V., Nichols, N.N., 2005. Fine grinding and air classification of field pea. Cereal Chem. 82, 341–344.

Yang, B., Zhaom, M., Shi, J., Yang, N., Jiang, Y., 2008. Effect of ultrasonic treatment on the recovery and DPPH radical scavenging activity of polysaccharides from longan fruit pericarp. Food Chem. 106, 685–690.

Zhang, R., Xu, Y., Shi, Y., 2003. The extracting technology of flavonoids compounds. Food Machin. 1, 21–22.

The nutritional value of whole pulses and pulse fractions

13

Emma Derbyshire
Manchester Food Research Centre, Manchester Metropolitan University, Manchester, UK

13.1 Introduction

Peas, beans, lentils, chickpeas and fava beans are all examples of pulses, defined as the edible seeds of legumes (usually grown in a pod) that can be used for human consumption (Rochfort and Panozzo, 2007). Pulse crops are produced on many continents worldwide, including North America, especially Canada, areas of Asia and the Middle East. Although consumed worldwide and to varying levels, pulse crops are imported to heavily populated areas such as India and Egypt, where they form a staple part of the diet (Roy et al., 2010).

Pulses have long been known for their nutritional and health-promoting properties. Pulses are a good source of fiber and proteins, are a low-glycemic index food and contain a number of important bioactive substances that may be important for health and

Pulse Foods: Processing, Quality and Nutraceutical Applications. DOI: 10.1016/B978-0-1238-2018-1.00007-0

well-being. Increasingly, scientists are seeing pulses as an abundant, under-utilized and nutrient-rich food source that has great potential within the food industry when used alone or combined with other food ingredients, i.e. cereal-based products to form foods with "added health benefits".

This chapter aims to discuss patterns of human consumption of pulses, their nutritional value and role in health and disease prevention. The potential to include pulses within food products and their promising role as functional foods and neutraceuticals will also be discussed in this chapter.

13.2 Pulses – intakes and trends

The human consumption of pulses varies geographically. As mentioned, areas such as India and Egypt consume some of the highest intakes of pulses while consumption rates are generally lower in the European Union. In these regions, pulses play a fundamental role in helping the population to consume suitable levels of several important nutrients, particularly protein while, in developed regions, the staple diet remains based largely on animal-derived protein (Rochfort and Panozzo, 2007). Within Europe, about 60% of pulses are consumed in Spain, France and the UK. It is also important to consider that the way in which pulses are eaten varies compared to other world regions. This can be attributed mainly to different food habits and traditions. In regions of Latin America, regular consumption of pulses is a tradition that is embedded in many people and communities from an early age. In these regions, constraints in terms of how beans are prepared and cooked limit the way in which these can be consumed (Leterme and Carmenza Munoz, 2002). In more developed regions, advances in food processing methods mean that the consumption of canned pulses is more of a common occurrence (Schneider, 2002).

Most countries use food models that have specific recommendations for subgroups of vegetables, including pulses. In the USA, scientists have compared the nutrient intakes of individuals consuming pulses (includes dry beans, peas and lentils) to non-consumers using data from the 1999–2002 National Health and Nutrition Examination Survey (NHANES). Unfortunately, only 7.9% of US adults consumed dry beans or peas on any given day. This is disappointing considering that those eating half a cup of dry beans or peas on a daily basis had lower intakes of total and saturated fat and significantly higher intakes of fiber, protein, folate, zinc, iron and magnesium (Mitchell et al., 2009).

Very few additional studies or surveys have accounted for intakes of pulses, or determined how they can contribute to habitual nutrient intakes. In the UK, the National Diet and Nutrition Survey (NDNS) includes pulses within the category of "fruit and vegetables" but specific intakes are not quantified. This unfortunately limits the conclusions that can be drawn linking patterns of pulse consumption to specific health outcomes. In future, surveys and studies need clearly to define and quantify what constitutes "a portion" of pulses as different countries may use different measures. In the UK NDNS, one portion of pulses has been defined as "three heaped teaspoons of pulses" (Hoare et al., 2004). Such surveys should also consider collecting data for pulses separately from fruits and vegetables as there is emerging evidence that these may have separate health benefits (Lanza et al., 2006; Sievenpiper et al., 2009).

13.3 Nutritional value of whole pulses

Pulses are unique in that they have a varied nutrition profile and one that is difficult to match from other food sources. As demonstrated in Table 13.1, pulses are a good source of protein, fiber, B vitamins, iron, zinc and copper, among other nutrients. Compared to other food sources they are also relatively low in energy, fat and sodium, making them a relatively nutrient-dense food source. In particular, pulses are a good source of protein, providing balanced amino acid profiles. It is thought that the functional properties of pulse proteins may be comparable to those such as soy and whey and, when blended with cereal proteins, could be a promising source of nutritional and functional proteins in the future (Boye et al., 2010).

Pulses are also foods with very low glycemic index values and may have an important role in helping to regulate blood sugar and lipid levels in both diabetic and healthy individuals (Rizkalla et al., 2002). Incorporating such foods into the daily diet may go some way to preventing the development of coronary heart disease and regulating body weight, also a risk factor for diabetics. A study undertaken by Wong et al. (2009) tested the theory that a diet containing

Table 13.1 Nutritional composition of pulses

Nutritional composition (g 100 g⁻¹)	Peas (boiled)	Chickpeas (soaked and boiled)	Lentils (dried and boiled)	Soya beans (dried and boiled)
Energy (kcal)	69	121	105	141
Carbohydrate (g)	9.7	18.2	16.9	5.1
Protein (g)	6.0	8.4	8.8	14.0
Fat (g)	0.9	2.1	0.7	7.3
Dietary fiber (NSP) (g)	5.1	4.3	3.8	6.1
Saturated fat (g)	0.2	0.2	0.1	0.9
Monounsaturated fat (g)	0.1	0.4	0.1	1.4
Polyunsaturated fat (g)	0.5	1.0	0.3	3.5
Sodium (mg)	Tr	5.0	3.0	1.0
Potassium (mg)	230	270	310	510
Calcium (mg)	19	46	22	83
Magnesium (mg)	29	37	34	63
Phosphorus (mg)	130	83	130	250
Iron (mg)	1.5	2.1	3.5	3.0
Copper (mg)	0.03	0.28	0.33	0.32
Zinc (mg)	1.0	1.2	1.4	0.9
Selenium (mg)	Tr	1.0	40	5.0
Iodine (mg)	2.0	–	–	2.0
Carotene (μg)	250	23	–	6.0
Vitamin E (mg)	0.21	1.10	–	1.13
Thiamin (mg)	0.70	1.10	0.14	0.12
Riboflavin (mg)	0.03	0.07	0.08	0.09
Niacin (mg)	1.8	0.7	0.6	0.5
Vitamin B₆ (mg)	0.09	0.14	0.28	0.23
Folate (μg)	27	66	30	54
Vitamin C (mg)	16	Tr	Tr	Tr

Tr, trace; –, negligible.
Source: Data extracted from Food Standards Agency (2006).

pulses may reduce the risk of overweight or obesity, possibly by improving satiety and lowering energy intake. In the research, young healthy-weight males (18–35 years) were randomized to eat three different diets containing pulses that had been subject to processing and consumed using different recipes and varieties of pulses. It was identified that the energy intake of the pulse meals was the strongest indicator of appetite and later energy intake. The composition, processing, recipe or variety of pulses had little effect on satiety or food intake. Similar studies are needed to see whether consistent findings would be yielded if this research were replicated.

With regard to micronutrients, in regions where malnutrition is prevalent, such as Asia and Africa (women and children in particular), integrating pulses, such as lentils, within the daily diet may help to improve diet quality and may go some way towards reducing the prevalence of micronutrient deficiencies. In one study, scientists measured the iron and zinc content of 19 different lentil genotypes. Although there was some variation in the levels of nutrients between samples, generally lentils were found to be a rich source of important micronutrients, with the iron content ranging between 73 and 90 mg kg^{-1} and zinc varying from 44 to 54 mg kg^{-1} (Thavarajah et al., 2009).

For vegetarians, in particular, pulses are an important source of protein and iron (FSA, 2010). Pulses can be easily added to soups, casseroles and meat sauces to add extra taste and flavor. They can also be used as a substitute for meat, which is often cheaper and generally makes dishes lower in fat. However, although the mineral content of pulses may be high, the bioavailability may be low due to the presence of phytate. Using enzymatic degradation during food processing, either by increasing the activity of the naturally occurring phytases (soaking, germination and fermentation) or by adding enzyme preparations, may go some way to improving the bioavailability (Sandberg, 2002).

Overall, pulses are an abundant source of important nutrients. However, the nutritional properties of pulses may need to be further communicated to government organizations, policy makers and health professionals. Better understanding of the nutritional value of pulses may lead to more prominent integration within both food models and dietary recommendations. This may be of benefit to both the lay population of healthy adults but also those with diagnosed metabolic medical conditions, such as individuals with diabetes mellitus (Leterme, 2002).

13.4 Nutritional value of pulse fractions

Pulses may also contain a range of bioactive components that may be extracted and used for their beneficial health properties (Table 13.2). There is an array of constituents that have the potential to be extracted from pulses and used as a condensed form, or integrated with other food components. However, this section will focus on the pulse fractions for which there is most scientific evidence; this includes the fiber and protein components and isoflavones which, to date,

Table 13.2 Potential health benefits of bioactive compounds found in pulses

Compound	Beneficial effects	Adverse effects
Protease inhibitors	Anticarcinogenic (?)	Increased carcinogenesis (?)
Amylase inhibitors	Therapeutic role for diabetes (?)	Reduced starch digestion
Lectins	Role in obesity treatment (?), reduced tumor growth (?)	Reduced nutrient absorption
Phytates	May help to reduce cholesterol (?), anticarcinogenic (?)	Reduced bioavailability of minerals
Oxalates		Reduced bioavailability of minerals
Phenolic compounds	Reduce risk factors for CHD (?)	
Flavonoids, isoflavones (phytoestrogens)	Reduced risk of hormone-dependent cancer?	Infertility syndrome in animals
Condensed tannins		Astringent taste
Ligans (phytoestrogens)	Reduced risk factors for menopause?	
Lignins		Reduced fermentability of dietary fibers, bitter taste, reduced food intake (animals)
Saponins	May help to reduce cholesterol (?), anticarcinogenic (?)	
Alkaloids		

Source: Adapted and reproduced with permission from Champ (2002).

have been extracted mainly from the soybean. The potential role of bioactive compounds will also be touched upon, very briefly.

13.4.1 Fiber

Today, dietary fiber is recognized as an important food constituent required for healthy nutrition (Mann and Cummings, 2009). At times, the definition of dietary fiber can be confusing and sometimes the term "non-starch polysaccharide (NSP)", defined as resistant starch and lignin, is used. NSP is usually determined using the Englyst method – when starch is removed completely from the fiber and NSP is measured by acid hydrolysis (Englyst et al., 1996). Englyst et al. (1987) believed that by defining and measuring dietary fiber as NSP this would give the analyst a clear task and yet retain the original concept of dietary fiber as cell wall material. NSP is therefore now regarded as being the focal point of interest when referring to dietary fiber.

The American Association of Cereal Chemists (AACC) (2001) has developed the most highly regarded definition of dietary fiber to date. This has been defined as:

the edible parts of plants or analogous carbohydrates that are resistant to digestion and absorption in the human small intestine with complete or partial fermentation in the large intestine. Dietary fiber includes polysaccharides, oligosaccharides, lignin, and associated plant substances. Dietary fibers promote beneficial physiological effects including laxation, and/or blood cholesterol attenuation, and/or blood glucose attenuation.

As demonstrated previously in Table 13.1, it can be seen that pulses are a rich source of fiber, with soy beans containing up to $6.1 \, \text{g} \, 100 \, \text{g}^{-1}$ NSP and chickpeas around $4.3 \, \text{g} \, 100 \, \text{g}^{-1}$ consumed (FSA, 2006). Although there is no UK legislation covering fiber claims, the FSA (2007) has issued guidance notes. It recommends that a "source" of fiber should contain $3 \, \text{g} \, 100 \, \text{g}^{-1}$ (or $100 \, \text{ml}^{-1}$) or at least $3 \, \text{g}$ in the "reasonable expected daily intake of the food", i.e. the amount that could reasonably be expected to be eaten in one day. If the food is naturally high in fiber, the claim must take the form "a high fiber food". For a claim of high fiber the food must contain at least $6 \, \text{g} \, 100 \, \text{g}^{-1}$ (or $100 \, \text{ml}^{-1}$) or at least $6 \, \text{g}$ in the reasonable expected daily intake of the food. When applying such guidelines, soy beans may be regarded as being a "high fiber food".

There is great potential for flours and fiber-rich fractions obtained from pulse crops to be incorporated into processed foods, as a measure to increase their dietary fiber content and serve as functional ingredients (Tosh and Yada, 2010). The health benefits of fiber-rich diets are well documented and total daily intakes have been associated with reduced risk of medical conditions, such as breast cancer risk (Cade et al., 2007) and cardiovascular disease (Estruch et al., 2009). However, while there is some evidence of the benefits of consuming extracted and synthetic fibers, epidemiological confirmation of clinical benefits and long-term safety is lacking. The food industry would benefit from further work within this area, linking the fiber content of these food components to specific health outcomes.

13.4.2 Protein

Pulses have a high protein content that is about twice that of cereals, although this may also be influenced by genetic and environmental factors (Rochfort and Panozzo, 2007). Chickpea, lentil and dry pea contain approximately 22%, 28.6% and 23.3% protein, respectively on a dry weight basis (DW) (Pulse Canada, 2004). Pulse seeds generally accumulate protein as they develop; hence mature pulse seeds are typically highest in protein, although this may depend on plant species, variety and growing conditions also (Pulse Canada, 2004). The storage proteins of pulses include albumins, globulins, prolamines, glutelins and residual proteins (Duranti, 2006).

Studies have shown that pulse protein fractions are generally lacking in the sulfur-containing amino acids and tryptophan but are rich in essential amino acids such as lysine, unlike cereals (Boye et al., 2010). Therefore, combining pulses with cereals is a good way to ensure that good balances of amino acids are consumed (Emami and Tabil, 2002).

13.4.3 Isoflavones

In terms of constituents that may be beneficial to health, soybeans, in particular, are an abundant source of flavones and isoflavones (Mateos-Aparicio et al., 2008). In terms of specific isoflavones, genistein and diadzein have been studied most to date with regard to their potential health benefits. According to the United States Department of Agriculture (USDA) (2002) survey on isoflavone content of pulses, lentils contain only marginal amounts of isoflavones while chickpeas contain 0.06 mg 100 g^{-1} genistein and 0.04 mg 100 g^{-1} daidzein and soybeans 74 mg 100 g^{-1} genistein and 47 mg 100 g^{-1} daidzein.

Within the last decade there has been emerging interest in the biological activity of theses fractions. The consumption of these soybean fractions has been linked to prevention and possible treatment of prostate cancer (Jian, 2009), improved immunity (Sakai and Kogiso, 2008), cardiovascular health (Rimbach et al., 2008) and prevention of bone loss associated with aging (Yamaguchi, 2006).

Further studies are now needed to continue studying the potential health benefits of these food components in isolation. By undertaking such research scientists can then determine how much of the individual fractions are needed to exert relative health benefits. Bioavailability studies also play an important role in determining the absorption and metabolism of isoflavones when provided as pure compound, as plant extract or as part of a food or beverage (Cassidy, 2006).

13.4.4 Bioactive compounds

Pulses also contain a range of bioactive compounds, some of which may have health benefits. There is not as much evidence for these as other constituents, but their role may become increasingly important in the future. A list of bioactive compounds found in pulses is given in Table 13.2. This section of the chapter will focus on the potential role and health benefits of the saponins and alkaloids.

In the past, saponins (plant glycosides) have been considered to be undesirable for consumption, mainly because of toxicity symptoms and their hemolytic ability to break down red blood cells. However, it is increasingly being documented that these may also have health benefits (Rochfort and Panozzo, 2007). Recent studies have suggested that legume saponins may possess anticancer activity (Chang et al., 2006), while earlier studies have documented that saponins may reduce cholesterol by forming an insoluble complex, preventing absorption in the intestine. Some saponins may also increase the excretion of bile acids – an indirect method of decreasing cholesterol (Sidhu and Oakenfull, 1986).

With regard to alkaloids (a type of plant phytochemical), the majority of alkaloids from edible legumes have been reported from lupins. Lupins have a relatively short history of use as a grain crop, and it is only recently (the past 20–30 years) that these have been studied (Hymowitz et al., 1990). It has been documented that lupin and quinolizidine alkaloids may play a role in helping to facilitate the secretion of insulin. Although this increased secretion has only been observed in participants with relatively high glucose levels, there may be a potential role for alkaloids in the management of type II diabetes (Lopez et al., 2004). More work is required within this field to study further the possible health benefits of alkaloids.

13.4.5 Antinutritional compounds (ANCs)

While pulses contain substances that may be beneficial to health, it is also important to consider that they can also be a source of antinutritional compounds (ANCs), components that may reduce their potential health benefits. ANCs may be defined as molecules that disrupt the processes of digestion, typically when consumed in a raw state (Domoney, 1999). For pulses this includes food components such as lectins (carbohydrate-binding proteins), protease inhibitors and the angiotensin I-converting enzyme (ACE) inhibitor, a compound widely distributed in human tissues (may contribute to high blood pressure when converted to angiotensin II, causing vasoconstriction). Such ANCs may cause unpleasant feelings, particularly after the consumption of raw pulses – bloating, vomiting and enlargement of the pancreas (Roy et al., 2010). Consequently, such ANCs should be taken into consideration when determining the nutritional value of such foods containing pulses.

In many instances the effects of ANCs may be diminished after cooking or may have multiple effects on health. For example, it is well known that phytic acid may reduce the bioavailability of some minerals, particularly iron. However, it also known that phytic acid has antioxidant properties and protects against DNA damage. Therefore, this compound has dual roles, one which may be injurious and the other beneficial to health (Campos-Vega et al., 2010).

13.5 Health benefits of eating pulses

There is considerable research studying the health benefits of pulses. The consumption of pulses has been linked to a range of health outcomes including a reduced risk of diabetes, cardiovascular disease and certain cancers, and may have an important role in weight management (Capos-Vega et al., 2010). Table 13.3 lists some examples of main potential positive and beneficial effects of legume consumption. This section aims to explain the physiological role of pulses in the prevention and treatment of such conditions.

13.5.1 Diabetes prevention

Around 135 million people worldwide suffer from type 2 diabetes, with prevalence rates varying from 1.1% in sub-Saharan Africa to 3.3% in developing countries and to 5.6% in the industrialized countries (Eschwège, 2000). The incidence of type 2 diabetes has been reported to be lower in Asian populations compared with Western countries, which has been attributed to higher intakes of fermented soybean products. Although the mechanism of action is yet to be confirmed it is thought that dietary phytoestrogens and soy peptides in fermented soybean foods may go some way towards preventing or delaying the progression of type 2 diabetes mellitus (Kwon et al., 2010).

Table 13.3 Pulses and their main potential positive and beneficial effects

Source	Involved metabolism	Beneficial effect	Reference
Legumes (including some pulses)	Cardiovascular	22% lower risk of coronary heart disease, and an 11% lower risk of cardiovascular disease	Bazzano et al. (2001)
	Cardiovascular and diabetes	Modulation of glucose, insulin and homocysteine concentrations and lipid peroxidation in coronary artery disease patients	Jang et al. (2001)
Azuki bean juice	Hypertriglyceridemia	Decreased triglyceride concentrations by inhibited pancreatic lipase activity	Maruyama et al. (2008)
Legumes	Type 2 diabetes mellitus	Risk reduction to develop type 2 diabetes in the order of 20–30%	Venn and Mann (2004)
Mung bean	Glucose and lipid metabolism	Modify glucose and lipid metabolism favorably in rats	Lerer-Metzger et al. (1996)
Legumes	Endometrial cancer	Low risk of endometrial cancer	Tao et al. (2005)
	Breast cancer	Low breast cancer risk	Velie et al. (2005)
	Colon cancer	Low risk of colorectal adenoma	Agurs-Collins et al. (2006)
Legumes and cereals	Obesity	Low average body mass index (BMI) and low risk of obesity	Greenwood et al. (2000)
Pulses	Obesity	Low waist-to-hip (WHR) ratio	Williams et al. (2000)
Whole grains, beans and legumes	Obesity	Low body mass index and waist circumference (WC)	Haveman-Nies et al. (2001)
Whole-grain bread and beans	Glycemia and obesity	Glycemic control and weight loss	Jimenes-Cruz et al. (2003)
Chickpeas	Skin and ear inflammation	Low risk of skin diseases and inflammation of the ear	Agharkar (1991); Warrier et al. (1995)
		Tonic, appetizer, stimulant and aphrodisiac, anthelmintic properties	Sastry and Kavathekar (1990)
	Hypertriglyceridemia	Reductions in serum total and low-density lipoprotein cholesterols	Pittaway et al. (2006)
Vegetables (including some seeds of pulses)	Lymphoblastic leukemia	Low risk of lymphoblastic leukemia	Petridou et al. (2005)
Common bean	Colon cancer	Inhibition of aberrant foci crypt development in rat colon	Feregrino-Perez et al. (2008)

Source: Capos-Vega et al. (2010).

Pulses are also a food with very low glycemic index values. There is accumulating evidence that the incorporation of such foods into the daily diet may play a fundamental role in helping to improve glycemic control and metabolic parameters, such as blood lipid levels, and be used as part of prevention and treatment strategies for medical conditions such as diabetes and coronary heart disease (Rizkalla et al., 2002). A meta-analysis of 41 trials (39 reports) identified that pulses alone helped to reduce both fasting blood glucose and insulin levels, both markers of long-term glycemic control (Sievenpiper et al., 2009). Work needs to be continued within this field in the form of larger, well-designed trials.

13.5.2 Cardiovascular health

Coronary heart disease (CHD) is the most common heart condition in the UK with around 2.6 million people living with CHD (British Heart Foundation, 2010). As discussed earlier, pulses are relatively low in fat, cholesterol free and a good source of dietary fiber and phytosterols which have been found to have cholesterol-lowering activity.

There is emerging evidence to suggest that eating pulses may reduce the risk of CHD. In a large US study of almost 10 000 men and women, findings showed that those eating pulses four or more times a week had a 22% lower risk of CHD and 11% lower risk of cardiovascular events compared to those eating pulses less than once a week (Bazanno et al., 2001). This finding was present even after adjusting for other factors that can also influence the risk of heart disease. Therefore, the cardiovascular health benefits of pulses were found to be independent of other factors that can increase the risk of heart disease.

Finley et al. (2007) also undertook a study to establish whether the consumption of pulses influenced cardiovascular well-being. Eighty volunteers (18–55 years), half are whom were healthy and half with metabolic syndrome (an indicator of cardiovascular disease risk), were randomized to eat either half a cup of cooked pinto beans daily alongside their regular diet or a replacement serving of chicken soup instead of pinto beans over a 12-week period. Study findings showed that cholesterol levels lowered among those eating the pinto beans.

Unfortunately, evidence does not appear be consistent between studies. A South African study also investigated the effects of dry bean consumption on serum lipid levels. In this randomized controlled trial, men followed their normal diets for 4 weeks and then ate 110 g per day of extruded dry beans as baked products or continued with their normal diet. After a 4-week wash-out period in between, the groups were crossed over. Results showed that mean intakes of pulses were slightly lower than the allocated target (91.9 g day^{-1}) and had very little effect on serum cholesterol, lipid or triglyceride levels. It was concluded that the incorporation of dry beans in the daily diet did not appear to lower lipid levels (Oosthuizen et al., 2000). Larger, longer studies are needed to ensure that findings are consistent across the board. A meta-analysis of studies would also be useful within this area of research.

13.5.3 Lowering cholesterol

There is an accumulation of evidence suggesting that soy consumption (namely from soybeans) may help to reduce plasma levels of total cholesterol, triglycerides and low-density lipoprotein cholesterol, although their biological effects are not yet fully understood (Song et al., 2007; Harland and Haffnner, 2008).

In one study, conducted in Japan, where the consumption of soybeans is one of the highest globally, 120 female students were asked to eat either:

- 6.26 g day^{-1} soybean protein and 50 mg day^{-1} isoflavone
- 1.36 g day^{-1} soybean protein and 50 mg day^{-1} isoflavone or
- wheat puff as a placebo over a 4-week study period.

Study findings showed that both total cholesterol and rates of hypercholesterolemia decreased significantly among those eating higher levels of soybean protein (Takahashi et al., 2004). It was concluded that soy protein, when consumed even over relatively short time periods, may help to reduce CHD risk among Japanese women, particularly those with higher cholesterol levels and most at risk of heart disease.

Other studies suggest that isoflavones, genistein and daidzein in particular, and certain peptides or protein fractions found in soybeans may help to protect against heart disease (Omoni and Aluko, 2005). Further work is needed to investigate the potential benefits of both pulse fractions and whole pulses with further detail. Although there may be potential in the future for isoflavone and soy protein extracts from soybeans to be used as functional ingredients within the food industry, more work is needed to decipher what levels of consumption are required to exert health benefits.

13.5.4 Cancer prevention

Pulses contain a range of bioactive constituents and are a low glycemic index food, both of which may play some role in reducing tumor risk. Most recently it has been suggested that soy isoflavones may have anticarcinogenic properties and reduce prostate cancer risk. Both genistein and daidzein have been found to induce cell death (apoptosis) in cells that may be carcinogenic (Hsu et al., 2010). Further understanding is needed to confirm the role of pulses in preventing cancer and diet–gene interactions which may modulate cancer risk (Mathers, 2002) but there are some preliminary findings from research studies.

In one study investigating the diet of patients with adenomatous polyps (benign tumors with a glandular structure) some unforeseen findings were observed. Scientists conducting the polyp prevention trial (PPT) were studying the effects of a high-fiber ($4.30\,g\ MJ^{-1}$ energy), high-fruit and vegetable (5–8 servings) and low-fat diet (less than 20% energy intake) on the reoccurrence of polyps in the large bowel. However, at the end of the 4-year intervention study it was found that habitual intakes of dry beans had tripled over the course of the study. Scientists quite unexpectedly found that individuals with the highest intakes of dry beans were 45% less likely to develop advanced adenomas (benign tumors) in the bowel (Lanza et al., 2006).

Although adenomas are not carcinogenic, they have the potential to become so. Therefore, pulse consumption may go some way to reducing the risk of colonic cancer. Further investigation is warranted.

13.5.5 Obesity management

Although few studies have investigated the role of pulses in weight management programs, or in relation to changes in body weight, it has been suggested that their low glycemic index or load diets may stimulate greater weight loss than higher glycemic index or load diets, or other weight reduction diets (Thomas et al., 2009).

There is some evidence that can be extracted from research papers indicating that pulses may help to regulate body weight. In the UK women's cohort study, seven clusters of food consumption were identified, three of which had high cereal levels: health conscious (high bran, wholemeal and pulses), low-diversity vegetarians (high wholemeal bread and pulses) and high-diversity vegetarians (high wholemeal bread, cereals, pasta and rice, and pulses) (Greenwood et al., 2000). Women with these food consumption patterns had significantly lower average BMI values as well as the lowest proportion of obese subjects (5–9% vs. 10–12% in the other four clusters). Another prospective study carried out in the UK identified four diet patterns and found the one with high intakes of rice, pasta and pulses was inversely related to waist-to-hip ratio (Williams et al., 2000). Research trials in the future could be designed specifically to study the effects of pulses or pulse components within functional foods and their impact on markers of obesity.

13.5.6 Osteoporosis

Osteoporosis is a major health problem in the developed world. Soy isoflavones are thought to preserve bone mineral density, increasing bone mineral density in postmenopausal women and reducing the risk of osteoporosis (Cassidy et al., 2006b).

Although there is a lack of human studies, animal studies have shown that the consumption of hull extract from pulses (400 mg or 800 mg kg^{-1} for 3 months) significantly helped to improve the bone mineral density in mice. It is possible that the antioxidants found in the bean hull may help to reduce levels of oxidative stress, which is associated with bone resorption and reduced bone density (Cao et al., 2010).

13.6 Conclusions

Overall, pulses are an important source of nutrients and bioactive components that may help to improve the diet quality and health of both children and adults. There is good evidence that consuming pulses may help to improve glycemic control and be used as part of prevention and treatment strategies for the management of diabetes and heart disease, and emerging evidence that regular consumption of pulses may help to regulate body weight and improve bone mineral density. It is important that the public is fully aware of the potential health benefits of pulses and are advised about how these could be included within their daily diets and the proportions that should be eaten.

There is also great potential for the food industry to develop tasty, appealing products containing pulses, particularly in developed regions where pulses are not consumed in the traditional manner. If publicity about the benefits of pulses is improved and if the food industrial and relevant organizations can incorporate pulses into contemporary, convenient and healthy food products, the trend in the consumption of pulses could change. This could be of great benefit to all sectors: public, health and private.

References

Agharkar, S.P., 1991. Medicinal plants of Bombay presidency. India Scientific Publishers, Jodhpur.

Agurs-Collins, T., Smoot, D., Afful, J., Makambi, K., Adams-Campbell, L., 2006. Legume intake and reduced colorectal adenoma risk in African-Americans. J. Nat. Black Nurses Assoc. 17, 6–12.

American Association of Cereal Chemists (AACC), 2001. The definition of dietary fibre. Cereal Foods World 46, 112–126.

Bazzano, L.H.J., Ogden, L.G., Loria, C., Vupputuri, S., Myers, L., Whelton, P.K., 2001. Legume consumption and risk of coronary heart disease in US men and women: NHANES I epidemiologic follow-up study. Arch. Intern. Med. 161, 2573–2578.

Boye, J., Zare, F., Pletch, A., 2010. Pulse proteins: processing, characterisation, functional properties and applications in food and feed. Food Res. Int. 43, 414–431.

British Heart Foundation, 2010. Beating heart disease together. Available at: http://www.bhf.org.uk/living_with_a_heart_condition/default. aspx (accessed May 2010).

Cade, J.E., Burley, V.J., Greenwood, D.C., UK Women's Cohort Study Steering Group, 2007. Dietary fibre and risk of breast cancer in the UK Women's Cohort Study. Int. J. Epidemiol. 36, 431–438.

Campos-Vega, R., Loarca-Pina, G., Oomah, D., 2010. Minor components of pulses and their potential impact on human health. Food Res. Int. 43, 461–482.

Cao, J.J., Gregoire, B.R., Sheng, X., Liuzzi, J.P., 2010. Pinto bean hull extract supplementation favorably affects markers of bone metabolism and bone structure in mice. Food Res. Int. 43, 560–566.

Cassidy, A., 2006. Factors affecting the bioavailability of soy isoflavones in humans. J. AOAC Int. 89, 1182–1188.

Cassidy, A., Albertazzi, P., Lise Nielsen, I., 2006. Critical review of health effects of soyabean phyto-oestrogens in postmenopausal women. Proc. Nutr. Soc. 65, 76–92.

Champ, M.M., 2002. Non-nutrient bioactive substances of pulses. Br. J. Nutr. 88 (3), S307–S319.

Chang, W.W., Yu, C.Y., Lin, T.W., Wang, P.H., Tsai, Y.C., 2006. Soyasaponin I decreases the expression of alpha2,3-linked sialic acid on the cell surface and suppresses the metastatic potential of B16F10 melanoma cells. Biochem. Biophys. Res. Commun. 341, 614–619.

Domoney, C., 1999. Inhibitor of legume seeds. In: Shewry, P.R., Casey, R. (Eds.), Seed protein. Kluwer Academic Publishers, Amsterdam, pp. 635–655.

Duranti, M., 2006. Grain legume proteins and nutraceutical properties. Fitoterapia 77, 67–82.

Emami, S., Tabil, L.G., 2002. Processing of starch-rich and protein-rich fractions from chickpeas – a review. ASAE/CSAE North-Central Intersectional Meeting 02-212, 1–21.

Englyst, H., Trowell, H., Southgate, D., Cummings, J., 1987. Dietary fiber and resistant starch. Am. J. Clin. Nutr. 46, 873–874.

Englyst, H., Quigley, E., Bravo, L., Hudson, G., 1996. Measurement by the Englyst NSP procedure. Measurement by the AOAC procedure. Explanation of differences. J. Assoc. Publ. Anal. 32, 1–38.

Eschwège, E., 2000. Epidemiology of type II diabetes, diagnosis, prevalence, risk factors, complications. Arch. Mal. Coeur. Vaiss. 93 (4), 13–17 [Abstract only].

Estruch, R., Martínez-González, M.A., Corella, D., PREDIMED Study Investigators, 2009. Effects of dietary fibre intake on risk factors for cardiovascular disease in subjects at high risk. J. Epidemiol. Commun. Health 63, 582–588.

Feregrino-Perez, A.A., Berumen, L.C., Garcia-Alcocer, G., 2008. Composition and chemopreventive effect of polysaccharides from common beans (*Phaseolus vulgaris* L.) on azoxymethane-induced colon cancer. J. Agric. Food Chem. 56, 8737–8744.

Finley, J.W., Burrell, J.B., Reeves, P.G., 2007. Pinto bean consumption changes SCFA profiles in fecal fermentations, bacterial populations of the lower bowel, and lipid profiles in blood of humans. J. Nutr. 137, 2391–2398.

FSA (Food Standards Agency), 2006. McCance and Widdowson's, the composition of foods, 6th edn. Royal Society of Chemistry, London.

FSA (Food Standards Agency), 2007. List of UK health claims carbohydrates, diets, fats, fibre, and foods and beverages. Available at: http://www.food.gov.uk/multimedia/pdfs/listofukhealthclaims01.pdf (accessed May 2010).

FSA (Food Standards Agency), 2010. Pulses, nuts and seeds. Available at: http://www.eatwell.gov.uk/healthydiet/nutritionessentials/eggsandpulses/pulses (accessed May 2010).

Greenwood, D., Cade, J., Draper, A., Barrett, J.H., Calvert, C., Greenhalgh, A., 2000. Seven unique food consumption patterns identified among women in the UK women's cohort study. Eur. J. Clin. Nutr. 54, 314–320.

Harland, J.I., Haffner, T.A., 2008. Systematic review, meta-analysis and regression of randomised controlled trials reporting an association between an intake of circa 25 g soya protein per day and blood cholesterol. Atherosclerosis 200, 13–27.

Haveman-Nies, A., Tucker, K., De Groot, L., Wilson, P.W.F., van Staveren, W.A., 2001. Evaluation of dietary quality in relationship to nutritional and lifestyle factors in elderly people in the US Framingham Heart Study and the European SENECA study. Eur. J. Clin. Nutr. 55, 870–880.

Hoare, J., Henderson, L., Bates, C.J., 2004. The National Diet and Nutrition Survey: adults aged 19 to 64 years Summary Report. The Stationery Office, London.

Hsu, A., Bray, T.M., Helferich, W.G., Doerge, D.R., Ho, E., 2010. Differential effects of whole soy extract and soy isoflavones on apoptosis in prostate cancer cells. Exp. Biol. Med. (Maywood) 235, 90–97.

Hymowitz, T., 1990. Grain legumes. In: Janick, J., Simon, J.E. (Eds.), Advances in new crops. Timber Press, Portland.

Jang, Y., Lee, J.H., Kim, O.Y., Park, H.Y., Lee, S.Y., 2001. Consumption of whole grain and legume powder reduces insulin demand, lipid peroxidation, and plasma homocysteine concentrations in patients with coronary artery disease: randomized controlled clinical trial. Arterioscler. Thromb. Vasc. Biol. 21, 2065–2071.

Jian, L., 2009. Soy, isoflavones, and prostate cancer. Mol. Nutr. Food Res. 53, 217–226.

Jimenez-Cruz, A., Bacardi-Gascon, M., Turnbull, W.H., Rosales-Garay, P., Severino Lugo, I., 2003. A flexible, low-glycemic index. Mexican-style diet in overweight and obese subjects with type 2 diabetes improves metabolic parameters during a 6-week treatment period. Diabetes Care 26, 1967–1970.

Kwon, D.Y., Daily, III, J.W., Kim, H.J., Park, S., 2010. Antidiabetic effects of fermented soybean products on type 2 diabetes. Nutr. Res. 30, 1–13.

Lanza, E., Hartman, T.J., Albert, P.S., 2006. High dry bean intake and reduced risk of advanced colorectal adenoma recurrence among participants in the polyp prevention trial. J. Nutr. 136, 1896–1903.

Lerer-Metzger, M., Rizkalla, S., Luo, J., 1996. Effects of long-term low-glycemic index starchy food on plasma glucose and lipid concentrations and adipose tissue cellularity in normal and diabetic rats. Br. J. Nutr. 75, 723–732.

Leterme, P., 2002. Recommendations by health organisations for pulse consumption. Br. J. Nutr. 88 (3), S239–S242.

Leterme, P., Carmenza Munoz, L., 2002. Factors influencing pulse consumption in Latin America. Br. J. Nutr. 88 (3), S251–S255.

Lopez, G., Pedro, M., Garzon de la Mora, P., 2004. Quinolizidine alkaloids isolated from *Lupinus* species enhance insulin secretion. Eur. J. Pharmacol. 504, 139–142.

Mann, J.I., Cummings, J.H., 2009. Possible implications for health of the different definitions of dietary fibre. Nutr. Metab. Cardiovasc. Dis. 19, 226–229.

Maruyama, C., Araki, R., Kawamura, M., 2008. Azuki bean juice lowers serum triglyceride concentrations in healthy young women. J. Clin. Biochem Nutr. 43, 19–25.

Mateos-Aparicio, I., Cuenca, A.R., Villanueva-Suarez, M.J., Zapata-Revilla, M.A., 2008. Soybean, a promising health source. Nutr. Hosp. 23, 305–312.

Mathers, J.C., 2002. Pulses and carcinogenesis for the prevention of colon, breast and other cancers. Br. J. Nutr. 88 (3), S273–S279.

Mitchell, D.C., Lawrence, F.R., Hartman, T.J., Curran, J.M., 2009. Consumption of dry beans, peas, and lentils could improve diet quality in the US population. J. Am. Diet. Assoc. 109, 909–913.

Omoni, A.O., Aluko, R.E., 2005. Soybean foods and their benefits: potential mechanisms of action. Nutr. Rev. 63, 272–283.

Oosthuizen, W., Scholtz, C.S., Vorster, H.H., Jerling, J.C., Vermaak, W.J., 2000. Extruded dry beans and serum lipoprotein and plasma haemostatic factors in hyperlipidaemic men. Eur. J. Clin. Nutr. 54, 373–379.

Petridou, E., Ntouvelis, E., Dessypris, N., Terzidis, A., Trichopoulos, D., 2005. Maternal diet and acute lymphoblastic leukemia in young children. Epidemiol. Biomark. Prevent. 14, 1935–1939.

Pittaway, J.K., Ahuja, K.D.K., Cehun, M., 2006. Dietary supplementation with chickpeas for at least 5 weeks results in small but significant reductions in serum total and lowdensity lipoprotein cholesterols in adult woman and men. Ann. Nutr. Metab. 50, 512–518.

Pulse Canada, 2004. Canadian dry peas. Available at: www.pulsecanada. com (accessed April 2010).

Rimbach, G., Boesch-Saadatmandi, C., Frank, J., 2008. Dietary isoflavones in the prevention of cardiovascular disease – a molecular perspective. Food Chem. Toxicol. 46, 1308–1319.

Rizkalla, S., Bellisle, F., Slama, G., 2002. Health benefits of low glycaemic index foods, such as pulses, in diabetic patients and healthy individuals. Br. J. Nutr. 88 (3), S255–S262.

Rochfort, S., Panozzo, J., 2007. Phytochemicals for health, the role of pulses. J. Agric. Food Chem. 55, 7981–7994.

Roy, F., Boye, J.I., Simpson, B.K., 2010. Bioactive proteins and peptides in pulse crops: pea, chickpea and lentil. Food Res. Int. 43, 432–422.

Sakai, T., Kogiso, M., 2008. Soy isoflavones and immunity. J. Med. Invest. 55, 167–173.

Sandberg, A.S., 2002. Bioavailability of minerals in legumes. Br. J. Nutr. 88, 281–285.

Sastry, C.S.T., Kavathekar, K.Y., 1990. Plants for reclamation of wastelands. Council of Scientific and Industrial Research, New Delhi, p. 684.

Schneider, A.V., 2002. Overview of the market and consumption of pulses in Europe. Br. J. Nutr. 88 (3), S243–S250.

Sidhu, G.S., Oakenfull, D.G., 1986. A mechanism for the hypocholesterolaemic activity of saponins. Br. J. Nutr. 55, 643–649.

Sievenpiper, J.L., Kendall, C.W., Esfahani, A., 2009. Effect of non-oilseed pulses on glycaemic control: a systematic review and metaanalysis of randomised controlled experimental trials in people with and without diabetes. Diabetologia 52, 1479–1495.

Song, W.O., Chun, O.K., Hwang, I., 2007. Soy isoflavones as safe functional ingredients. J. Med. Food 10, 571–580.

Takahashi, K., Kamada, Y., Hiraoka-Yamamoto, J., 2004. Effect of a soybean product on serum lipid levels in female university students. Clin. Exp. Pharmacol. Physiol. 31 (2), S42–S43.

Tao, M.H., Xu, W.H., Zheng, W., 2005. A case-control study in Shanghai of fruit and vegetable intake and endometrial cancer. Br. J. Cancer 92, 2059–2064.

Thavarajah, D., Thavarajah, P., Sarker, A., Vandenberg, A., 2009. Lentils (*Lens culinaris* Medikus subspecies *culinaris*): a whole food for increased iron and zinc intake. J. Agric. Food Chem. 57, 5413–5419.

Thomas, D., Elliott, E.J., Baur, L., 2009. Low glycaemic index or low glycaemic load diets for overweight and obesity (review). Cochrane Database Syst Rev. 21 (1), CD006296.

Tosh, S.M., Yada, S., 2010. Dietary fibres in pulse seeds and fractions: characterisation, functional attributes, and applications. Food Res. Int. 43, 450–460.

USDA, 2002. USDA-Iowa State University database on the Isoflavone Content of Foods. Available at: http://www.nal.usda.gov/fnic/foodcomp/Data/isoflav/isoflav.html (accessed May 2010).

Velie, E.M., Schairer, C., Flood, A., He, J.-P., Khattree, R., Schatzkin, A., 2005. Empirically derived dietary patterns and risk of postmenopausal breast cancer in a large prospective cohort study. Am. J. Clin. Nutr. 82, 1308–1319.

Venn, B.J., Mann, J.I., 2004. Cereal grains. Legumes and diabetes. Eur. J. Clin. Nutr. 58, 1443–1461.

Warrier, P.K.W., Nambiar, V.P.K., Remankutty, C., 1995. Indian medicinal plants. Orient Longman, Chennai.

Williams, D.E., Prevost, A.T., Whichelow, M.J., Cox, B.D., Day, N.E., Wareham, N.J., 2000. A cross-sectional study of dietary patterns with glucose intolerance and other features of the metabolic syndrome. Br. J. Nutr. 83, 257–266.

Wong, C.L., Mollard, R.C., Zafar, T.A., Luhovyy, B.L., Anderson, G.H., 2009. Food intake and satiety following a serving of pulses in young men: effect of processing, recipe, and pulse variety. J. Am. Coll. Nutr. 28, 543–552.

Yamaguchi, M., 2006. Regulatory mechanism of food factors in bone metabolism and prevention of osteoporosis. Yakugaku Zasshi 126, 1117–1137.

Role of pulses in nutraceuticals

14

Marina Carbonaro
Istituto Nazionale di Ricerca per gli Alimenti e la Nutrizione (INRAN),
Rome, Italy

14.1 Introduction

The word "nutraceutical" (a combination of nutrition and pharmaceutical) indicates "a food, or components of a food, that provides health benefits, including the prevention and treatment of diseases" (Diplock et al., 1999). The terms "nutraceutical" and "functional food" are often used interchangeably. However, nutraceutical is used more often to describe an isolated molecular extract, whereas functional foods describe whole foods or their concentrates. Many food components, especially those of plant origin, have been claimed to exert beneficial effects on human health: dietary fiber, phenolics, vitamins, minerals, lipids, proteins and peptides are some examples, even though the molecular mechanisms responsible for the observed effects are only partially understood (Milner, 2004).

Many of the bioactive compounds described so far are present in significant amounts in pulse seeds that are, therefore, functional foods. The different functional components in pulses may be used in the prevention and management of cardiovascular diseases, cancer, obesity, osteoporosis and degenerative diseases that are caused by oxidative stress.

Pulse Foods: Processing, Quality and Nutraceutical Applications. DOI: 10.1016/B978-0-1238-2018-1.00007-0

In the following sections, pulse compounds with biological beneficial activity, but also with potential antinutritional undesired effects, such as allergenicity, together with their use in nutraceutical applications, are discussed.

14.2 Nutritional benefits of pulses

Besides being excellent sources of protein, pulses are rich in minerals and trace elements (potassium, calcium, magnesium, iron, zinc, copper, manganese and selenium). Because they are rich in potassium and magnesium, they may decrease the risk of cardiovascular disease by helping to lower blood pressure. Consumption of pulse proteins, due to their low methionine content, reduces serum homocysteine concentration, another important factor in the control of the risk of cardiovascular diseases.

Notably, all pulses, especially dry beans (*Phaseolus* spp. L.), are a major source of dietary fiber (Trinidad et al., 2010). The soluble fiber fraction, although accounting for a small part of total fiber, is the fraction of fiber that helps to lower cholesterol levels and the risk of heart disease. In addition, soluble fiber helps to regulate blood glucose levels. Having a low glycemic index (a measure of how much blood glucose rises after a food is consumed), pulses are useful in the diet of people with diabetes. Some evidence has also indicated a protective effect of pulse fiber on the risk of development of colon cancer (Van Loo et al., 1995; Trinidad et al., 2010).

Recent research findings have highlighted several health benefits of pulses specifically related to minor bioactive substances present in their seed: considerable research is being performed on non-nutrient compounds that may exert significant metabolic and physiological effects (phytochemicals) (Rochfort and Panozzo, 2007). However, threshold concentrations at which beneficial effects can be exerted have not been assessed in most cases. Pulse phytochemicals include simple phenolics, especially flavonoids (flavonols, anthocyanidins, phytoestrogens and catechins) and phenolic acids, complex polyphenols (tannins), phytates, saponins, alkaloids and sterols. Many of these compounds have been reported to be able to reduce the growth of different type of cancer cells and to lower cholesterol levels.

More recently, beneficial health properties similar to those found for phytochemicals have been described for undigested pulse proteins (enzymatic inhibitors and lectins) and for peptides derived from specific protein fractions.

A number of research findings have shed light about the real risk of adverse health effects from the so-called "antinutritional factors" present in the seed. This has been mostly limited to high content of heat-stable (non-protein) compounds (such as tannins and phytic acid) (Gilani et al., 2005), besides the well-known hazards presented by specific compounds, e.g. vicine and convicine. Nonetheless, concern for people with precarious health, as well as for malnourished subjects, needs to be taken into consideration when developing nutraceutical foods based on pulses.

Plant breeding methods have the potential of changing pulse seed composition towards reduction of specific antinutritional factors. However, the use of processing procedures – thermal treatments, fermentation, germination, soaking – that are effective in reducing antinutrient levels remains an important strategy.

Specific bioactivities of pulses with beneficial or detrimental effects and related compounds are discussed below.

14.3 Antinutritional factors of pulses

A number of protein and non-protein compounds with antinutritional effects have been found to be present in pulse seeds. In the active form, these compounds are toxic or deleterious for the growth of most animal species. Major bioactive compounds of pulses and their biological effects are presented in Table 14.1. For a long time, bioactive compounds have been considered essentially in relation to their adverse effect on nutrient bioavailability (Muzquiz, 2004). However, the current literature contains substantial evidence that they may exert beneficial effects at low concentrations, protecting against cardiovascular diseases, cancer, diabetes, osteoporosis and hypertension. Therefore, both adverse and potential beneficial effects are reported in Table 14.1.

Unlike antinutritional factors of protein nature, such as enzyme inhibitors and lectins, that can be inactivated by processing preceding consumption (soaking, thermal treatment), non-protein compounds are often heat-stable and, therefore, represent the most harmful antinutrients. Diets high in phytic acid have been shown to have a negative effect on mineral absorption, an effect also exerted by oxalic acid.

Table 14.1 Major bioactive compounds of pulses, their biological activities and treatments for inactivation

Bioactive compound	Chemical nature	Sources	Biological activity	Inactivation
Enzymatic inhibitors	Proteins MW 600–24 000	Soybean, common bean, chickpea, faba bean, pea, lentil, peanut	Pancreatic hyperthrophy, impaired growth and protein utilization, inhibition of proteolytic enzymes and α-amylase. Decrease of blood glucose (amylase inhibitor). Anticancer and antiviral activity	Thermal treatment, germination
Lectins	Proteins or glycoproteins MW 10 000–124 000	Soybean, common bean, jack bean, chickpea, faba bean, pea, lentil, peanut	Reduced intestinal absorption of nutrients, growth inhibition, hyperproliferation of intestinal villi, pancreatic hypersecretion. *In vitro* hemagglutinating activity. Anticancer properties. Decrease of blood glucose	Thermal treatment
Seed globulins	Storage proteins	Peanut, soybean, lupin	Allergenic properties (susceptible subjects). Reduction of blood cholesterol and glycemia. Anticancer activity	Thermal treatment, high pressure
Polyphenols	Tannins, flavonoids, phenolic acids, stilbenes, lignins	Colored beans, black gram, faba bean, colored pea, lentil, soybean	Reduced food intake, protein and mineral utilization (condensed tannins), inhibition of digestive enzymes. Antioxidant and anticancer activity.	Soaking, dehulling, thermal treatment
Phytic acid	Myo-inositol hexaphosphate	Soybean, common bean, chickpea, faba bean, pea, lentil	Reduced mineral (Ca, Zn, Fe) absorption. Antioxidant and anticancer activity	Soaking, germination, fermentation
Favism factors	Pyrimidine glucosides (vicine, convicine)	Faba bean major, faba bean minor	Acute hemolytic anemia (susceptible subjects)	Germination
Flatus-inducing factors	α-Galactosides (raffinose, stachyose, verbascose)	Tropical pulses, soybean, common bean, chickpea, faba bean, pea, lentil	Intestinal discomfort. Possible prebiotic properties	Soaking, fermentation, germination

Bioactive compound	Chemical nature	Sources	Biological activity	Inactivation
Saponins	Steroidal or triperpenic glycosides	Soybean, chickpea, faba bean, pea, lentil, peanut	Growth-retarding effects, decrease of enzymatic activity, cardiovascular and hemolytic activity. Anticancer properties, immune-modulatory effects. Reduction of blood cholesterol	Soaking, thermal treatment
Alkaloids	Amino acids/ purine/ pyrimidine derivatives	Lupin	Inhibition of digestion, adverse effects on cardiovascular and central nervous system circulation	Soaking in saline solution
Cyanogens	Cyanogenic glycosides	Lima bean, common bean, chickpea	Respiratory distress	Soaking, thermal treatment
Lathyrogens	Amino acid derivatives	Grass pea, faba bean	Effect on nervous system, spastic paralysis and skeletal deformities	Dehulling, thermal treatment

Concentrations of trypsin inhibitor, lectins, phytic acid, total polyphenols, tannins, α-galactosides and saponins in the seed of the main pulse species are shown in Table 14.2. The common bean contains the highest level of lectins, whereas soybean has a high amount of trypsin inhibitor and phytic acid (Valdebouze et al., 1980; Gueguen, 1983; Carbonaro et al., 2001c; Champ, 2002). Irrespective of the species, concentration of antinutrients in pulse seeds appears to be significant.

Among polyphenols, high molecular weight tannins (M_r 500–5000), especially condensed tannins (proanthocyanidins), form strong insoluble complexes with proteins, thus lowering their digestibility. However, when the molecular weight is very large (>5000 Da), tannins become insoluble, and lose their protein precipitating capacity, fermentability in the colon and bioavailability (Serrano et al., 2009).

An adverse effect of tannins on trace element absorption, especially of iron (but also of zinc and copper), as a consequence of metal insolubilization, has also been observed in cooked pulses (Carbonaro et al., 2001a). In addition, binding of tannins to digestive enzymes and to proteins of the gut wall, with interference on

Table 14.2 Content of bioactive compounds in main pulse species (% dry weight, unless otherwise indicated)

Species	Trypsin inhibitor (TI unit mg⁻¹ d.w.)[a]	Lectins (Hemagglutinating activity mg⁻¹ d.w.)[a]	Phytic acid[b]	Polyphenols (tannic acid equivalents)[b]	Tannins (catechin equivalents)[b]	α-Galactosides[c]	Saponins[d]
Common bean (*Phaseolus vulgaris* L.)	9.6	8200	1.2	0.5	0.1	3.1	0.03
Faba bean (*Vicia faba* L.)	6.7	50	1.1	1.2	0.5	2.9	0.4
Lentil (*Lens culinaris* L.)	8.4	640	0.6	1.0	0.1	5.6	0.5
Chickpea (*Cicer arietinum* L.)	7.5	nd[e]	0.5	0.6	nd	3.8	0.4
Pea (*Pisum sativum* L.)	6.6	250	0.9	0.2	0.1	5.9	0.2
Soybean (*Glycine max* L.)	50	2400	1.8	0.4	0.1	4.0	0.6

[a]Valdebouze et al. (1980); Gueguen (1983); Champ (2002).
[b]Carbonaro et al. (2001c); Carnovale et al. (2001).
[c]Gulewicz et al. (2000); Carnovale et al. (2001).
[d]Sodipo and Arinze (1985); Savage and Deo (1989).
[e]Not detectable.

gut function may also occur (Carbonaro et al., 2001b). Although secretion of salivary-rich proteins in the mouth has been described as a protective mechanism against antinutritional effects of tannins, much evidence for their potential deleterious effects has been collected and strategies are being developed for decreasing their content in pulse seeds. Processing techniques (soaking, boiling, steaming) have been found to decrease the condensed tannin content of pulses. However, these treatments also cause a reduction in low molecular weight phenolics (anthocyanins, phenolic acids, isoflavones) with beneficial effects (Xu and Chang, 2008).

Pulse seeds contain high amounts of α-galactosides: raffinose, stachyose and verbascose (2.9–5.9 g $100 \, g^{-1}$) (Gulewicz et al., 2000; Carnovale et al., 2001). These oligosaccharides can be fermented in the colon and are responsible, at least in part, for the problem of flatus consequent to pulse consumption.

Saponins (triterpene or steroid derivatives attached to one or more sugar moieties) of different types are also found in many pulses (common bean, lentil, chickpea, pea, faba bean, soybean). They have a bitter taste and foaming properties. Saponin content of pulses ranges from 0.03 g $100 \, g^{-1}$ (common bean) to 0.6 g $100 \, g^{-1}$ (soybean) (Sodipo and Arinze, 1985; Savage and Deo, 1989) (see Table 14.2). Saponins affect lipid membranes and cause hemolysis *in vitro* by a mechanism, involving cholesterol binding, that is still under study (Fu and Renxiao, 2010). Once hydrolyzed in the intestinal tract, some saponins may result in systemic toxic effects (e.g. weight loss, gastroenteritis) in ruminants, because of inhibition of microbial fermentation and synthesis in the rumen (Lu and Jorgensen, 1987). In non-ruminants (e.g. chickens, pigs), the main antinutritional concern is probably a decrease in growth rate due to reduction of food intake.

Potential harmful effects for susceptible subjects (with glucose-6-phosphate dehydrogenase deficient erythrocytes) resulting from vicine and convicine from faba bean (*V. faba* L.) have been well documented. Hemolytic anemia (favism) is caused by oxidant aglycones (divicine and isouramil) liberated upon glucoside hydrolysis in the gut (Arese et al., 1981). High amounts of vicine and convicine in the seed of faba bean varieties can be present (up to 6.7 g $100 \, g^{-1}$). Italian varieties of faba bean (*V. faba* L., major) and field bean (*V. faba* L., minor) showed variability in glucoside content and different vicine/convicine ratios. Mean glucoside concentration was 0.6 g $100 \, g^{-1}$ for fresh faba bean (corresponding to 0.3 g $100 \, g^{-1}$ aglycone content) and

$0.95\,g\ 100\,g^{-1}$ for dry field bean ($0.48\,g\ 100\,g^{-1}$ aglycone content) (Carnovale et al., 2001).

Alkaloids of lupins (*L. albus*, *L. angustifolius*, *L. luteus* and *L. mutabilis* L.) are amino acid (or purine and pyrimidine) derivatives that can inhibit the central nervous system, circulation, digestion, reproduction and the immune system. More than 150 different alkaloids have been identified as toxic, bitter lupins containing up to $1-4\,g\ 100\,g^{-1}$ seed. On the other hand, modern sweet lupin cultivars usually have an alkaloid content of less than $0.02\,g\ 100\,g^{-1}$ seed (Lallès and Jansman, 1993).

Other deleterious effects have been linked to the presence of cyanogens, glycosides of a sugar and cyanide. They can be hydrolyzed by enzymes, especially in the rumen, with release of hydrogen cyanide (HCN) and consequent respiratory distress when ingested in large amounts. Cyanogenic glucosides have been found in high concentration in lima beans (*P. lunatus* L.), especially in black varieties ($14\,mg$ HCN $100\,g^{-1}$), while low concentrations (about $2\,mg$ HCN $100\,g^{-1}$) can be present in *P. vulgaris* and *C. arietinum* seeds (Liener, 1979).

Toxic amino acids, known as lathyrogens, are present in the genera *Lathyrus* and *Vicia*. Neurolathyrogens (β-*N*-oxalyl-α,β-diamino-propionic acid, ODAP) cause spastic paralysis of legs in humans and animals, while osteolathyrogens (β-aminopropionitrile) induce skeletal deformities by interfering with the growth of cartilage and bones. Seeds of *Lathyrus* cultivated in temperate climatic regions have been found to contain lower levels of lathyrogens than those from subtropical and tropical areas (Grela et al., 2000). Values of ODAP of $0.2-1.0\,g\ 100\,g^{-1}$ and of $0.5-1.1\,g\ 100\,g^{-1}$ have been reported for *Lathyrus* seeds from India and Ethiopia, respectively (Rao, 1978; Urga et al., 2005). Low toxic varieties with less than $0.1\,g\ 100\,g^{-1}$ β-ODAP have also been described (Rotter et al., 1990; Grela et al., 2001).

Elimination or reduction of antinutritional factors can be achieved by classical breeding, molecular biology techniques and by several technological treatments, often used in combination. Most commonly employed post-harvesting treatments are: dry and moist thermal treatment, extrusion cooking, steaming, soaking, germination, fermentation, dehulling, enzymatic treatment (Champ, 2002). In particular, germination can be used to reduce the content of phytates, tannins and protease inhibitors, while dehulling is effective in decreasing the levels of tannins. A phytate-degrading enzyme has been isolated and characterized from faba bean seeds and used for phytate removal during processing (Greiner et al., 2001). Enzymatic treatment or fermentation of legume proteins, such as globulins, with microbial proteases has been proposed as a tool to increase digestibility and reduce allergenicity, as well as to improve their functional properties (Schwenke, 1997; Frokiaer et al., 2001). Most of these post-harvesting processing techniques have proven to be successful in achieving threshold levels of antinutritional factors that did not interfere with optimal human and animal performance. Although the presence of unusually heat-stable antinutritional compounds in legume seed has been described, it should be taken into consideration that severe thermal treatment often required for their elimination is not recommended because it can result in protein modification and reduction in protein availability, as well as in undesidered interaction among nutrients (Carbonaro et al., 2000, 2005).

In recent years, there has been general agreement that antinutritional factors in diets should not necessarily be reduced to zero levels. In fact, it has been demonstrated that biological activities described for many of these compounds have health-promoting properties when present in low concentration in food and feed. Therefore, threshold levels of each antinutritional compound to be safely included in the diet need to be established.

14.4 Prebiotic properties of pulses

A prebiotic is a non-digestible food ingredient that beneficially affects the host by selectively stimulating the growth of beneficial commensal bacteria in the colon, especially *Bifidobacteria* (Gibson and Roberfroid, 1995). Because humans lack the enzymes capable of digesting raffinose-like sugars in beans, bacterial fermentation in the colon may cause intestinal discomfort. On the other hand, non-digestible α-galactosides – raffinose and other carbohydrates with a degree of polymerization of up to 60 units – have recently been hypothesized to have prebiotic properties, similar to those ascribed to inulin and other fructo-oligosaccharides of cereals, as well as to soluble dietary fiber (pectins, hemicellulose) (Martinez-Villaluenga et al., 2008a). Functional effects linked with prebiotics include improvement of glycemic control, modulation of the metabolism of lipids, hypocholesterolemic effects, anticancer properties, anti-inflammatory and other immune effects (Flight and Clifton, 2006; Rideout et al., 2008; Sievenpiper et al., 2009). Moreover, a possible enhancing effect of non-digestible carbohydrates on mineral (calcium, magnesium, iron) absorption has been reported (Macfarlane et al., 2008).

Clinical trials have been shown that soluble dietary fiber of (non-soybean) pulses has the highest hypocholesterolemic effect, followed by that of vegetable protein (Anderson and Major, 2002). Soluble fiber absorbs water to become a viscous substance that is readily fermented by bacteria in the digestive tract. Fermentation in the colon produces short-chain fatty acids (acetate, butyrate, propionate), that are likely responsible for the observed physiological effects (Van Loo et al., 1995; Marcil et al., 2002). A high concentration (20.9–46.9 g $100\,g^{-1}$) and fermentability of dietary fiber from several legumes have recently been reported (Mallillin et al., 2008).

In Table 14.3 the concentrations of total, soluble and insoluble fractions of dietary fiber from common bean, faba bean, lentil, chickpea, pea and soybean are reported. Common bean was found to contain the highest total fiber content, whereas faba bean showed the highest amount of the soluble fiber fraction (Hidalgo et al., 1997; Carnovale et al., 2001; Carbonaro et al., 2001c). Due to their high fiber content, all pulses can contribute to the recommended high intake of dietary fiber (25–30 g day^{-1}), that is seldom achieved in Western country diets.

14.5 Antioxidant compounds of pulses

Among phytochemicals, antioxidant compounds with free radical scavenging activity, such as phenolic compounds, deserve a special mention, being likely implicated in a number of beneficial health

Table 14.3 Dietary fiber (total, soluble and insoluble fractions) content of main pulse species (% dry weight)[a]

Species	Total	Soluble	Insoluble
Common bean (*Phaseolus vulgaris* L.)	21.6	1.2	20.4
Faba bean (*Vicia faba* L.)	21.1	1.6	19.5
Lentil (*Lens culinaris* L.)	14.9	1.2	13.7
Chickpea (*Cicer arietinum* L.)	14.0	1.3	12.7
Pea (*Pisum sativum* L.)	15.7	1.3	14.4
Soybean (*Glycine max* L.)	11.9	0.9	11.0

[a]Hidalgo et al. (1997); Carnovale et al. (2001); Carbonaro et al. (2001c).

effects. Indeed, they have been reported to prevent chronic degenerative diseases including cancer, diabetes, osteoporosis, cardiovascular, autoimmune, inflammatory and neurodegenerative diseases (Alzheimer's and Parkinson's diseases, multiple sclerosis) (Scalbert et al., 2005).

Phenolic compounds, including their major subcategory, flavonoids, have been extensively studied in pulses. Both highly polymerized polyphenols, that is tannins (M_r 500–5000), and low molecular weight phenolics (flavonoids, phenolic acids) have been found to be present (0.01–$4.0\,g\ 100\,g^{-1}$ of dry weight). Pulses with the highest polyphenolic content are the dark varieties (i.e. red common bean, black gram). Condensed tannins (proanthocyanidins) have been found in several varieties of field bean (*V. faba* L., minor) and in colored common bean (*P. vulgaris* L.). Pulse phenolics with a low molecular weight are predominantly of flavonoid origin, although the concentration varies widely among the different pulse species. Flavonoid chemistry of the Leguminosae family includes the isoflavonoids, 5-deoxyflavonoids and many unusual glycosides and dimers. Flavonoids are commonly found as glycoside derivatives.

Oligomeric proanthocyanidins (2–10 catechin units), compounds with a wide range of pharmacological activity including protection of collagen distruction, antimicrobial and ulcer activity, have recently been found in significant amounts in the lentil seed coat (*L. culinaris* L.) and in broad bean (*V. faba* L.) ($0.16\,g\ 100\,g^{-1}$) (Auger et al., 2004; Amarowicz et al., 2009).

Phenolic acids are derivatives of benzoic acid (gallic, syringic, vanillic acid) or of cinnamic acid (caffeic and quinic acids). Ferulic, p-coumaric, caffeic and vanillic acids have been quantified in different pulse species. The total phenolic acid content of common bean (*P. vulgaris* L.) has been found to be $30\,mg\ 100\,g^{-1}$, with ferulic acid as the prevalent compound, followed by p-coumaric acid (Luthria and Pastor-Corrales, 2006; Carbonaro et al., 2008b). Phenolic acids are expected to have a positive effect on prevention of low-density lipoprotein cholesterol oxidation. They may also inhibit carcinogenesis in breast and liver.

A study on 24 commercial US common bean samples has indicated hydroxycinnamic acid derivatives as the main phenolic component (Lin et al., 2008). The flavonoid content was more variable and absent in some varieties. Analysis of glycosylated flavonoid components indicated the presence of 3-*O*-glucosides of delphinidin, petunidin and malvidin in black bean, while pink and dark red bean contained the diglycosides of quercetin and kaempferol

(Lin et al., 2008). Quantitative analysis of flavonols in several accessions of three Italian *P. vulgaris* L. ecotypes showed a very variable content, from 19 to 840 mg $100 g^{-1}$, and kaempferol 3-O-glucoside as the most abundant compound (Dinelli et al., 2006).

Soybean has been demonstrated to represent a unique source of isoflavones: genistein, daidzein and glycitein. Indeed, the concentration of isoflavones in seeds of most pulses is much lower than that of soybean (Bravo, 1998). These compounds have shown antioxidant activity and antiproliferative properties towards cancer cells. Because of their structure, similar to that of steroid hormones, they also have estrogenic activity and may lower the risk of heart disease and bone loss. On the other hand, contrasting data on their potential in preventing hormone-dependent cancers (e.g. breast and prostate) have been provided. Genistein have been found to be effective in inhibiting carcinogenesis in animal models by binding to tyrosine kinase and disrupting cell growth signals (Banerjee et al., 2008). However, especially in a concentrated form, it may act as a natural estrogen and promote cancer development (Taylor et al., 2009).

Each of these compounds can exist as an aglycone, a glycoside or an acetylated or malonated glycoside, metabolism affecting the rate of transport, the biological properties and the stability to processing of isoflavones in a complex way.

Bioavailability assessment is key to claim bioactivity of pulse phenolics (Manach et al., 2005; Ferruzzi et al., 2009). The soy isoflavones – genistein, daidzein and glycitein – appear to be sufficiently bioavailable to humans to act *in vivo*. Most studies have indicated that glycosylated forms of isoflavones are absorbed at a lower rate in humans than their respective aglycones (Izumi et al., 2000).

Also gallic acid and catechin are well absorbed, whereas no sufficient data have been collected on absorption and bioefficacy of most flavonoids from pulses. Polymerization degree, galloylation, glycosylation and methylation affect the antioxidant properties and bioavailability of catechin and proanthocyanidins (Landis-Piwowar and Dou, 2008). Their absorption through the gut barrier is probably limited to the molecules of low polymerization degree and to metabolites (partially unknown) formed by the colonic microflora. Moreover, they have a high affinity for proteins. Interaction with proteins has been reported to impair bioavailability of phenolics (catechin, tannic acid) in the small intestine (Carbonaro et al., 2001b). A significant part of pulse phenolics has been found to be bound to proteins in the seed (Carbonaro et al., 2000, 2008b).

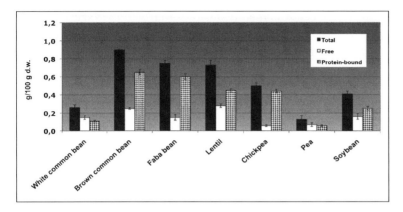

Figure 14.1 Concentration of total, free and protein-bound polyphenols in common bean, faba bean, lentil, chickpea, pea and soybean (Carbonaro, personal communication).

Total, free and protein-bound polyphenol content of the main pulse species is reported in Figure 14.1. Percentage of protein-bound polyphenols was lower in white common bean, pea, soybean and lentil (42, 46, 61 and 62% of total polyphenols, respectively) than in brown common bean, faba bean and chickpea (72, 80 and 88%, respectively). In the latter seeds, most of polyphenols appeared to be bound to proteins.

Further studies on the bioavailability and bioefficacy of phenolic compounds of pulses, and related influencing factors, are still required. The possibility for systemic effects or local effects in the gut needs to be ascertained too.

Antioxidant activity and anticancer effects have also been ascribed to phytic acid. Phytic acid (myo-inositolhexaphosphate) is very reactive with positively charged ions, especially Zn, Ca and Fe. Chelating activity may lower iron-mediated colon cancer in experimental animals and humans. Anticancer activity towards human prostate and mammary carcinoma cells has also been reported (Reddy et al., 2000). A further positive effect concerns control of serum cholesterol and tryglicerides by phytic acid.

It should be considered that diets high in phytic acid may interfere with mineral absorption. However, it has been shown that levels of phytic acid in the diet as low as 0.013% are effective in reducing serum triglycerides and cholesterol, and that antinutritional effects occurred only at 10-fold higher levels (Onomi et al., 2004).

Many new areas for phytic acid utilization including medical and industrial applications, based on the antioxidant properties of phytic acid, have been proposed (Oatway et al., 2001). Medical applications include prevention of urinary and nephritic calculi, prevention of colonic cancer and suppression of bacterial plaque formation. For industrial use, phytic acid is widely employed as a preservative for fishpaste products and pasta, in the reduction of fermentation time, in taste enrichment of bean paste and soy sauce, and as an oxidation inhibitor for oils.

14.6 Pulse and soybean bioactive peptides and proteins

Pulse seeds are known to contain high levels of proteins, generally 20–30% w/w. Among oilseed pulses, protein content of soybean (40% w/w) is the highest and considerably higher than that of cereals, eggs and milk. Therefore, soybean is an excellent source not only of high-quality vegetable proteins, but also of significant amounts of unique biologically active components, such as trypsin inhibitor, lectins, and the most recently discovered peptide called lunasin.

Pulse bioactive proteins – protease inhibitors, α-amylase inhibitor and lectins – have long been considered as antinutrients. However, there is substantial evidence that these proteins, similarly to other non-protein bioactive compounds, exert beneficial effects at low concentrations. Therefore, they can be used in the formulation of therapeutic products for treatment and prevention of human diseases (cardiovascular diseases, cancer, diabetes, osteoporosis and hypertension) (Barac et al., 2005; Duranti, 2006; Aluko, 2008).

An antidiabetic and antiobesity role for α-amylase inhibitor from kidney bean (*P. vulgaris* L.) has been claimed (MacCarty, 2005; Obiro et al., 2008). Although much evidence has been presented that indicates the reduction of glucose, total and low-density lipoprotein (LDL) blood cholesterol by soy proteins, the actual mechanism of action is still under study (Dewell et al., 2006). However, the body of evidence was such that the American Food and Drug Administration authorized a specific health claim regarding soy consumption and prevention of heart disease in 1999 (FDA, 1999).

Soy 7S globulin – β-conglycinin – as well as its closely related protein conglutin γ from lupin, have been shown to have *in vitro* kinase activity and to bind peptide hormones and mammalian insulin, likely representing the active molecule for the regulatory effect on glycemia and on lowering cholesterol and triglyceride levels (Lovati et al., 1998; Duranti et al., 2004; Scarafoni et al., 2007). Active peptides embedded in β-conglycinin have been found to inhibit lipid accumulation in adipocytes *in vivo*, thus representing interesting components for the control of obesity (Martinez-Villaluenga et al., 2008b).

Possible effects of individual amino acids (methionine, cysteine, lysine, leucine, aspartate, glutamate) on cholesterol metabolism have recently been demonstrated in animal models (Blachier et al., 2010). In particular, methionine increased high-density lipoprotein cholesterol and apolipoprotein A-I, while cysteine decreased very low-density lipoprotein cholesterol in blood.

Anticarcinogenic effects of the trypsin and chymotrypsin inhibitors of the Bowman-Birk class have been described not only from soybean, but also for pea seeds, soybean inhibitors being active towards colon, liver, lung, esophagus and breast cancers, and pea inhibitors presenting very high antiproliferative activity on human colon cancer (Clemente et al., 2005; Clemente and Domoney, 2006). Animal models have clearly shown that these inhibitors are absorbed after oral administration and are found in liver, kidney and lung in an intact form (Wan et al., 2002).

An inhibitory activity on tumor growth has been demonstrated for lectins from several plant sources (Gonzalez and Prisecaru, 2005; Pryme et al., 2006). In particular, concanavalin A, a lectin from jack bean seeds (*Canavalia ensiformis* L.), has been shown to exert a potent antihepatoma effect, being a potential therapeutic agent for liver tumor (Lei and Chang, 2009).

Binding and internalization of trypsin inhibitor and lectins in the small intestinal villi of rat have previously been demonstrated by Pusztai et al. (1997) and by Ewen et al. (1998). Stability to processing and during gastrointestinal digestion of bioactive proteins is likely to be a prerequisite for the preventive activity attributed to a regular pulse intake.

Bioactive peptides from soybean may exist naturally or be derived from protein hydrolysates. They have been shown to have common structural properties such as a short peptide length (2–9 amino acid residues) and hydrophobic amino acid residues in addition to proline, lysine or arginine (Kim et al., 2000). Bioactive peptides are resistant to the action of peptidase and act as physiological modulators during gastrointestinal digestion of soy products. They stimulate superoxide anions, which trigger non-specific immune defense systems (Kitts and Weiler, 2003) and show antioxidant, antiobesity and anticancer activities (lunasin and hydrophobic peptides) (Pena-Ramos and Xiong, 2002; Nakamori, 2002).

14.6.1 Lunasin

Lunasin is the most recently isolated bioactive polypeptide from soybean with cancer-preventive effects (Hernandez-Ledesma et al., 2009). It has subsequently been isolated from wheat, barley and rice (Jeong et al., 2009). It consists of a unique 43 amino acid peptide, whose carboxyl end contains nine residues of aspartic acid, an Arg-Gly-Asp cell adhesion motif, and a helix with structural homology to a conserved region of chromatin-binding proteins. It is now known that lunasin is a linker peptide of Bowman-Birk inhibitors with molecular mass of 5.45 kDa. The peptide is heat stable, surviving temperatures up to 100 °C for 10 min (Jeong et al., 2003). The content of lunasin in soybean seed is dependent on varieties and environment.

Lunasin from soybean is a promising chemopreventive peptide, because of its high bioavailability after oral administration, as shown in studies using synthetic lunasin. *In vitro* and *in vivo* chemopreventive properties of lunasin against chemical carcinogenesis have been reported. According to Galvez et al. (2001), transfection of mammalian cells with the lunasin gene leads to mitotic arrest and cell death attributed to binding of its poly-aspartyl carbonyl end to regions of hypoacetylated chromatin in centromeres. Also, animal studies showed that lunasin inhibited skin tumorigenesis when applied topically in a mouse skin cancer model.

14.6.2 Hydrophobic peptides

During processing of soy products, especially during fermentation or heat processing, many low molecular weight peptides can be formed that exhibit cytotoxicities on several cell tumor lines (Kitts and Weiler, 2003). In particular, a hydrophobic nonapeptide with high cancer lowering activity was isolated and characterized from hydrolyzed defatted soybean by Kim et al. (2000). The molecular weight of this peptide was 1.16 kDa and the amino acid sequence was X-Met-Leu-Pro-Ser-Tyr-Ser-Pro-Tyr.

14.7 Structural aspects and bioactivity of pulse proteins

The structural properties and amino acid sequence of pulse proteins have been reported to affect their digestibility, amino acid availability and bioactivity (Begbie and Pusztai, 1989; Carbonaro, 2006). Structural aspects of pulse proteins in comparison with animal (meat and milk) proteins are summarized in Table 14.4.

The role of disulfides in stabilizing protein structure and hindering susceptible sites to specific proteases is well known: improvement in the susceptibility to proteolysis of food proteins by chemical cleavage – sulfytolysis, S-carboxymethylation – of disulfides has been demonstrated for both plant and animal proteins (Reddy et al., 1988; Deshpande and Damodaran, 1989).

The protein fraction of pulses with a relatively high cysteine content, that is albumin (according to classical Osborne classification), has been found to be resistant to proteolytic digestion by trypsin, as a consequence of high stability conferred by a number of disulfide bonds in low molecular weight proteins (Marletta et al., 1992; Clemente et al., 2000; Faris et al., 2008).

Maintainance of the native protein conformation (especially the inhibitory domain) after cooking and during gastrointestinal digestion may explain the protective antiproliferative effect of Bowman-Birk inhibitors of pulses towards colon cancer. Bowman-Birk inhibitors

Table 14.4 Structural properties of pulse proteins in comparison with meat and milk proteins

Protein	MW (kDa)	pI	No. of SS	No. of SH	Structure type[a]	α-helix (%)	β-sheet (%)
Glycinin	340	4.6	6	2	G	15	36
Conglycinin	200	5	0	1	G	15	31
7S protein	150	5	0	1	G	16	40
Concanavalin A	25.5	5.5	0	0	G	4	47
Myoglobin	16.7	6.7	0	1	G	88	0
Collagen	300	9	0	0	F	Collagen helix	0
α-Casein	23.5	5.1	0	0	RC	0	0
β-Lactoglobulin	18.4	5.2	2	1	G	15	41
α-Lactalbumin	14.2	5.1	4	0	G	43	10
BSA	66	4.8	17	1	G	66	4

[a]G, globular; F, fibrous; RC, random coil.

have been found in an intact form in several organs after ingestion (Wan et al., 2002), but the mechanism responsible for their bioactivity needs to be further elucidated.

Intrinsic structural features, such as stabilization by disulfide bridges and presence of hydrophobic patches, have long been claimed to be responsible for low digestibility of seed storage proteins from both pulses and cereals (Semino et al., 1985; Nielsen et al., 1988; Deshpande and Damodaran, 1989; Shewry and Halford, 2002). Moreover, structural constraints that exist in the native proteins have been shown not to be completely overcome by heating: oligomeric plant proteins undergo dissociation and subsequent reaggregation upon processing, with formation of highly stable complexes at high temperature. These latter have been found to have reduced functionality, digestibility and amino acid availability (Utsumi et al., 1984; Carbonaro et al., 1993; Duodu et al., 2003; Carbonaro, 2006).

Previous results on *in vivo* gastrointestinal digestion of native and denatured 7S globulin indicated that high amounts of intermediate proteolytic fragments of phaseolin (MW 22–27 kDa) are present in the small intestine after digestion of the autoclaved 7S globulin (Carbonaro et al., 2005). Differences between the proteolytic pattern of 7S vs. 11S globulins could be evidenced and related to the *in vivo* digestibility behavior of native and denatured proteins.

The role of secondary and quaternary structures of pulse proteins has been examined in relation to their digestibility. Native phaseolin, the 7S storage protein of common bean (*P. vulgaris* L.), has been found to be more resistant to proteolysis than vicilin, the storage protein of green pea (P. *sativum* L.). Far-UV and circular dichroism spectroscopy indicated that the native conformation of vicilin was more flexible than that of phaseolin upon heating. However, heating induced undesiderable changes in vicilin structure so as to confer resistance to proteolysis (Deshpande and Damodaran, 1989).

Studies on the effect of dry heating and autoclaving on the secondary structure of pulse proteins by analysis of the amide I band in the mid-infrared spectrum have recently been performed (Carbonaro et al., 2008a). A high amount of β-sheet conformation was measured for proteins in both *P. vulgaris* L. and *L. culinaris* L. flours (32–47%). Dry heating did not change the secondary structure of lentil proteins, while it greatly decreased β-sheet content of common bean proteins and increased random coil percentage. Upon autoclaving, up to 60% of inter- and intramolecular beta aggregates were observed in proteins from both pulses (Carbonaro et al., 2008a). However, β-sheet structure was still detectable after autoclaving of lentil, but not common bean, proteins. A high amount of β-sheet structure and stabilization of the 11S hexamer of lentil by disulfide bonds are believed to be responsible for the different behavior and may explain the lower small intestinal digestibility that was measured for 11S than for 7S globulins from pulses (Carbonaro et al., 2000, 2005).

Fourier transform infrared studies on soy and buckwheat proteins have indicated an increase in antiparallel β-sheet aggregates upon thermal processing (Prudencio-Ferreira and Areas, 1993; Choi and Ma, 2005). Similar changes in secondary structure have also been observed upon thermal treatment of wheat gliadins (Secundo and Guerrieri, 2005).

Understanding of the relationship between structural properties of pulse proteins and release of bioactive sequences will make it possible to optimize their biological effects, especially from techno-logically processed foods.

14.7.1 Allergenicity of pulse proteins

Although pulse proteins have been recognized to be important food ingredients because of their nutraceutical properties, at the same time, they represent potential food allergens. Indeed, many pulse proteins (storage globulins, 2S albumins, α-amylase inhibitors) have been recognized to have allergenic properties and high stability towards severe treatments aimed at their inactivation (Cuadrado et al., 2009).

Structure and properties common to plant allergens facilitate their grouping into four classes:

- Bet v 1 homologues responsible for cross-reacting pollen–fruit allergy syndromes
- Class I chitinases of fruit and vegetables responsible for cross-reacting latex–fruit allergy syndromes
- A-class proteins: 2S albumins, α-amylase inhibitors, lipid trans-fer proteins
- Seed storage globulins.

Most of these proteins are glycoproteins of low molecular weight (10–20 kDa), with a compact structure stabilized by disulfide bonds. They are thermostable, resistant to proteolysis and show biological activities conferring resistance to pests and pathogens.

Seed storage globulins, such as vicilin from peanut (Ara h 1), lentil (Len c 1) and pea (Pis s 1), belong to the cupin superfamily. They are characterized by a cupin motif that consists of a β-barrel core domain built from two walls of antiparallel β-sheet associated with a loop domain which mainly contains α-helices (Barre et al., 2005). Three cupin motifs are arranged in a homotrimeric structure in a similar way in the three major vicilin allergens Ara h 1, Lens c 1 and Pis s 1. A similar structure has been observed in other storage vicilins: β-conglycinin, phaseolin, canavalin.

Peanut Ara h 1 trimers have been recently described to interact with proanthocyanidins to build up high molecular weight complexes of low digestibility and high allergenicity (Van Boxtel et al., 2007). Non-covalent interaction between procyanidins and Ara h 1 trimers has been explained by the distribution of proline-rich regions on both sides of the proteins.

In vivo gastrointestinal digestion protocols and immunological assays have been used to clarify the role of large digestion-resistant fragments and of the food matrix on the stimulation of the immune system (Moreno, 2007). Although most food allergens maintain their activity after passing through the digestive tract, the relationship between food allergenicity and digestibility is still unclear. Stability to denaturation and proteolysis, and potential for aggregation seem to be important factors, as well as allergen concentration (Lee et al., 2007; Van Boxtel et al., 2008).

Besides conformational epitopes, linear epitopes of 12–14 amino acids that are located in the hydrophobic interior of the protein and become available after digestion in the gut may contribute to the allergenicity of the whole protein.

14.8 Pulse components as nutraceutical ingredients

The bioactive components of pulses have been exploited in the preparation of new or improved food products, in particular bakery products, soups, extruded products, ready-to-eat snacks, infant formulas and dietetic foods. Examples of use of pulse flours, extracts or isolated components as ingredients in pulse-fortified products are given in Table 14.5.

Table 14.5 Examples of use of pulse flours, extracts or isolated components in food formulations

Application study	Reference
Pasta with faba bean protein concentrate	Carnovale and Lombardi (1979)
Gluten-free crackers with pulse flours	Han et al. (2010)
Spaghetti with pea and faba bean flour	Petitot et al. (2010)
Bread and cookies with pea and soy flour	Raidl and Klein (1983); Sadowska et al. (2001); Kamaljit et al. (2010)
Biscuits with defatted soybean	Singh et al. (2000)
Snack bars with chickpea and lentil flours	Meng et al. (2010); Ryland et al. (2010)
Meat products with chickpea and pea flours	Thushan Sanjeewa et al. (2010); Pietrasik and Janz (2010)
Egg albumen with basic soy proteins	Wang and Wang (2009)
Bakery products with lupin galactans	Pfoertiner and Fisher (2001)
White common bean extract (phaseolamin) with high glycemic foods (white bread)	Udani et al. (2009)
Soy protein isolate and zein microspheres for nutraceutical delivery	Chen and Subirade (2009)

Pulse proteins have good functional properties (solubility, foaming, water- and fat-binding capacity). They provide well-balanced essential amino acids when consumed with cereals and food rich in S-amino acids and tryptophan. Protein concentrates (>65% protein) and isolates (>90% protein) can be easily obtained from pulse flours (Fig. 14.2) and then used as ingredients, especially in combination with cereals (Carnovale and Lombardi, 1979). Gluten-free cracker model formulations with good acceptability, physical and nutritional properties have been prepared with chickpea, lentil, pea and navy bean flours and with their protein or fiber isolates (Han et al., 2010). Fortified pasta (spaghetti) has been produced by adding high amounts of pea and faba bean flour (35%) to wheat semolina (Petitot et al., 2010). However, decreases in some quality attributes (cooking loss) have been observed and strategies for improving the products are being developed.

Pea flour (5%) has been successfully incorporated into bread and cookies (Sadowska et al., 2001; Kamaljit et al., 2010), while defatted soy and soy-fortified biscuits could be prepared with standardized levels of ingredients and emulsifiers (Singh et al., 2000). Both field pea and soy flours have been used in chemically leavened quick bread (Raidl and Klein, 1983). Snack bar formulations with desirable texture properties were obtained with extruded chickpea flour and micronized flaked lentils (Meng et al., 2010; Ryland et al., 2010).

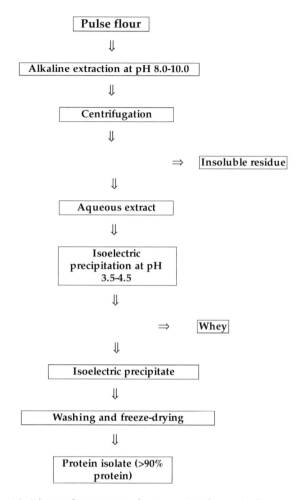

Figure 14.2 Scheme of preparation of protein isolate from pulse flour.

Chickpea and pea flours have also been found to be useful in low-fat emulsion-type meat products because of their high technological functionality (Thushan Sanjeewa et al., 2010; Pietrasik and Janz, 2010). Similarly, soybean basic proteins have been added to egg albumen to improve foaming properties (Wang and Wang, 2009). Biological effects of some of these novel food preparations are currently being tested. Fiber fractions have also often been used as additives in foods in order to decrease their glycemic index, thus improving glucose tolerance, and help in controlling blood cholesterol and triglyceride levels. They have good water- and oil-binding capacity and can be used as a fat replacer. Galactans from lupin have been employed as emulsifying agents in baking (Pfoertner and Fisher, 2001). Soy and other pulse prebiotics (fiber components, β-galacto-sides), being relatively cheap to manufacture or extract, may be used as functional ingredients, as an alternative to probiotics that are difficult to handle in some foodstuffs (Tenorio et al., 2010). However, some studies have indicated no long-term evidence of benefit (lower risk of cardiovascular disease, glycemic control) when isolated dietary fiber components from several sources, pulses included, have been added to functional or manufactured foods (Mann, 2007). On the other hand, a recent study has shown that whole yellow pea flour added to functional foods (banana bread, biscotti and spaghetti) reduced the postprandial glycemic response (Marinangeli et al., 2009).

A standardized protein extract from white common bean (phase-olamin) with α-amylase inhibiting activity has recently been proposed as a novel method for reducing the glycemic index of food (Udani et al., 2009). α-Amylase inhibitor, the active compound in the extract from white bean, acts as a carbohydrate (starch) absorption blocker. It has been demonstrated to work *in vivo* and may have clinical utility as a nutraceutical additive in the therapy of type 2 diabetes and obesity.

In an interesting application, microspheres of soy protein isolate and zein blends have been produced by a cold gelation method to be tested as a possible delivery system for nutraceutical products (Chen and Subirade, 2009). These microspheres have then been demonstrated to provide efficient release of nutraceuticals, such as riboflavin, in a simulated gastrointestinal digestion system (Chen et al., 2010).

When pulse flours, or their extracts, are prepared to be used in food formulations, their composition and safety must be carefully tested. Indeed, analysis of antinutritional factors of chickpea protein concentrates obtained by different procedures (isoelectric precipitation, ultrafiltration) showed that, while phytates and polyphenols were decreased, for both processes the trypsin inhibitor content of the concentrates remained high (Mondor et al., 2009). In addition, potential risks for suceptible subjects are related to the allergenic properties of soybean and other pulse proteins, especially from lentil and chickpea (Pascual et al., 1999). Therefore, both characterization of whole flour or extracts containing the nutraceutical components and assessment of their stability to conventional food processes are recommended, besides testing the chemical, physical and sensory properties of the resulting products.

It is also worth considering that only a few studies have been concerned with the consequence of the interaction of the various bioactive components in a single food or with a comparison between effects of dietary exposure to purified components and whole foods. Available results indicate that the activity of a single compound or the sum of activities of several compounds, often tested in cell cultures, do not reflect the bioactivity of the whole food provided in the diet (Jeffery, 2005; Serra et al., 2009).

14.9 Conclusions

Pulse seeds have been unequivocally proved to represent key functional foods because they contain a number of compounds with potential beneficial health effects; therefore, their regular dietary intake is highly recommended. However, current research needs to be focused on the bioavailability and bioefficacy of specific ingredients in novel pulse-based foods for their efficient use in nutraceutical applications.

Both the food matrix and technological processing affect the absorption and metabolism of phytochemicals by humans in a complex way. These aspects require further studies to plan suitable food formulations for delivering bioactive compounds to humans.

References

Aluko, R.E., 2008. Determination of nutritional and bioactive properties of peptides in enzymatic pea, chickpea, and mung bean protein hydrolysates. J. AOAC Int. 91, 947–956.

Amarowicz, R., Estrella, I., Hernandez, T., 2009. Antioxidant activity of a red lentil extract and its fractions. Int. J. Mol. Sci. 10, 5513–5527.

Anderson, J.W., Major, A.W., 2002. Pulses and lipaemia, short- and long-term effect: potential in the prevention of cardiovascular disease. Br. J. Nutr. 88, S263–S271.

Arese, P., Bosia, A., Naitana, A., Gaetani, S., D'Aquino, M., Gaetani, G.F., 1981. Effect of divicine and isouramil on red cell metabolism in normal and G6PD-deficient (Mediterranean variant) subjects.

Possible role in the genesis of favism. Prog. Clin. Biol. Res. 55, 725–746.

Auger, C., Al-Awwadi, N., Bornet, A., et al., 2004. Catechins and procyanidins in Mediterranean diets. Food Res. Int. 37, 233–245.

Banerjee, S., Wang, L.Y., Sarkar, F.H., 2008. Multi-targeted therapy of cancer by genistein. Cancer Lett. 269, 226–242.

Barac, M.B., Stanojevic, S.P., Pesic, M.B., 2005. Biologically active components of soybean and soy protein products – a review. Acta Period. Technol. APTEF 36, 155–168.

Barre, A., Borges, J.-P., Rougé, P., 2005. Molecular modelling of the major peanut allergen Ara h 1 and other homotrimeric allergens of the cupin superfamily: a structural basis for their IgE-binding cross-reactivity. Biochimie 87, 499–506.

Begbie, R., Pusztai, A., 1989. The resistance to proteolytic breakdown of some plant (seed) proteins and their effects on nutrient utilization and gut metabolism. In: M. Friedman (Ed.), Absorption and utilization of amino acids, Vol. III. CRC Press, Boca Raton, pp. 243–263.

Blachier, F., Lancha, Jr., A.H., Boutry, C., Tomé, D., 2010. Alimentary proteins, amino acids and cholesterolemia. Amino Acids 38, 15–22.

Bravo, L., 1998. Polyphenols: chemistry, dietary sources, metabolism, and nutritional significance. Nutr. Rev. 56, 317–333.

Carbonaro, M., 2006. 7S globulin from *Phaseolus vulgaris* L.: impact of structural aspects on the nutritional quality. Biosci. Biotechnol. Biochem. 70, 2620–2626.

Carbonaro, M., Vecchini, P., Carnovale, E., 1993. Protein solubility of raw and cooked bean (*Phaseolus vulgaris*): role of the basic residues. J. Agric. Food Chem. 41, 1169–1175.

Carbonaro, M., Grant, G., Cappelloni, M., Pusztai, A., 2000. Perspectives into factors limiting *in vivo* digestion of legume proteins: antinutritional compounds or storage proteins? J. Agric. Food Chem. 48, 742–749.

Carbonaro, M., Grant, G., Mattera, M., Aguzzi, A., Pusztai, A., 2001a. Investigation of the mechanisms affecting Cu and Fe bioavailability from legumes. Biol. Trace Elem. Res. 84, 181–196.

Carbonaro, M., Grant, G., Pusztai, A., 2001b. Evaluation of polyphenol bioavailability in rat small intestine. Eur. J. Nutr. 40, 84–90.

Carbonaro M., Mattera M., Cappelloni, M., 2001c. Effect of processing on antinutritional compounds of common bean, faba bean, lentil, chickpea and pea. In: AEP, Proceedings of 4th International Conference on Grain Legumes, Cracow, Poland, pp. 418–419.

Carbonaro, M., Grant, G., Cappelloni, M., 2005. Heat-induced denaturation impairs digestibility of legume (*Phaseolus vulgaris* L. and *Vicia faba* L.) 7S and 11S globulins in the small intestine of rat. J. Sci. Food Agric. 85, 65–72.

Carbonaro, M., Maselli, P., Dore, P., Nucara, A., 2008a. Application of Fourier transform infrared spectroscopy to legume seed flour analysis. Food Chem. 108, 361–368.

Carbonaro, M., Viglianti, A., Forte, M., Nardini, M., 2008b. Characterization of nutritionally active fractions of polyphenols in different legume species. In: Escribano-Bailon, M.T. (Ed.), Polyphenols communications 2008, Vol. 2. Globalia Artes Grafica, Salamanca, pp. 465–466.

Carnovale, E., Lombardi, M., 1979. Emploi d'un concentrat proteique de féverole dans la formulation de pates alimentaires. In: Fabriani, G., Lintas, C. (Eds.), Symposium international sur matières premières et pates alimentaires. CNR, Italy, pp. 127–140.

Carnovale, E., Marconi, E., Carbonaro, M., Marletta, L., 2001. Qualità nutrizionale ed utilizzazione delle leguminose da granella nell'alimentazione umana. In: Ranalli, P. (Ed.), Leguminose e agricoltura sostenibile. Calderini Edagricole, Bologna, pp. 339–379.

Champ, M.-J., 2002. Non-nutrient bioactive substances of pulses. Br. J. Nutr. 88, S307–S319.

Chen, L., Subirade, M., 2009. Elaboration and characterization of soy/zein protein microspheres for controlled nutraceutical delivery. Biomacromolecules 10, 3327–3334.

Chen, L., Hebrard, G., Beyssac, E., Denis, S., Subirade, M., 2010. *In vitro* study of the release properties of soy-zein protein microspheres with a dynamic artificial digestive system. J. Agric. Food Chem. 58, 9861–9867.

Choi, S.M., Ma, C.Y., 2005. Conformational study of globulin from common buckwheat (*Fagopyrum esculentum* Moench) by Fourier transform infrared spectroscopy and differential scanning calorimetry. J. Agric. Food Chem. 53, 8046–8053.

Clemente, A., Domoney, C., 2006. Biological significance of polymorphism in legume protease inhibitors from the Bowman-Birk family. Curr. Prot. Pept. Sci. 7, 201–216.

Clemente, A., Vloque, J., Vloque, R.S., Pedroche, J., Bautista, J., Millan, F., 2000. Factors affecting the *in vitro* protein digestibility of chickpea albumins. J. Sci. Food Agric. 80, 79–84.

Clemente, A., Gee, J.M., Johnson, I.T., Mackenzie, D.A., Domoney, C., 2005. Pea (*Pisum sativum* L.) protease inhibitors from the Bowman-Birk class influence the growth of human colorectal adenocarcinoma HT29 cells *in vitro*. J. Agric. Food Chem. 53, 8979–8986.

Cuadrado, C., Cabanillas, B., Pedrosa, M.M., et al. 2009. Influence of thermal processing on IgE reactivity to lentil and chickpea proteins. Mol. Nutr. Food Res. 53, 1462–1468.

Deshpande, S.S., Damodaran, S.D., 1989. Structure–digestibility relationship of legume 7S proteins. J. Food Sci. 54, 108–113.

Dewell, A., Hollenbeck, P.L.W., Hollenbeck, C.B., 2006. Clinical review: a critical evaluation of the role of soy protein and isoflavone supplementation in the control of plasma cholesterol concentrations. J. Clin. Endocrinol. Metab. 91, 772–780.

Dinelli, G., Bonetti, A., Minelli, M., Marotti, I., Catizone, P., Mazzanti, A., 2006. Content of flavonols in Italian bean (*Phaseolus vulgaris* L.) ecotypes. Food Chem. 99, 105–114.

Diplock, A.T., Aggett, P.J., Ashwell, M., Bornet, F., Fern, E.B., Roberfroid, M.B., 1999. Scientific concepts of functional foods in Europe: consensus document. Br. J. Nutr. 81, S1–S27.

Duodu, K.G., Taylor, J.R.N., Belton, P.S., Hamaker, B.R., 2003. Factors affecting sorghum protein digestibility. J. Cereal Sci. 38, 117–113.

Duranti, M., 2006. Grain legume proteins and nutraceutical properties. Fitoterapia 77, 67–82.

Duranti, M., Lovati, M.R., Dani, V., Barbiroli, A., Scarafoni, A., Castiglioni, S., 2004. The α′ subunit from soybean 7S globulin lowers plasma lipids and upregulates liver β-VLDL receptors in rats fed a hypercholesterolemic diet. J. Nutr. 134, 1334–1339.

Ewen, S.W.B., Bardocz, S., Grant, G., Pryme, I.F., Pusztai, A., 1998. The effects of PHA and misletoe lectin binding to epithelium of rat and mouse gut. In: Bardocz, S., Pfuller, U., Pusztai, A. (Eds.), COST 98 – Effects of antinutrients in the nutritional value of legume diets, Vol. 5. Office Offic. Publ. EC, Luxembourg, pp. 221–225.

Faris, R.J., Wang, H., Wang, T., 2008. Improving digestibility of soy flour by reducing disulfide bonds with thioredoxin. J. Agric. Food Chem. 56, 7146–7150.

Ferruzzi, M.G., Lobo, J.K., Janle, E.M., et al., 2009. Bioavailability of gallic acid and catechins from grape seed polyphenol extract is improved by repeated dosing in rats: implications for treatment in Alzheimer's disease. J. Alzheimers Dis. 18, 113–124.

Flight, I., Clifton, P., 2006. Cereals grains and legumes in the prevention of coronary heart disease and stroke: a review of the literature. Eur. J. Clin. Nutr. 60, 1145–1159.

Food and Drug Administration, 1999. Federal Register 64 (206), 57699–57733.

Frokiaer, H., Barkholt, V., Bagger, C.L., 2001. Processing: impact on seed nutritive value: scientific, technical and economic aspects. In: AEP, Proceedings 4th European Conference on Grain Legumes, Cracow (Poland), pp. 127–131.

Fu, L., Renxiao, W., 2010. Hemolytic mechanism of dioscin by molecular dynamics simulations. J. Mol. Model. 16, 107–118.

Galvez, A., Chen, N., Macasieb, J., de Lumen, B.O., 2001. Chemopreventive property of a soybean peptide (lunasin) that binds to deacetylated histones and inhibits acetylation. Cancer Res. 61, 7473–7478.

Gibson, G.R., Roberfroid, M.B., 1995. Dietary modulation of the human colonic microflora. Introducing the concept of prebiotics. J. Nutr. 125, 1401–1412.

Gilani, G.S., Cockell, K.A., Sepehr, E., 2005. Effects of antinutritional factors on protein digestibility and amino acid availability in foods. J. AOAC Int. 88, 967–987.

Gonzalez, D.E., Prisecaru, I.V., 2005. Lectins and bioactive plant proteins: a potential in cancer treatment. Crit. Rev. Food Sci. Nutr. 45, 425–445.

Greiner, R., Muzquiz, M., Burbano, C., Cuadrado, C., Pedrosa, M.M., Goyaga, C., 2001. Purification and characterization of a phytate-degrading enzyme from germinated faba bean (*Vicia faba* var. Almeda). J. Agric. Food Chem. 49, 2234–2240.

Grela, E.R., Studziaeski, T., Winiarska, A., 2000. Lathyrism in people and animals. Med. Wet. 56, 558–562.

Grela, E.R., Studziaeski, T., Matras, J., 2001. Antinutritional factors in seeds of *Lathyrus sativus* cultivated in Poland. Lath. Lathyr. Newslett. 2, 101–104.

Gueguen, J., 1983. Legume seed protein extraction, processing and end product characteristics. Plant Food Hum. Nutr. 32, 267–303.

Gulewicz, P., Ciesiolka, D., Frias, J., 2000. Simple method of isolation and purification of alpha-galactosides from legumes. J. Agric. Food Chem. 48, 3120–3123.

Han, J., Janz, J.A.M., Gerlat, M., 2010. Development of gluten-free cracker snacks using pulse flours and fractions. Food Res. Int. 43, 627–633.

Hernandez-Ledesma, B., Hsieh, C.C., de Lumen, B.O., 2009. Lunasin, a novel seed peptide for cancer prevention. Peptides 30, 426–430.

Hidalgo, P., Guerra-Hernandez, E., Garcia-Villanova, B., 1997. Determination of insoluble dietary fiber compounds: cellulose, hemicellulose and lignin in legumes. Ars Pharmac. 38, 357–364.

Izumi, T., Piskula, M.K., Osawa, S., 2000. Soy isoflavone aglycones are absorbed faster and in higher amounts than their glucosides in humans. J. Nutr. 130, 1659–1699.

Jeffery, E., 2005. Component interactions for efficacy of functional foods. J. Nutr. 135, 1223–1225.

Jeong, H.J., Park, H.J., Lam, Y., de Lumen, B.O., 2003. Characterization of lunasin isolated from soybean. J. Agric. Food Chem. 51, 7901–7906.

Jeong, H.J., Lee, J.R., Jeong, J.B., Park, J.H., Cheong, Y.K., de Lumen, B.O., 2009. The cancer preventive seed peptide lunasin from rye is bioavailable and bioactive. Nutr. Cancer 61, 680–686.

Kamaljit, K., Baljeet, S., Amarjeet, K., 2010. Preparation of bakery products by incorporating pea flour as a functional ingredient. Am. J. Food Technol. 5, 130–135.

Kim, S.E., Kim, H.H., Kim, J.Y., Woo, H.J., Lee, H.J., 2000. Anticancer activity of hydrophobic peptides from soy proteins. BioFactors 12, 151–155.

Kitts, D.D., Weiler, K., 2003. Bioactive proteins and peptides from food sources: applications of bioprocesses used in isolation and recovery. Curr. Pharm. Des. 9, 1309–1323.

Lallès, J.P., Jansman, A.J.M., 1993. Recent progress in the understanding of the mode of action and effects of antinutritional factors from legume seeds in non-ruminant farm animals. In: Huisman, J., Van der Poel, A.F.B., Liener, I.E. (Eds.), Recent advances of research in antinutritional factors in legume seeds. Pudoc, Wageningen, pp. 219–232.

Landis-Piwowar, K.R., Dou, Q.P., 2008. Polyphenols: biological activities, molecular targets, and the effect of methylation. Curr. Mol. Pharmacol. 1, 233–243.

Lee, H.W., Keum, E.H., Lee, S.J., 2007. Allergenicity of proteolytic hydrolysates of the soybean 11S globulin. J. Food Sci. 72, 168–172.

Lei, H.Y., Chang, C.P., 2009. Lectin of concanavalin A as an anti-hepatoma therapeutic agent. J. Biomed. Sci. 16, 10.

Liener, I.E., 1979. Significance for humans of biologically active factors in soybeans and other food legumes. J. Am. Oil Chem. Soc. 56, 121–129.

Lin, L., Harnly, J.M., Pastor-Corrales, M.A., Luthria, D.L., 2008. The poliphenolic profiles of common bean (*Phaseolus vulgaris* L.). Food Chem. 107, 399–410.

Lovati, M.R., Manzoni, C., Gianazza, E., Sirtori, C., 1998. Soybean protein products as regulators of liver low-density lipoprotein receptors. J. Agric. Food Chem. 46, 2472–2480.

Lu, C.D., Jorgensen, N.A., 1987. Alfalfa saponins effect: site and extent of nutrient digestion in ruminants. J. Nutr. 117, 919–927.

Luthria, D.L., Pastor-Corrales, M.A., 2006. Phenolic acid content of fifteen dry edible beans (*Phaseolus vulgaris* L.) varieties. J. Food Comp. Anal. 19, 205–211.

MacCarty, M.F., 2005. Nutraceutical resources for diabetes prevention. Med. Hypotheses 64, 151–158.

Macfarlane, G.T., Steed, H., Macfarlane, S., 2008. Bacterial metabolism and health-related effects of galacto-oligosaccharides and other prebiotics. J. Appl. Microbiol. 104, 305–344.

Mallillin, A.C., Trinidad, T.P., Raterta, R., Dagbay, K., Loyola, A.S., 2008. Dietary fiber and fermentability characteristics of root crops and legumes. Br. J. Nutr. 100, 485–488.

Manach, C., Williamson, G., Morand, C., Scalbert, A., Rémésy, C., 2005. Bioavailability and bioefficacy of polyphenols in humans. I. Review of 97 bioavailability studies. Am. J. Clin. Nutr. 81, 230S–242S.

Mann, J., 2007. Dietary carbohydrate: relationship to cardiovascular disease and disorders of carbohydrate metabolism. Eur. J. Clin. Nutr. 61, S100–S111.

Marcil, V., Delvin, E., Seidman, E., 2002. Modulation of lipid synthesis, apolipoprotein biogenesis, and lipoprotein assembly by butyrate. Am. J. Physiol. Gastrointest. Liver Physiol. 283, G340–G346.

Marinangeli, C.P.F., Cassis, A.N., Jones, P.J.H., 2009. Glycemic responses and sensory characteristics of whole yellow pea flour added to novel functional foods. J. Food Sci. 74, S385–S389.

Marletta, L., Carbonaro, M., Carnovale, E., 1992. *In vitro* protein and sulphur amino acid availability as a measure of bean protein quality. J. Sci. Food Agric. 59, 497–504.

Martinez-Villaluenga, C., Frias, J., Vidal-Valverde, C., 2008a. Alpha-galactosides: antinutritional factors or functional ingredients? Crit. Rev. Food Sci. Nutr. 48, 301–316.

Martinez-Villaluenga, C., Bringe, N.A., Berhow, M.A., Gonzalez de Mejia, E., 2008b. Beta-conglycinin embeds active peptides that inhibit lipid accumulation in 3T3-L1 adipocytes *in vitro*. J. Agric. Food Chem. 56, 10533–10543.

Meng, X., Threinen, D., Hansen, M., Driedger, D., 2010. Effect of extrusion conditions on system parameters and physical properties of a chickpea flour-based snack. Food Res. Int. 43, 650–658.

Milner, J.A., 2004. Molecular targets for bioactive food components. J. Nutr. 134, 2492S–2498S.

Mondor, M., Aksay, S., Drolet, H., Roufik, E., Boye, J.I., 2009. Influence of processing on composition and antinutritional factors of chickpea protein concentrates produced by isoelectric precipitation and ultrafiltration. Innovat. Food Sci. Emerg. Technol. 10, 342–347.

Moreno, F.J., 2007. Gastrointestinal digestion of food allergens. Effect on their allergenicity. Biomed. Pharmacother. 87, 1576S–1581S.

Muzquiz, M., 2004. Recent advances in antinutritional factors in legume seeds and oilseeds. Chipsbooks.

Nakamori, T., 2002. Antiobesity effects of soy proteins and soy peptides. Food Style 21, 86–88.

Nielsen, S.S., Deshpande, S.S., Hermodson, M.A., Scott, M.P., 1988. Comparative digestibility of legume storage proteins. J. Agric. Food Chem. 36, 896–902.

Oatway, L., Vasanthan, T., Helm, J.H., 2001. Phytic acid. Food Rev. Int. 17, 419–431.

Obiro, W.C., Zhang, T., Jiang, B., 2008. The nutraceutical role of *Phaseolus vulgaris* alpha-amylase inhibitors. Br. J. Nutr. 100, 1–12.

Onomi, S., Okazaki, Y., Katayama, T., 2004. Effect of dietary level of phytic acid on hepatic and serum lipid status in rats fed a high-sucrose diet. Biosci. Biotechnol. Biochem. 68, 1379–1381.

Pascual, C.Y., Fernandez-Crespo, J., Sanchez-Pastor, S., 1999. Allergy to lentil in Mediterranean pediatric patients. J. Allergy Clin. Immunol. 103, 154–158.

Pena-Ramos, E.A., Xiong, Y.L., 2002. Antioxidant activity of soy protein hydrolysates in a liposomal system. J. Food Sci. 76, 2952–2956.

Petitot, M., Boyer, L., Minier, C., Micard, V., 2010. Fortification of pasta with split pea and faba bean flours: pasta processing and quality evaluation. Food Res. Int. 43, 634–641.

Pfoertner, H.P., Fisher, J., 2001. Dietary fibres of lupins and other grain legumes. In: McCleary, B.V., Prosky, L. (Eds.), Advanced dietary fiber technology. Blackwell Science, Oxford, pp. 361–366.

Pietrasik, Z., Janz, J.A.M., 2010. Utilization of pea flour, starch-rich and fiber-rich fractions in low fat bologna. Food Res. Int. 43, 602–608.

Prudencio-Ferreira, S.H., Areas, J.A.G., 1993. Protein–protein interactions in the extrusion of soy at various temperatures and moisture contents. J. Food Sci. 58, 378–384.

Pryme, I.F., Bardocz, S., Pusztai, A., Ewen, S.W.B., 2006. Suppression of growth of tumour cell lines *in vitro* and tumours *in vivo* by mistletoe lectins. Histol. Histopathol. 21, 285–299.

Pusztai, A., Grant, G., Bardocz, S., Bainter, K., Gelencser, E., Ewen, S.W.B., 1997. Both free and complexed trypsin inhibitors stimulate pancreatic seretion and change duodenal enzyme levels. Am. J. Physiol. 272, G340–G350.

Raidl, M.A., Klein, B.P., 1983. Effect of soy or field pea flour substitution on physical and sensory characteristics of chemically leavened quick bread. Cereal Chem. 60, 367–370.

Rao, S.L.N., 1978. A sensitive and specific colorimetric method for the determination of α,β-diaminopropionic acid and the *Lathyrus sativus* neurotoxin. Anal. Biochem. 86, 386–395.

Reddy, B.S., Hirose, Y., Cohen, L.A., Simi, B., Cooma, I., Rao, C.V., 2000. Preventive potential of wheat bean fractions against experimental colon carcinogenesis: implications for human colon cancer prevention. Cancer Res. 60, 4792–4797.

Reddy, I.M., Kella, N.K.D., Kinsella, J.E., 1988. Structural and conformational basis of the resistance of β-lactoglobulin to peptic and chymotryptic digestion. J. Agric. Food Chem. 36, 737–741.

Rideout, T.C., Harding, S.V., Jones, P.J., Fan, M.Z., 2008. Guar gum and similar soluble fibers in the regulation of cholesterol metabolism: current understandings and future research priorities. Vasc. Health Risk Manag. 4, 1023–1033.

Rochfort, S., Panozzo, J., 2007. Phytochemicals for health, the role of pulses. J. Agric. Food Chem. 55, 7981–7994.

Rotter, R.G., Marquardt, R.R., Low, R.K.-C., Briggs, C.J., 1990. Influence of autoclaving on the effects of *Lathyrus sativus* feed to chicks. Can. J. Anim. Sci. 70, 739–741.

Ryland, D., Vaisey-Genser, M., Arntfield, S.D., Malcolmson, L.J., 2010. Development of a nutritious acceptable snack bar using micronized flaked lentils. Food Res. Int. 43, 642–649.

Sadowska, J., Blaszczak, W., Fornal, J., Vidal-Valverde, C., Frias, J., 2001. Effect of pea flour addition on bread quality and structure. In: AEP, Proceedings of 4th European conference on grain legumes, Cracow (Poland), pp. 142–143.

Savage, G.P., Deo, S., 1989. The nutritional value of peas (*Pisum sativum*). A literature review. Nutr. Abs. Rev. 59, 66–83.

Scalbert, A., Manach, C., Morand, C., Rémésy, C., Jiménez, L., 2005. Dietary polyphenols and the prevention of diseases. Crit. Rev. Food Sci. Nutr. 45, 287–306.

Scarafoni, A., Magni, C., Duranti, M., 2007. Molecular nutraceutics as a mean to investigate the positive effects of legume seed proteins on human health. Trends Food Sci. Technol. 18, 454–463.

Schwenke, K.D., 1997. Enzyme and chemical modification of proteins. In: Damodaran, S., Paraf, A. (Eds.), Food proteins and their applications. Marcel Dekker, New York, pp. 393–423.

Secundo, F., Guerrieri, N., 2005. ATR-FT/IR study on the interactions between gliadins and dextrins and their effects on protein secondary structure. J. Agric. Food Chem. 53, 1757–1764.

Semino, G.A., Restani, P., Cerletti, P., 1985. Effect of bound carbohydrates on the action of trypsin on lupin seed glycoprotein. J. Agric. Food Chem. 33, 196–199.

Serra, A., Macia, A., Romero, M.P., 2009. Bioavailability of procyanidin dimers and trimers and matrix food effects in *in vitro* and *in vivo* models. Br. J. Nutr. 14, 1–9.

Serrano, J., Puupponen-Pimia, R., Dauer, A., Aura, A.-M., Saura-Calixto, F., 2009. Tannins: current knowledge of food sources, intake, bioavailability and biological effects. Mol. Nutr. Food Res. 53, S310–S329.

Shewry, P.R., Halford, N.G., 2002. Cereal seed storage proteins: structures, properties and role in grain utilization. J. Exp. Bot. 53, 947–958.

Sievenpiper, J.L., Kendall, C.W., Esfahani, A., 2009. Effect of non-oilseed pulses on glycaemic control: a systematic review and meta-analysis of randomised controlled experimental trials in people with and without diabetes. Diabetologia 52, 1479–1495.

Singh, R., Singh, G., Chauhan, G.S., 2000. Nutritional evaluation of soy fortified biscuits. J. Food Sci. Technol. 37, 162–164.

Sodipo, O.A., Arinze, H.U., 1985. Saponin content of some Nigerian foods. J. Sci. Food Agric. 36, 407–408.

Taylor, C.K., Levy, R.M., Elliott, J.C., Burnett, B.P., 2009. The effect of genistein aglycone on cancer and cancer risk: a review of *in vitro*, preclinical, and clinical studies. Nutr. Rev. 67, 398–415.

Tenorio, M.D., Espinosa-Martos, I., Préstamo, G., Rupérez, P., 2010. Soybean whey enhance mineral balance and caecal fermentation in rats. Eur. J. Nutr. 49, 155–163.

Thushan Sanjeewa, W.G., Wanasundara, P.D., Pietrasik, Z., Shand, P.J., 2010. Characterization of chickpea (*Cicer arietinum* L.) flours and application in low-fat pork bologna as a model system. Food Res. Int. 43, 617–626.

Trinidad, T.P., Mallillin, A.C., Loyola, A.S., Sagum, R.S., Encabo, R.R., 2010. The potential health benefits of legumes as a good source of dietary fibre. Br. J. Nutr. 103, 569–574.

Udani, J.K., Singh, B.B., Barrett, M.L., Preuss, H.G., 2009. Lowering the glycemic index of white bread using a white bean extract. Nutr. J. 8, 52–56.

Urga, K., Fufa, H., Biratu, E., Husain, A., 2005. Evaluation of *Lathyrus sativus* cultivated in Ethiopia for proximate composition, minerals, β-ODAP and anti-nutritional components. Afr. J. Food Agric. Nutr. Dev. 5, 1–15.

Utsumi, S., Damodaran, S., Kinsella, J.E., 1984. Heat-induced interactions between soybean proteins: preferential association of 11S basic subunits and β subunits of 7S. J. Agric. Food Chem. 32, 1406–1412.

Valdebouze, P., Bergeron, E., Gaborit, T., Delort-Laval, J., 1980. Content and distribution of trypsin inhibitors and haemagglutinins in some legume seeds. Can. J. Plant Sci. 60, 695–701.

Van Boxtel, E.L., Van den Broek, L.A.M., Koppelman, S.J., Vincken, J.-P., Gruppen, H., 2007. Peanut allergen Ara h 1 interacts with proanthocyanidins into higher molecular weight complexes. J. Agric. Food Chem. 55, 8772–8778.

Van Boxtel, E.L., Van den Broek, L.A.M., Koppelman, S.J., Gruppen, H., 2008. Legumin allergens from peanuts and soybeans: effects of denaturation and aggregation on allergenicity. Mol. Nutr. Food Res. 52, 674–682.

Van Loo, J., Cummings, J., Delzenne, N., 1995. Functional food properties of non-digestible oligosaccharides: a consensus report from the ENDO project (DGXII AIRII-CT94-1095). Br. J. Nutr. 81, 121–132.

Wan, X.S., Serota, D.G., Ware, J.H., Crowell, J.A., Kennedy, A.R., 2002. Detection of Bowman-Birk inhibitor antibodies in sera of humans and animals treated with Bowman-Birk inhibitor concentrate. Nutr. Cancer 43, 167–173.

Wang, G., Wang, T., 2009. Improving foaming properties of yolk-contaminated egg albumen by basic soy protein. J. Food Sci. 74, C581–C587.

Xu, B.J., Chang, S.K.C., 2008. Effect of soaking, boiling, and steaming on total phenolic content and antioxidant activities of cool season food legumes. Food Chem. 110, 1–13.

Quality standards and evaluation of pulses

15

Mahesh Gupta[1], Brijesh Tiwari[2], Amarinder Singh Bawa[3]
[1]School of Food Science and Environmental Health, Dublin Institute of Technology, Dublin, Ireland
[2]Department of Food and Tourism, Manchester Metropolitan University, Manchester, UK
[3]Defence Food Research Laboratory, Siddartha Nagar, Mysore, India

15.1 Introduction

The mature grains of grain legumes are marketed as dry products usually called pulses in the trade and industry sectors. Pulses may be defined as the dried edible seeds of cultivated legumes. Production of and international trade in pulse crops, particularly dry peas, lentils, chickpeas and beans, have increased dramatically over the past 15 years (Hofer, 2004). The principal marketing countries have been Australia, Canada, the USA and the European Union. Pulses provide protein, fiber, complex carbohydrates, vitamins and minerals to the diet and are an important part of vegetarian diets. While standards for nomenclature and quality measurement for pulses exist between some buyers and sellers of pulses, these do not have universal recognition. It is possible that in as many as 30% of trades involving pulses, differences in methodology and nomenclature between buyers and sellers may result in disputes requiring arbitration. This suggests a critical need

Pulse Foods: Processing, Quality and Nutraceutical Applications. DOI: 10.1016/B978-0-1238-2018-1.00007-0

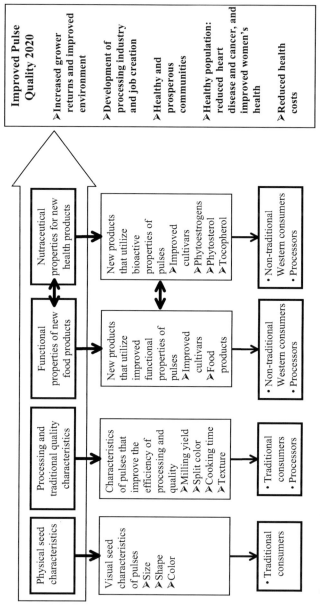

Figure 15.1 Developing quality pulses for a sustainable environment, population and community (Panozzo and Materne, 2007).

for developing standard methods for testing the quality characteristics of pulses and a common international language for trade. Figure 15.1 represents the changing emphasis for plant breeding and consumer demand for developing quality pulses for a sustainable environment, population and community (Jenson, 2002). Providing more definitive quality assessment methods would reduce trade disputes. Pulse grain quality attributes can be broadly classified into two categories: (1) extrinsic and (2) intrinsic quality parameters (Table 15.1). These quality attributes play an important role in pulse production and processing chain. In general, initial quality assessment is common for most of the grains, including cereals. Pulse quality definition around the word has been confusing until now because of several reasons such as:

1. The problem of language
2. Variation in consumption pattern
3. Differences in national specification and
4. Import and export specifications. Some national standards for quality of pulses exist in certain countries such as the UK (British Edible Pulse Association), Australia (Pulse Australia) and Canada (Pulse Canada).

In developing countries, legumes are the most important high-protein foods and play the role that is played in rich countries by meat and other animal products (Singh and Singh, 1992). Pulses provide 67% calories and a very substantial portion of protein to the daily requirement of the human body. In general, the role of pulses in the human diet is attributed to both the quality and quantity of proteins.

Table 15.1 Quality parameter indices for pulse commercial quality

Extrinsic quality parameters (wholesomeness, soundness and safety)			Intrinsic quality parameters (technological, physicochemical and nutritional quality)	
Presence of impurities	Grain defects	Contaminants	Technological	Physicochemical criteria
Broken kernels	Moldy grains	Pesticide residues	Moisture	Protein content
Sprouted kernels	Insect damaged	Mycotoxins	Test weight	Starch content
Foreign matter impurities	Sprout damaged	Heavy metals	1000-kernel weight	Starch quality
Inorganic matter	Shriveled kernels	Noxious seeds	Kernel size	Lipids
Weed seeds	Diseased kernels	Radioactivity	Color	Fiber (cellulosic)
Other grains	Heat damaged		Hardness	Total ash content
Dead insects	Visual defects			Enzyme activity

Adapted from Fleurat-Lessard (2002).

According to the Food and Agriculture Organization (FAO) of the United Nations pulses shall meet the following general requirements/limits as determined using the relevant standards listed below:

1. Grain shall be the dried mature seeds of pulse.
2. Grain shall be sweet, well-filled, smooth, hard, clean, wholesome, uniform in size, shape, color and in sound merchantable condition.
3. Grain shall be free from a substance which renders them unfit for human or animal consumption or processing into or utilization thereof as food or feed.
4. Grain shall be free of pests, live animals, animal carcasses, animal droppings, fungus infestation, added coloring matter, molds, weevils, obnoxious substances, glass, metal, coal, dung, discoloration and all other impurities except to the extent indicated in this standard and must meet any other phytosanitary requirements specified by the importing country authority.
5. Grain shall be free from filth (impurities of plant and animal origin including insects, rodent hair and excreta) in amounts that represent a hazard to human health.
6. Grain shall be free from toxic or noxious seeds that are commonly recognized as harmful to health.
7. Grain shall be free from abnormal flavors, musty, sour or other undesirable odor, obnoxious smell and discoloration.
8. Grain shall be free from microorganisms and substances originating from microorganisms, fungi or other poisonous or deleterious substances in amounts that may constitute a hazard to human health.
9. Grain shall contain no chemical residues, which exceed the prescribed maximum residue limit, provided that:
   (a) The prescribed maximum residue limit of the importing country shall be complied with; and
   (b) The export documents are accordingly endorsed with the name of the importing country.
10. Grain shall contain not more than 10 microgram per kilogram aflatoxin of which not more than 5 microgram per kilogram may be aflatoxin B1, provided that:
    (a) The prescribed maximum aflatoxin limit of the importing country shall be complied with
    (b) The export documents are accordingly endorsed with the name of the importing country; and
    (c) An inspector shall verify compliance to the levels of aflatoxin by sampling and submitting samples for analysis of only certain consignments according to a risk-based plan.

11. Grain shall comply with the requirements for declared plant injurious organisms of phytosanitary importance as determined by the competitive authority.

15.2 Pulse quality standard parameters

Unlike cereals, there are no specific quality standards for pulses, with a lack of common nomenclature and various types of market grades of pulses. A common approach for measuring quality is also lacking. There are certain initiatives for the development of pulse quality standards by the International Pulse Quality Committee and Codex Alimentarius.

15.2.1 International Pulse Quality Committee (IPQC)

The International Pulse Quality Committee (IPQC) was created in 2000 with a vision to facilitate the marketing of pulses through the development of standardized nomenclature and methods of testing quality parameters in order to meet customer needs. Members of the IPQC include Canada, the USA, Australia and Europe, representing the scientific community, and growers and traders of the world's leading pulse exporting countries. One goal of the IPQC was to assign nomenclature for each type and class of pulse crop to end the confusion that exists due to the multitude of names currently in use. In addition, there were very few current internationally accepted methods for evaluating pulse quality. Pulse quality standard parameters identified by IPQC are listed in Table 15.2 and specifications for common pulses by the Codex Alimentarius commission are listed in Table 15.3. The IPQC identified 13 quality parameters aimed at creating common measurements for everything from seed size, moisture and crude protein, to dehulling efficiency, and cooking and canning qualities (Table 15.2). The IPQC is currently working on several projects that will provide replicable testing methods for quality parameters, such as color, texture, taste, cooking time, and splitting and milling ability of the commonly traded pulses. The type of work required ranges from complete development of new methods to verification and

Table 15.2 Quality parameters identified by the International Pulse Quality Committee (IPQC)

S. no.	Parameters
1	Seed size
2	Moisture
3	Crude protein
4	Fiber
5	Starch
6	Water absorption
7	Split yield and dehulling efficiency
8	Cooking quality and time
9	Trypsin inhibitor activity (TIA)
10	Seed coat integrity
11	Tannins

development of internationally recognized precision and accuracy for established methods.

15.2.2 Codex Alimentarius Commission (CAC)

The Codex Alimentarius Commission is an intergovernmental body with over 170 members, within the framework of the Joint FAO/WHO Food Standards Program established by the FAO and the World Health Organization (WHO), with the purpose of protecting the health of consumers and ensuring fair practices in the food trade. The Commission also promotes coordination of all food standards work undertaken by international governmental and non-governmental organizations. The CAC standard applies to the whole, shelled or split pulses defined below which are intended for direct human consumption. The standard does not apply to pulses intended for factory grading and packaging, industrial processing, or to those pulses intended for use in the feeding of animals. It does not apply to fragmented pulses when sold as such, or to other legumes for which separate standards may be elaborated.

15.2.2.1 CAC pulses standard

According to CAC, pulses are dry seeds of leguminous plants which are distinguished from leguminous oil seeds by their low fat content. The pulses covered by Codex Standard (CODEX STAN 171-1989) are the following:

Table 15.3 Specifications for pulses

Parameter	Requirements			Method of test
	Grade 1	Grade 2	Grade 3	
Physical characteristics	The pulse grains shall be dried and mature. They shall be sweet, clean, wholesome, uniform in size, shape, color and in sound merchantable conditions			ISO 605
Purity, % min by wt Whole, defective whole pulses other than specified type	99.0	99.0	99.0	ISO 605
Moisture, % max by wt	10.0	12.0	12.0	ISO 24557
Size grading 98% must be within a 2 mm range based on slotted sieves and within this range 75% must be within 0.8 mm. (Seed size will be recorded on the certificate)	98% 75% must be in 0.8 mm range	9% 75% must be in 0.8 mm range	n/a	EAS 46; 4.3.2
Defective, % max by wt Whole pulses grain not of the specified variety. Grains that are bin burnt, broken, caked, chipped, total damaged, heat damaged, insect damaged, shriveled, split, sprouted, weather damaged, total foreign matter, contrasting classes, wrinkled and affected by mold (field or storage). Includes pods that contain green grams, whether broken or unbroken, and loose seed coat	2.0% max by wt	4.0% max by wt	6.0% max by wt	
Immature grain	2.0% max by wt	3.0% max by wt	4.0% max by wt	
Contrasting classes Beans with more than 2.0% contrasting classes are graded as mixed beans	0.5% max by wt	1.0% max by wt	2.0% max by wt	
Classes that blend Beans with more than 15.0% classes that blend are graded as mixed beans	5.0% max by wt	10.0% max by wt	15.0% max by wt	
Germination excluding hard seeds	90%	n/a	n/a	
Sprout test for sprouting green grams	Suitable	n/a	n/a	
Poor color Seed coat or kernel which is distinctly off color from the characteristic color of the predominating class of the specified type	1% max by wt	1% max by wt	3% max by wt	

(Continued)

Table 15.3 Specifications for pulses – continued

Parameter	Requirements			Method of test
	Grade 1	Grade 2	Grade 3	
Foreign material Unmillable material and all vegetable matter other than green gram seed material	0.5% max by wt, includes 0.3% max by wt unmillable material	1.0% max by wt, includes 0.5% max by wt unmillable material	1.5% max by wt, includes 0.75% max by wt unmillable material	
Other edible grains Any edible grains (including oil seeds) other than the one which is under consideration	0.1% max by wt	0.5% max by wt	3.0% max by wt	
Weevilled grains Grains per cent by count clean-cut weevil bored	0.1% max by wt	0.2% max by wt	0.5% max by wt	
Unmillable material Soil, stones, metals and non-vegetable matter	0.3% max by wt (of which 0.2% soil, stones)	0.5% max by wt (of which 0.4% soil, stones)	0.75% max by wt (of which 0.6% soil, stones)	EAS 46; 4.3.3
Snails Dead or alive. Whole or substantially whole (more than half) including bodies 200 g^{-1} sample	Nil tolerance	Nil tolerance	Nil tolerance	
Field insects Dead or alive 200 g^{-1} sample	Nil tolerance	Nil tolerance	Nil tolerance	EAS 46
Foreign seeds	0.3% max by wt	0.5% max by wt	0.5% max by wt	EAS 46
Objectionable material Includes objectionable odor, taste	Nil tolerance	Nil tolerance	Nil tolerance	EAS 46
Ryegrass ergot Pieces laid end to end 200 g^{-1} sample	Nil tolerance	Nil tolerance	Nil tolerance	
Uric acid Maximum kg^{-1} sample	100 ml	100 ml	100 ml	
Aflatoxin – Total ppb (Total aflatoxin including (AFB1 + AFB2 + AFG1 + AFG2))	10	10	10	ISO 16050
Aflatoxin B1 only, ppb	5	5	5	ISO 16050
Fumonison – Total ppb (Total fumonison including (FB1 + FB2 + FB3))	5	5	5	ISO 16050

NOTE Foreign matter is mineral or organic matter (dust, twigs, seed coats, seeds of other species, dead insects, fragments, or remains of insects, other impurities of animal origin). Green grams shall have not more than 1% extraneous matter of which not more than 0.25% shall be mineral matter and not more than 0.10% shall be dead insects, fragments or remains of insects, and/or other impurities of animal origin.

Codex Alimentarius and EAC standards for pulses.

- Beans of *Phaseolus* spp. (except *Phaseolus mungo* L. syn. *Vigna mungo* (L.))
- Hepper and *Phaseolus aureus* Roxb. syn. *Phaseolus radiatur* L., *Vigna radiata* (L.) Wilczek
- Lentils of *Lens culinaris* Medic. syn. *Lens esculenta* Moench.
- Peas of *Pisum sativum* L.
- Chickpeas of *Cicer arientinum* L.
- Field beans of *Vicia faba* L.
- Cowpeas of *Vigna unguiculata* (L.) Walp., syn. *Vigna sesquipedalis* Fruhw., *Vigna sinensis* (L.) Savi exd Hassk.

15.2.2.2 Moisture content

CAC allows two maximum moisture levels considering variations in climatic conditions and marketing practices. Lower levels of moisture are suggested for tropical climates or when long-term storage is a normal commercial practice. Higher moisture levels are suggested for more moderate climates or when other short-term storage is the normal commercial practice as given in Table 15.4. These moisture levels are for whole pulses (unhulled); however, in the case of dehulled pulses (without seed coat) moisture content shall be 2% (absolute).

15.2.2.3 Extraneous matter

Extraneous matter is mineral or organic matter (dust, twigs, seed coats, seeds of other species, dead insects, fragments, or remains of insects, other impurities of animal origin). Pulses shall have not more than 1% extraneous matter of which not more than 0.25% shall be mineral matter and not more than 0.10% shall be dead insects, fragments or remains of insects, and/or other impurities of animal origin.

Table 15.4 Moisture levels (%) for certain pulses

Pulses	Lower level	Higher level
Beans, field beans	15	19
Lentils	15	16
Peas, cowpeas	15	18
Chickpeas	14	16

15.2.2.4 Toxic or noxious seeds

The products covered by the provisions of this standard shall be free from the following toxic or noxious seeds in amounts which may represent a hazard to human health:

- Crotolaria (*Crotalaria* spp.), corn cockle (*Agrostemma githago* L.), castor bean (*Ricinus communis* L.), Jimson weed (*Datura* spp.) and other seeds that are commonly recognized as harmful to health.

15.2.2.5 Contaminants

15.2.2.5.1. Heavy metals Pulses shall be free from heavy metals in amounts which may represent a hazard to health. Table 15.5 shows limits for heavy metal contaminants based on the parameter limit test method.

15.2.2.5.2. Pesticide residues Pulses shall comply with those maximum residue limits established by the CAC for this commodity. Table 15.6 lists the maximum pesticide residue limits and extraneous maximum residue limits in pulses.

15.2.2.5.3. Mycotoxins Pulses shall comply with those maximum mycotoxin limits established by the CAC for this commodity.

15.2.2.6 Hygiene

It is recommended that the products covered by the provisions of this standard be prepared and handled in accordance with the appropriate sections of the *Recommended International Code of Practice – General Principles of Food Hygiene* (CAC/RCP 1-1969) and other Codes of Practice recommended by the CAC which are relevant to these products.

Table 15.5 Heavy metal contaminant limits (parameter limit test method)

S. no.	Parameter	Limit	Test method
1	Arsenic (As), ppm max.	0.10	
2	Copper (Cu), ppm max.	5.0	
3	Lead (Pb), ppm max.	0.20	EAS 41
4	Cadmium (Cd), ppm max.	0.02	
5	Mercury (Hg), ppm max.	0.01	

Codex Alimentarius and EAC standards for pulses.

Table 15.6 Maximum pesticide residue limits and extraneous maximum residue limits in pulses (current as at 2009-06-09)

S. no.	Type	Unit symbol	Limit
1	Diquat	mg kg^{-1}	0.2
2	Fludioxonil	Undef	0.07
3	Glufosinate-ammonium	mg kg^{-1}	3
4	Glyphosate	Undef	5
5	Methidathion	mg kg^{-1}	0.1
6	Methiocarb	mg kg^{-1}	0.1
7	Parathion-methyl	mg kg^{-1}	0.3
8	Pyraclostrobin	Undef	0.3
9	Quintozene	mg kg^{-1}	0.01

Codex Alimentarius and EAC standards for pulses.

To the extent possible in good manufacturing practice, the products shall be free from objectionable matter. When tested by appropriate methods of sampling and examination, the products:

1. Shall be free from microorganisms in amounts which may represent a hazard to health
2. Shall be free from parasites which may represent a hazard to health; and
3. Shall not contain any substance originating from microorganisms in amounts which may represent a hazard to health. Microbiological limits for pulses are listed in Table 15.7.

15.2.2.7 Packaging
1. Pulses shall be packaged in containers which will safeguard the hygienic, nutritional, technological and organoleptic qualities of the product.

Table 15.7 Microbiological limits for pulses

S. no.	Type of microorganism	Limits	Test method
1	Yeasts and molds, max. g^{-1}	102	
2	S. aureus 25 g^{-1}	Nil	EAS 217
3	E. coli, max. g^{-1}	Nil	
4	Salmonella, max. 25 g^{-1}	Nil	

Codex Alimentarius and EAC standards for pulses.

2. The containers, including packaging material, shall be made of substances which are safe and suitable for their intended use. They should not impart any toxic substance or undesirable odor or flavor to the product.
3. When the product is packaged in sacks, these must be clean, sturdy and strongly sewn or sealed.

15.2.2.8 Labeling

In addition to the requirements of the Codex *General Standard for the Labeling of Prepackaged Foods* (CODEX STAN 1-1985), the following specific provisions apply:

- *Name of the product:* the name of the product to be shown on the label shall be the commercial type of the pulse.
- *Labeling of non-retail containers:* information for non-retail containers shall either be given on the container or in accompanying documents, except that the name of the product, lot identification and the name and address of the manufacturer or packer shall appear on the container. However, lot identification and the name and address of the manufacturer or packer may be replaced by an identification mark, provided that such a mark is clearly identifiable with the accompanying documents.

15.2.3 The AACC Pulse and Grain Legume Technical Committee (TC)

The American Association of Cereal Chemists (AACC) is a non-profit international organization of nearly 4000 members who are specialists in the use of cereal grains in foods. The AACC has been an innovative leader in gathering and disseminating scientific and technical information to professionals in the grain-based foods industry worldwide for over 85 years. AACC has modified its mandate to include pulses and was very interested in developing a committee to promote pulse-oriented methodology. In 2002, an AACC Pulse and Grain Legume Technical Committee (TC) was established. One of the goals for the Pulse and Grain Legume TC is to act on developing new methods and revising existing methods for pulses. Another goal is to organize technical sessions on pulses and grain legumes at the AACC annual meetings. Methods developed by the IPQC will be tested and verified by the AACC Pulse and Grain Legume TC to ensure that the methods are valid internationally. The AACC will also publish the methods so that they are accredited and available for laboratories around the world.

15.2.4 Standardized names and seed testing

The IPQC is creating a standardized naming system (nomenclature) for pulse crops. While local names may vary from region to region, the pulse industry will have an international standardized system that can be used for each type of pulse crop, based on seed shape and color, seed coat characteristics, and the color of split seed. These internationally recognized names for each of the major pulse crops will be available on the Internet and this international naming system will help all countries communicate for trade. There has been good progress on developing standard testing methods for evaluating pulse quality. Common testing methods for measuring quality will help define the characteristics that each country requires in its pulses. Pulse crops contain moisture that must be estimated in order to apply the appropriate premium or discount when trading these crops. However, there was no published international standard method for the determination of moisture content in pulses. Through collaboration with researchers from Canada, the USA and Australia, a gravimetric procedure for determining the moisture content of pulses has been developed and subjected to an inter-laboratory study. The method has been accepted and published by the AACC as a standard reference method (AACC, 2004). Water hydration capacity of pulses is another important quality parameter and is defined as the amount of water that whole seeds absorb after soaking in excess water for 16 h at room temperature ($22 \pm 2\,°C$). Pulses are generally soaked before cooking to ensure uniform expansion of the seed coat and cotyledon for uniform cooking and to ensure their tenderness. The ability of seeds to hydrate has been linked to cooking or canning quality. Several methods for measuring the water hydration capacity of pulses have been reported, but no universally accepted methods exist. A collaborative study on a method for determining the water hydration capacity of pulses was carried out in the summer of 2004 and the precision of the method was established. The AACC Pulse and Grain Legume TC have approved the method. A method for determining the crude protein content of pulses and other agricultural commodities has been developed and is at the Draft International Standard stage in ISO (the International Organization for Standardization).

15.3 Techniques for quality evaluation

Legumes are important sources of macro- and micronutrients and are consumed by humans in many forms. The nutrient bioavailability from legumes depends on the nutrient content (Bressani and Elfas, 1980) and factors such as post-harvest handling, processing methods and conditions. In addition to traditional food and other uses, legumes can be milled into flour, used to make bread, doughnuts, tortillas, chips, spreads, and extruded snacks or used in liquid form to produce milks, yogurt and infant formula (Garcia et al., 1998). Pop beans (Popenoe et al., 1989), licorice (*Glycyrrhiza glabra*) and soybean candy (Genta et al., 2002) provide novel uses for specific legumes. For all uses, quality evaluation of pulses is most important for milling and its application in different food products.

Physical properties as applicable to cereal grains are also important for pulses. These properties provide essential engineering data required in design of various processing machines, structures, processes and controls; in analysis and determining the efficiency of a machine or an operation; in developing new consumer products; and in evaluating and retaining the quality of the final product. The physical properties such as grain characteristics (e.g. seed dimensions, grain size, sphericity), grain volume, thousand grain mass, bulk density, true density, bulk porosity, terminal velocity, angle of repose and static coefficient of friction have been reported for faba beans (Fraser et al., 1978), pigeon pea (Shepherd and Bhardwaj, 1986), chickpea (Dutta et al., 1988; Gupta and Prakash, 1990), soybean (Kulkarni et al., 1993; Deshpande et al., 1993; Unde and More, 1996), lentil (Carman, 1996), black gram (Munde, 1999; Nimkar et al., 2004), green gram (Nimkar and Chattopadhyay, 2001) and moth bean (Nimkar et al., 2005). These physical properties are known to vary depending on various factors including genotypic and extrinsic conditions such as moisture content.

The color of pulses as a quality parameter is assessed after the removal of damaged grains by using standard color charts or guides for picking out grains of other colors. For example, in Canada, the color of peas is considered as good if they are bright, normal color, lightly earth tagged or lightly stained and considered as fair if the color is moderately immature, moderately earth tagged or stained (Pulse Canada, 2010). Other methods for measuring color include the use of colorimeters. For example, the color of dehulled grains or pulse flour is measured using a Hunter color lab XE spectrocolorimeter with the

CIE (1976) L^* a^* and b^* color scale. Illuminant D65 is used, where the following parameters are measured: L^*=darkness to brightness; a^*=greenness (−) to redness (+); b^*=blueness (−) to yellowness (+). Nutritional quality parameters such as protein, starch and other nutrients can be assessed by using internationally accepted approved methods of the American Association of Cereal Chemists (AACC), official methods of the Association of Official Analytical Chemists (AOAC) or methods developed by the International Organization for Standardization (ISO). Table 15.8 lists some of the standard methods applicable to pulses.

Table 15.8 List of quality standards for pulses

CAC/RCP 1, Recommended international code of practice – General principles of food hygiene
EAS 38, Labelling of prepackaged foods – Specification
EAS 79, Cereals and pulses as grain – Methods of sampling
EAS 217, Methods for the microbiological examination of foods
ISO 520, Cereals and pulses – Determination of the mass of 1000 grains
ISO 605, Pulses – Determination of impurities, size, foreign odors, insects, and species and variety

Test methods
ISO 2164, Pulses – Determination of glycosidic hydrocyanic acid
ISO 2171, Cereals, pulses and by-products – Determination of ash yield by incineration
ISO 4112, Cereals and pulses – Guidance on measurement of the temperature of grain stored in bulk
ISO 4174, Cereals, oilseeds and pulses – Measurement of unit pressure loss in one-dimensional airflow through bulk grain
ISO 5223, Test sieves for cereals
ISO 5526, Cereals, pulses and other food grains – Nomenclature
ISO 5527, Cereals – Vocabulary
ISO 6322-1, Storage of cereals and pulses – Part 1: General recommendations for the keeping of cereals
ISO 6322-2, Storage of cereals and pulses – Part 2: Practical recommendations
ISO 6322-3, Storage of cereals and pulses – Part 3: Control of attack by pests
ISO 6639-1, Cereals and pulses – Determination of hidden insect infestation – Part 1: General principles
ISO 6639-2, Cereals and pulses – Determination of hidden insect infestation – Part 2: Sampling
ISO 6639-3, Cereals and pulses – Determination of hidden insect infestation – Part 3: Reference method
ISO 6639-4, Cereals and pulses – Determination of hidden insect infestation – Part 4: Rapid methods
ISO 13690, Cereals, pulses and milled products – sampling of static batches
ISO 16002:2004, Stored cereal grains and pulses – Guidance on the detection of infestation by live invertebrates by trapping
ISO 16050, Foodstuffs – Determination of aflatoxin B1, and the total content of aflatoxin B1, B2, G1 and G2 in cereals, nuts and derived products – High performance liquid chromatographic method
ISO/TS 16634-2, Food products – Determination of the total nitrogen content by combustion according to the Dumas principle and calculation of the crude protein content – Part 2: Cereals, pulses and milled cereal products
ISO 20483, Cereals and pulses – Determination of the nitrogen content and calculation of the crude protein content – Kjeldahl method
ISO 22000, Food safety management systems – Requirements for any organization in the food chain
ISO 24557, Pulses – Determination of moisture content – Air-oven method
OIML R87, Quantity of product in prepackages

15.4 Conclusions

Pulses are consumed in various countries in different forms. Internationally recognized quality standards and common nomenclature of pulses will help pulse grain processors, product developers and traders through to consumers in making the right choice. Current efforts for developing uniform standards are sporadic and limited in comparison to cereals and oil seeds. This chapter highlights the need for internationally accepted quality standards. Various agencies such as AACC, IPQC and CAC are involved in developing new or improved methods for measuring and assessing pulse and end-use quality. Such initiatives will provide new and valuable tools which will assist all sectors of the pulse processing industry. International cooperation in developing quality standards would channel new and exciting developments for the world pulse industry into the future.

References

AACC, 2004. AACC Method 44–17. Approved methods of the American Association of Cereal Chemists, 10th edn. AACC, St. Paul.

AOAC, 1995. Crude protein in cereal grains and oilseeds. AOAC Official Methods of Analysis, 16th edn, Suppl. AOAC, Arlington.

Bressani, R., Elfas, L.G., 1980. Nutritional values of legume crops for humans and animals. In: Summerfield, R.J., Bunting, A.H. (Eds.), Advances in legume science. FAO, London, pp. 57–66.

Carman, K., 1996. Some physical properties of lentil seed. J. Agric. Eng. Res. 63, 87–92.

Codex Standard For Certain Pulses (CODEX STAN 171-1989, Rev. 1-1995).

Deshpande, S.D., Bal, S., Ojha, T.P., 1993. Physical properties of soya bean. J. Agric. Eng. Res. 56 (2), 89–98.

Dutta, S.K., Nema, V.K., Bhardwaj, R.K., 1988. Physical properties of gram. J. Agric. Eng. Res. 39, 259–268.

FAO/WHO (Food and Agriculture Organization of the United Nations and the World Health Organization) Food Standards Codex Alimentarius https://www.codexalimentarius.net (accessed on April 23, 2010).

Fleurat–Lessard, F., 2002. Qualitative reasoning and integrated management of the quality of stored grain: a promising new approach. J. Stored Prod. Res. 38, 191–218.

Fraser, B.M., Verma, S.S., Muir, W.E., 1978. Some physical properties of fababeans. J. Agric. Eng. Res. 23, 53–57.

Garcia, M.C., Marina, M.L., Laborda, F., Torre, M., 1998. Chemical characterization of commercial soybean products. Food Chem. 62, 325–331.

Genta, H.D., Genta, M.L., Alvarez, N.V., Santana, M.S., 2002. Production and acceptance of a soy candy. J. Food Eng. 53, 199–202.

Gupta, R.K., Prakash, S., 1990. Effect of moisture content on some engineering properties of pulses. Paper presented at XXVI Annual convention of Indian Society of Agricultural Engineering held at Hisar, 7–9 February 1990.

Hofer, J., 2004. Legumes crop and its origin. Grain Legumes 40, 9–40.

Jensen, E.S., 2002. The contribution of grain legumes, currently under-utilised in the EU, to a more environmentally-friendly and sustainable European. LINK. Grain legumes for Sustainable Agriculture, Strasbourg, 26 September 2002. Org Print 1852.

Kulkarni, S.G., Bhole, N.G., Sawarkar, S.K., 1993. Spatial dimensions and their dependence on grain moisture condition. J. Food Sci. Technol. 30, 335–338.

Munde, A.V., 1999. Effect of moisture content on gravimetric properties of black gram. J. Maharashtra Agric. Univ. 22, 833–835.

Nimkar, P.M., Chattopadhyay, P.K., 2001. Some physical properties of green gram. J. Agric. Eng. Res. 80, 183–189.

Nimkar, P.M., Guleria Nisha, S., Bute, A.N., 2004. Physical properties of black gram (*Phaseolus mungo*). J. Food Sci. Technol. 41, 326–329.

Nimkar, P.M., Mandwe, D.S., Dudhe, R.M., 2005. Physical properties of moth gram. Biosyst. Eng. 91, 183–189.

Panozzo, J., Materne, M., 2007. Developing quality pulses for a sustainable environment, population and community. Personal Communication.

Popenoe, H., King, S.R., Leon, J., Kalinowski, L.S., 1989. Lost crops of the Incas. National Academy Press, Washington, DC.

Pulse Canada: http://www.pulsecanada.com.

Shepherd, H., Bhardwaj, R.K., 1986. Moisture dependent physical properties of pigeonpea. J. Agric. Eng. Res. 35, 227–234.

Singh, U., Singh, B., 1992. Tropical grain legumes as important human foods. Econ. Bot. 46, 310–321.

Unde, P.A., More, H.G., 1996. Effect of grain size and moisture content on dehulling of soyabean. J. Maharashtra Agric. Univ. 21 (3), 456–459.

USDA, 2002. USDA-Iowa state university database on the isoflavone content of foods. Agricultural Research Service, US Department of Agriculture, Washington, DC.

Global pulse industry: state of production, consumption and trade; marketing challenges and opportunities

16

Peter Watts
Pulse Canada, Winnipeg, MB, Canada

16.1 Introduction

The first section of this chapter will focus on global pulse production, consumption and trade, highlighting major statistics and trends from the last four decades based on the United Nations' Food and Agriculture Organization (FAO) figures which start in 1961 and have been

Pulse Foods: Processing, Quality and Nutraceutical Applications. DOI: 10.1016/B978-0-1238-2018-1.00007-0

updated to 2008 for production, and to 2007 for trade and consumption. This will be a high-level overview of the data, looking primarily at developments that have occurred since 1961, with only cursory observations regarding the drivers behind the changes. The second section of this chapter will focus on opportunities for the sector, and related challenges, primarily from a Canadian perspective. While India is far and away the world's largest producer, consumer and importer of pulses, Canada exports 90% of its production and accounts for 35–40% of global pulse trade and therefore developing new market opportunities and addressing barriers to consumption and trade are of paramount importance. This section will look at the development of new markets particularly in the context of consumer drivers such as nutrition and health as well as environmental sustainability.

As indicated, the statistics used in this document are taken primarily from the FAO. On the FAO website, pulses are defined as the "dry seeds of leguminous plants which are distinguished from leguminous oil seeds by their low fat content." Within the FAO statistical data base, pulses are broken down into the following major classifications: dry beans, peas, chickpeas, cowpeas, pigeon peas, broad/horse/fava beans, lentils, vetches and lupins. It is of note that classifying the different genera, species and types of pulses is not a straightforward task. The job is further complicated by the fact that similar pulses have different names in different countries. In the case of chickpeas, there are two main types: desi and kabuli, which make this a less complicated example among the pulses, although chickpeas have a few different names, including garbanzo bean and Bengal gram for example. Dry beans offer a more complex example. *Phaseolus vulgaris*, or common beans, that include white or navy, pinto, kidney, black, pink, romano, yellow and several other types are captured under "dry beans" within the FAO statistics. And while mung beans (*Vicia radiata*) and black matpe or urad beans (*Vicia mungo*) are a different genus than common beans, they are also considered dry beans within the FAO statistical database. For the purposes of the statistics cited in this paper, only the major FAO classifications noted above will be covered.

16.2 Global pulse production, consumption and trade

Evidence of the cultivation, preparation and consumption of pulse crops can be traced back millennia to the tombs of ancient Egypt, to Greece in the time of Homer and even to the Old Testament.

The use of legumes as a basic dietary staple dates back more than 20 000 years in some Eastern cultures. These ancient civilizations recognized the value of legumes for their diversity and tremendous nutritional value. The Italian writer and academic Umberto Eco suggested the cultivation of beans in Europe during the Middle Ages was of immense importance, "saving" the Europeans from malnutrition and possible extinction (http://www.legumechef.com/English/historia_en.htm).

Today, pulses are produced all over the world, although 75% of global output is concentrated among 15 countries (India, Canada, Myanmar, Brazil, China, Nigeria, the USA, Russia, Ethiopia, Australia, Mexico, Turkey, Niger, Pakistan and France). Production patterns have shifted over the last 40 years, with a significant increase in output in Africa and countries like Myanmar and Brazil. Canada and Australia are both major producers today compared with very little cultivation in the early 1960s. In other countries output has dropped. The most significant example is China where production has fallen by 60% since the early 1960s largely due to the substitution of bean and pea production for crops such as corn and wheat. Even since 2005, China's production, which is predominantly broad beans, dry beans and peas, has declined by some 30%.

From a consumption perspective, in addition to human food, pulses are used in animal feed, pet food, aquaculture and certain bio-industrial applications, although to date pulses are rarely used in the production of biofuels. According to the FAO, about 71% of global pulse consumption is for food, while 18% goes to feed, with most of the balance used as seed. Generally speaking, lupins and vetches have been used primarily as animal feed, while a portion of other pulses is also used in the feed sector when prices and supply permit. The consumption section below will only focus on the use of pulses in the human food sector.

While total human food consumption of pulses globally has risen over the last four decades, this has been driven primarily by population growth. Global average per capita consumption of pulses has actually declined since the 1960s as people increasingly rely on other sources of food such as meat, fish, oils and sugars driven by factors such as rises in income levels and urbanization. Unfortunately, this shift in consumption patterns has been linked to the rise in disease among people in some traditional pulse-consuming countries, particularly in cardiovascular disease and diabetes.

No study of pulses would be complete without a discussion of India. With its huge population and the position of pulses as a staple food in daily life, India is the world's largest overall consumer of pulses. Its per capita consumption at around 13 kilograms (kg) is on the higher end of the scale, although there are many countries in Africa and Latin America where consumption in considerably higher (e.g. Burundi at 34 kg and Brazil at 17 kg). Globally, per capita consumption in 2008 was 6.5 kg. India is also by far the world's largest producer with output ranging from 13 to 15 million tonnes (Mt) in recent years, accounting for 21–26% of global output. However, over the years, with its dramatic population growth, India has gone from a significant net exporter of pulses to a major net importer. This has had important implications from a trade perspective, benefiting exporting nations such as the USA, Australia, Canada, Ukraine and others.

16.2.1 Production statistics

In 2008, the most recent year of complete FAO statistics, world production of pulses amounted to 60.9 Mt. Dry beans accounted for one-third of global pulse production, followed by peas at 16% and chickpeas or garbanzo beans at 14%. The remaining third of production is made up of cowpeas, pigeon peas, broad beans (or faba bean, horse bean), lentils, vetches and lupins (Fig. 16.1).

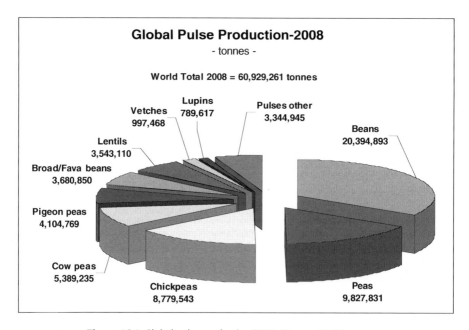

Figure 16.1 Global pulse production 2008. (Source: FAO.)

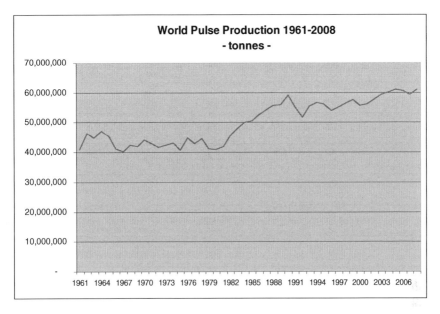

Figure 16.2 Global pulse production 1961–2008. (Source: FAO.)

Total world pulse production has increased over the last four decades from 40.8 Mt in 1961 to 60.9 Mt by 2008, a 49% increase (Fig. 16.2).

The growth has been predominantly in beans which increased by 9.2 Mt, followed by cowpeas at 4.5 Mt, lentils at 2.8 Mt, peas at 2.5 Mt and chickpeas at 1.1 Mt. Some production has dropped, such as broad beans, which fell by 1.2 Mt (Table 16.1).

In terms of global production by country, India remains the world's largest producer of pulse crops, accounting for 21–26% of global output between 2004 and 2008. Canada has emerged as the second largest producer of pulses in recent years, at around 5 Mt of production, followed Myanmar, Brazil and China (Fig. 16.3).

India's largest crop is chickpeas, followed by dry beans and pigeon peas. In Canada, the largest crop is peas, primarily yellow field peas, followed by lentils and beans (Table 16.2).

There have been some significant changes in production over the last four decades in terms of geography. For example, China's pulse production dropped from 8.5 to 3.5 Mt between 1961 and 2008, while in Africa output grew from 3.6 to 12.2 Mt. In India, while pulse production increased between 1961 and 2008 from 12.9 to 15.3 Mt, per capita output dropped by more than half from 28 to 13 kg.

Table 16.1 Global pulse production 1961–2008 changes

	1961	%	2008	%	Change	%
Beans	11 228 313	28	20 394 893	33	9 166 580	82
Peas	7 346 299	18	9 827 831	16	2 481 532	34
Chickpeas	7 681 851	19	8 779 543	14	1 097 692	14
Cowpeas	870 114	2	5 389 235	9	4 519 121	519
Pigeon peas	2 227 995	5	4 104 769	7	1 876 814	84
Broad/horse beans	4 842 684	12	3 680 850	6	(1 161 834)	−24
Lentils	854 877	2	3 543 110	6	2 688 233	314
Vetches	1 871 348	5	997 468	2	(873 880)	−47
Lupins	631 322	2	789 617	1	158 295	25
Other	3 228 720	8	3 421 945	6	193 225	6
Total	**40 783 483**		**60 929 261**		**20 145 778**	**49**

Source: FAO.

In contrast, pulse production in North America has increased from 1.2 to 6.9 Mt over the same period (most of the original production was in the USA). In Australia, production has grown from virtually nothing to become one of the world's top ten producers. In North America and Australia, new varieties of pulse crops were developed that were adapted to local climate, soils and diseases enabling farmers to compete from both a quality and yield standpoint (Table 16.3).

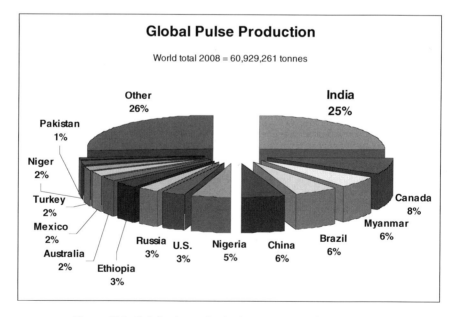

Figure 16.3 Global pulse production by country 2008. (Source: FAO.)

Table 16.2 Top five global pulse producers 2008

	India	Canada	Myanmar	Brazil	China
Chickpeas	5 748 600	67 000	260 000	N/a	6000
Beans	3 930 000	266 200	2 500 000	3 460 867	1 121 151
Pigeon peas	3 075 900	N/a	600 000	N/a	N/a
Lentils	910 000	1 043 200	1 400	N/a	130 000
Peas	800 000	3 571 300	40 000	4000	900 000
Broad beans	N/a	10 900	N/a	16 000	1 200 000
Cow peas	N/a	N/a	150 000	N/a	N/a
Other pulses	800 000	10 900	N/a	N/a	100 000
Total	15 264 500	4 969 500	3 551 400	3 480 867	3 457 151

Source: FAO.

16.2.2 Area and yield

The increase in global production since the 1960s has been a combination of both area and yield, although yield has had a greater impact than area. Since 1961, pulse harvested area increased 16% from 64 to 74 million hectares. In comparison, pulse yields increased 29% from 0.64 to 0.82 tonnes per hectare (Table 16.4). Area and yield figures will be looked at in more detail below in relation to individual crops.

Table 16.3 Global pulse production 1961–2008 changes in top 15 producers

		1961	%	2008	%	Change	%
1	India	12 859 873	32	15 264 500	25	2 404 627	19
2	Canada	60 960	0	4 958 600	8	4 897 640	8034
3	Myanmar	222 330	1	3 551 400	6	3 329 070	1497
4	Brazil	1 800 809	4	3 480 867	6	1 680 058	93
5	China	8 520 593	21	3 457 151	6	(5 063 442)	−59
6	Nigeria	472 000	1	2 969 000	5	2 497 000	529
7	USA	1 100 775	3	1 904 820	3	804 045	73
8	Russia	N/a	N/a	1 812 140	3	N/a	N/a
9	Ethiopia	0	0	1 774 338	3	1 774 338	N/a
10	Australia	22 644	0	1 478 000	2	1 455 356	6427
11	Mexico	890 329	2	1 412 841	2	522 512	59
12	Turkey	592 006	1	1 369 140	2	777 134	131
13	Niger	68 000	0	1 288 009	2	1 220 009	1794
14	Pakistan	934 400	2	845 100	1	(89 300)	−10
15	France	175 120	0	791 379	1	616 259	352
	Total	40 783 483	60 929 261	20 145 778	49		

Source: FAO.

Table 16.4 World pulse harvested area, yield and production comparison 1961 vs. 2008

World	1961	2008	% Change
Harvested area (ha)	64 008 312	74 129 733	16
Yield (t ha^{-1})	0.64	0.82	29
Production (tonnes)	40 783 483	60 929 261	49
Source: FAO.			

Today, peas yield the most of any pulse crop on average, followed by broad beans, lupins and lentils. Chickpeas and beans yields are approximately 50% of those of peas (Fig. 16.4).

Note that in the detailed production section below, we have excluded data on pigeon peas, cowpeas and vetches given these crops are not traded significantly on the global market.

16.2.3 Dry bean production

Dry beans come in a wide range of types including mung, white (or navy), pinto, kidney, black, romano (cranberry, borlotti), pink, yellow and many others. The FAO does not carry data on the production

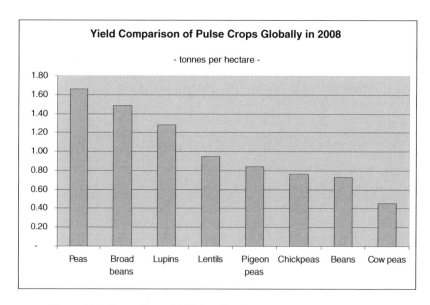

Figure 16.4 Comparison of yields by pulse 2008. (Source: FAO.)

Table 16.5 World bean harvested area, yield and production comparison 1961 vs. 2008

Beans	1961	2008	% Change
Area (ha)	22 766 818	27 988 440	23
Yield (t ha^{-1})	0.49	0.73	48
Production (tonnes)	11 228 313	20 394 893	82

Source: FAO.

of individual types of beans on a global basis. These data must therefore be collected on a country by country basis (Table 16.5). For the purposes of this chapter, only aggregated figures of all types of beans, peas, lentils and chickpeas will be discussed.

Dry bean production rose from 11.2 Mt in 1961 to 20.4 Mt in 2008, an increase of 9.2 Mt or 82% over 47 years. Rising global bean output has been the result of a combination of both area and yield increases at 23% and 48%, respectively. Area sown to beans has increased significantly in India, Myanmar and Brazil over the years, accounting for the majority of the rise in area globally and offsetting a major drop in dry bean plantings in China (Fig. 16.5). Today, India is the world's largest producer of dry beans, followed by Brazil, Myanmar, the USA and Mexico.

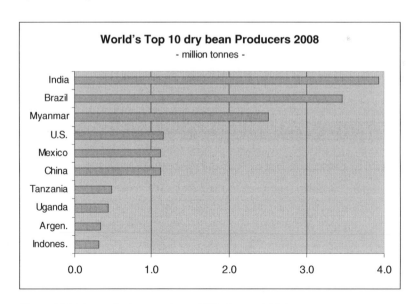

Figure 16.5 Top ten dry bean producers 2008. (Source: FAO.)

Table 16.6 World pea harvested area, yield and production comparison 1961 vs. 2008

Peas	1961	2008	% Change
Area (ha)	7 549 841	5 926 087	−22
Yield (t ha⁻¹)	0.97	1.66	70
Production (tonnes)	7 346 299	9 827 831	34

Source: FAO.

16.2.4 Dry pea production

Dry pea production has been quite variable over the years, primarily due to changes in the former Soviet Union, where output rose significantly through the 1980s and then dropped back in the 1990s. Total world pea production rose from 7.3 Mt in 1961 to 9.8 Mt in 2008, although production peaked at 16.6 Mt in 1990. Over this period, the most dramatic changes occurred in China, where pea production dropped from 2.9 Mt to 0.9 Mt, and in Canada, where production rose from a mere 30 000 tonnes to 3.5 Mt. The vast majority of the world's production of peas are yellow field peas, followed by green field peas and a few other types produced only in smaller quantities. Pea area has actually declined since 1961 by 22% while yields have increased by 70% (Table 16.6). The drop in area is largely the result of reduced pea seedings in China while yield improvements have occurred in all pea-producing countries.

Today, Canada is the world's largest pea producer (Fig. 16.6), followed by Russia, China, India and the USA.

16.2.5 Chickpea production

Chickpea production is made up primarily of desi and kabuli type chickpeas. Similar to peas, production has been more variable than most other pulse crops but, in this case, the variability is driven by India, which produces 65% of the world's chickpeas. Total world chickpea production has been gradually rising, up from 7.7 Mt in 1961 to 8.8 Mt in 2008, an increase of 15% (Table 16.7). Interestingly, chickpea area actually dropped between 1961 and 2008 by 2%, so the 15% increase in global production since 1961 has been entirely driven by increased yields.

Turkey, Pakistan, Australia and Iran are the next largest producers of chickpeas after India (Fig. 16.7).

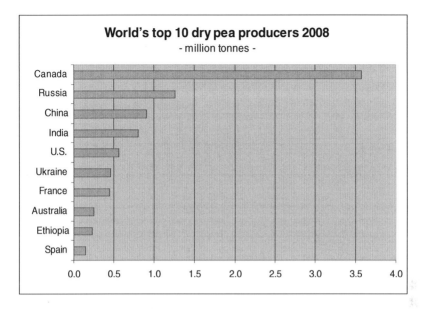

Figure 16.6 Top ten pea producers 2008. (Source: FAO.)

16.2.6 Lentil production

Lentil production has more than tripled since 1961, rising from 0.9 Mt to 3.5 Mt in 2008. The growth in production is largely due to increased output in Canada, India and Turkey. Red and green lentil production make up the lion's share of global production, although there are some other types of lentils produced such as French green and, on a very small scale, black lentils. More than any of the other pulse crops, lentil area has grown significantly over the last 40 years, more than doubling from 1.6 to 3.8 million hectares. Yields have risen 79% over the same period from 0.53 to 0.94 tonnes per hectare, contributing to the 314% increase in lentil output between 1961 and 2008 (Table 16.8).

Table 16.7 World chickpea harvested area, yield and production comparison 1961 vs. 2008

Chickpeas	1961	2008	% Change
Area (ha)	11 836 682	11 556 744	−2
Yield (t ha^{-1})	0.65	0.76	17
Production (tonnes)	7 681 851	8 779 543	14

Source: FAO.

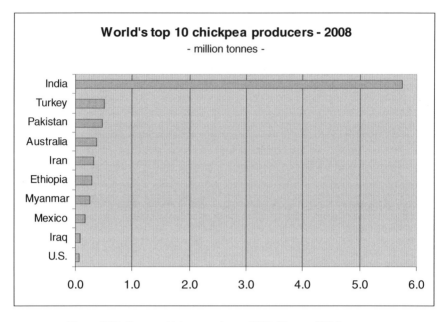

Figure 16.7 Top ten chickpea producers 2008. (Source: FAO.)

Canada, now the world's largest lentil producer (1.5 Mt of production in 2009), had been predominantly a green lentil producer until around 2004 when red lentil production began to take off. By 2009, green and red lentil area in Canada was almost evenly split. Almost all of Canada's lentils are grown in the province of Saskatchewan. In India and Turkey, the majority of production is composed of red lentils (Fig. 16.8).

16.2.7 Broad bean production

Broad bean production has declined from an average of between 5 and 6 Mt in the early 1960s to current levels of around 4 Mt. The decline has been primarily driven by lower plantings in China,

Table 16.8 World lentil harvested area, yield and production comparison 1961 vs. 2008

Lentils	1961	2008	% Change
Area (ha)	1 619 653	3 751 682	132
Yield (t ha^{-1})	0.53	0.94	79
Production (tonnes)	854 877	3 543 110	314

Source: FAO.

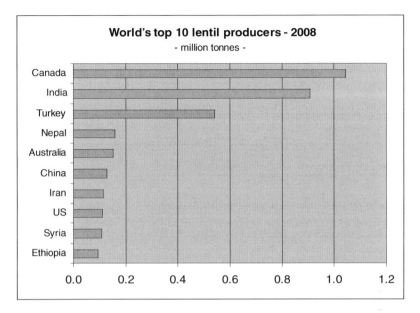

Figure 16.8 Top ten lentil producers 2008. (Source: FAO.)

although it remains the largest broad bean producer in the world with about 1.5 Mt of output annually (Fig. 16.9). Overall, global broad bean area has fallen by 54%. Yields meanwhile have increased quite substantially, rising from 0.9 to 1.48 t ha^{-1}, a 66% increase (Table 16.9). Other large producers include Ethiopia, France, Egypt and Australia.

16.2.8 Lupin production

World lupin production increased from about 600 000 tonnes in the early 1960s to more than 2 Mt in 1999, with the vast majority of this increase coming from Australia. Australia is by far the world's largest lupin producer, accounting for about 60% of global output. Most of the world's lupin production is used for feed for the livestock sector. Less than 5% of global lupin production is used in the human food sector (http://www.abare.gov.au/publications_html/ac/ac_07/a2_june. pdf). Lupin area declined since 1961 by 43% in spite of the emergence of Australia as a major producer from virtually no lupin production in the early 1960s (Table 16.10). Most of the drop in areas seeded to lupins is the result of reductions in countries like Russia, Ukraine, Poland and South Africa that used to be large producing countries (Fig. 16.10). Yields have improved quite significantly largely due to new varieties adopted in Australia with better disease resistance.

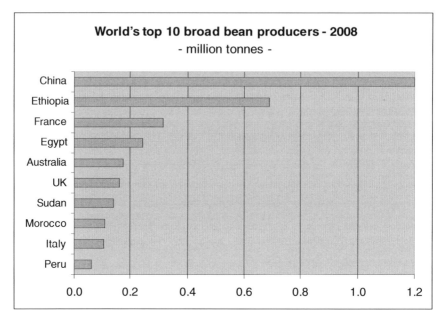

Figure 16.9 Top ten broad bean producers 2008. (Source: FAO.)

Table 16.9 World broad bean harvested area, yield and production comparison 1961 vs. 2008

Broad beans	1961	2008	% Change
Area (ha)	5 402 824	2 480 445	−54
Yield (t ha^1)	0.90	1.48	66
Production (tonnes)	4 842 684	3 680 850	−24

Source: FAO.

Table 16.10 World lupin harvested area, yield and production comparison 1961 vs. 2008

Lupins	1961	2008	% Change
Area (ha)	1 088 851	616 713	−43
Yield (t ha^1)	0.58	1.28	121
Production (tonnes)	631 322	789 617	25

Source: FAO.

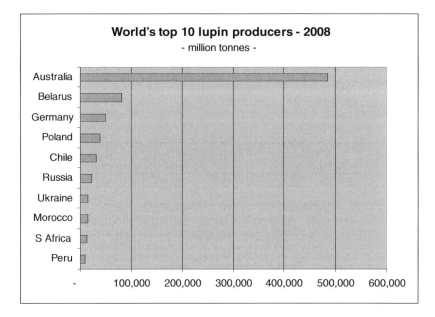

Figure 16.10 Top ten lupin producers 2008. (Source: FAO.)

16.2.9 Consumption statistics

The majority of pulses consumed as food are used in traditional pulse-consuming countries in South Asia, such as India, Pakistan, Bangladesh and Myanmar, in Latin America such as Brazil and Mexico, and in the Middle East and Africa in countries such as Turkey, Iran, Egypt, Ethiopia and Nigeria. These countries have high per capita consumption, combined with large populations and together account 26.6 Mt or 62% of global pulse consumption for food (India alone accounts for 35%). But other countries, not necessarily considered "traditional" consumers, also account for significant use of pulses. The USA consumes about 1.5 Mt of pulses annually, as does China, although this means that China's per capita consumption is very low given its population is over four times that of the USA (Fig. 16.11). Italy consumes over 300 000 tonnes per year, putting its per capita consumption at about 6 kg.

Globally, total consumption of pulses as food is rising. In 2007, the FAO estimates global consumption of pulses for food at 42.7 Mt, up 48% from 29.0 Mt in 1961. However, per capita consumption declined for most of the last 40 years, reaching 6.5 kg in 2007 down from 9.5 kg in 1961, a drop of 46%. In India alone, per capita consumption has dropped from 22.8 to 12.8 kg over this period (Fig. 16.12). Previously

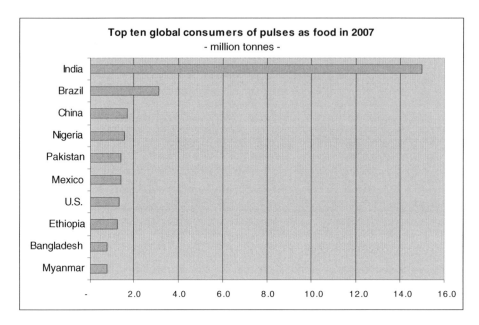

Figure 16.11 Top ten global consumers of pulses as food 2007. (Source: FAO.)

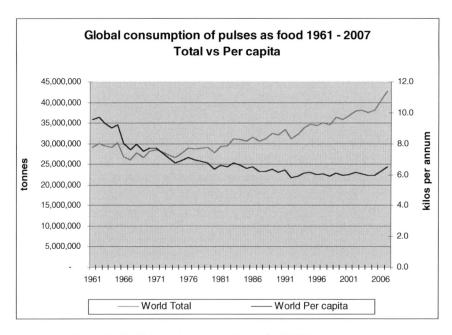

Figure 16.12 Global pulse consumption as food 2007: total vs. per capita. (Source: FAO.)

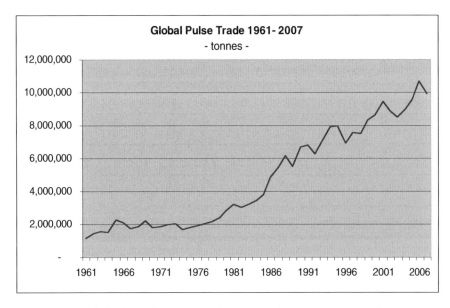

Figure 16.13 Global pulse trade 1961–2007. (Source: FAO.)

very dependent on pulses, the green revolution saw a rapid increase in domestic production of other crops such as wheat and rice, reducing India's reliance on pulses and therefore per capita availability and consumption.

16.2.10 Trade statistics

Global pulse trade was 9.9 Mt in 2007, just shy of the record 10.7 Mt in 2006, and up from 1.1 Mt in 1961, a tenfold increase (Fig. 16.13). It is of note that, after 1999, the FAO stopped publishing Australian lupin export data. Therefore, lupins are not included in the trade data. Global pulse trade has risen steadily since 1961, driven by increased demand in major importing countries such as India, Bangladesh, Pakistan, Egypt and China. In 1961, these five countries imported a combined 26 000 t compared with 4.8 Mt in 2007 or almost 50% of global trade. India alone imported 2.9 Mt in 2007 (Fig. 16.14).

Peas are the most traded crop at 3.7 Mt in 2007, followed by beans, lentils, chickpeas and broad beans. Of India's 2.9 Mt of pulse imports in 2007, 1.7 Mt were peas, 486 000 were beans, 230 000 were lentils and 145 000 were chickpeas (Fig. 16.15). Interestingly, India also exported 162 000 tonnes of chickpeas in 2007.

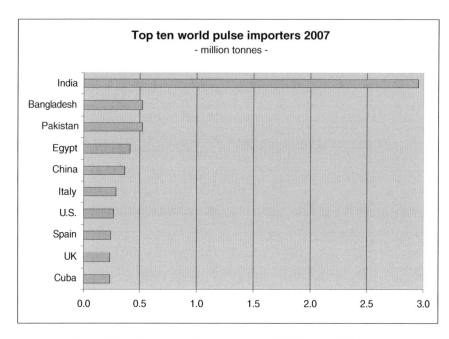

Figure 16.14 Top ten global pulse importers 2007. (Source: FAO.)

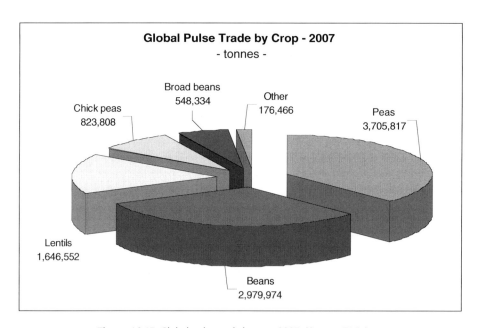

Figure 16.15 Global pulse trade by crop 2007. (Source: FAO.)

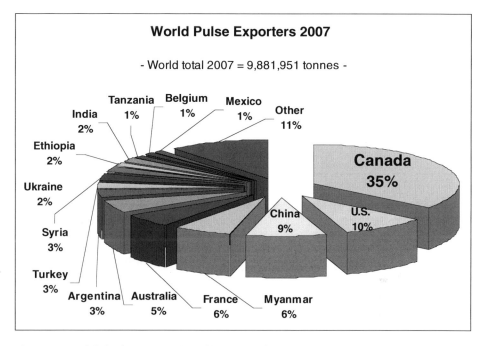

Figure 16.16 Global pulse exporters 2007. (Source: FAO.)

Today, Canada is the world's largest exporter of pulses, account-ing for 35% of global trade in 2007. The USA is the second largest exporter globally at 10%, followed by China, Myanmar, France and Australia (Fig. 16.16).

While Canada dominates pea and lentil exports in global trade, China and Myanmar are the largest bean exporters. Australia is the world's largest chickpea exporter while France exports the most broad beans (Table 16.11).

Table 16.11 Major pulse exporters 2007

2007	Canada	USA	China	Myanmar	France	Australia
Peas (tonnes)	2 188 418	482 861	2 384	–	350 364	133 856
Lentils (tonnes)	921 394	122 774	12 058	–	2 881	101 611
Beans (tonnes)	325 171	309 331	794 740	547 989	4 531	27 331
Chickpeas (tonnes)	70 778	20 908	770	82 393	951	159 584
Broad beans (tonnes)	3 845	3 072	40 519	–	196 027	99 403
Total (tonnes)	3 509 606	938 946	850 471	630 382	554 754	521 785

Source: FAO.

On the import side, India tops the list for four of the five major categories of globally traded pulses. Bangladesh and Pakistan are next with significant imports of peas, lentils and chickpeas. China is the world's fourth largest pulse buyer due to its imports of nearly 300 000 tonnes of peas each year, primarily to make starch for the vermicelli noodle market (Table 16.12).

16.3 Challenges and opportunities: perspective from Canada

The following section will provide some perspectives on new market opportunities for the pulse sector looking outward from Canada, the country that is the most heavily reliant on pulse exports. As noted above, Canada is the world's largest producer and exporter of both peas and lentils, and is also a significant producer and exporter of dry beans and chickpeas. Approximately 80% of pulse production is exported, with the balance used domestically as food (3%), seed (7%) and feed/other (10%). Currently, the largest markets for Canadian pulses are markets in the Indian subcontinent, the Mediterranean basin and Latin America where pulses are primarily imported in bulk, unprocessed form. However, in recent years, Canada has also ramped up exports of processed pulses including split peas and lentils (now the world's largest splitter and exporter of red lentils) and also has important exports markets in the USA, the UK and Italy for beans, lentils and chickpeas destined for the food-processing sector. In addition, Canada produces pulse flours and fractions, including pea protein, starch and fiber. With new, improved varieties of pulses leading to higher yields and production, the development of new markets is increasingly important for continued profitable growth of the sector.

This section will begin with a brief environmental scan and then examine some of the market opportunities that are ripe for development for the pulse sector.

16.3.1 Environmental scan

As indicated in the last section of this chapter, global pulse consumption is increasing at a slower rate than population growth, which implies a drop in per capita consumption. In fact, per capita pulse

Table 16.12 Major pulse importers 2007

Pea imports (tonnes)		Lentil imports (tonnes)		Bean imports (tonnes)		Chickpea imports (tonnes)		Broad bean imports (tonnes)	
India	1 738 283	India	230 557	India	486 159	India	145 605	Egypt	301 433
Bangladesh	317 455	Bangladesh	138 471	USA	171 151	Pakistan	136 677	Italy	67 413
China	282 193	Sri Lanka	102 505	Cuba	127 162	Bangladesh	63 077	Sudan	42 672
Belgium	159 644	Pakistan	88 610	UK	122 920	Spain	61 362	Norway	19 265
Pakistan	120 900	Egypt	84 463	Japan	122 838	Algeria	48 639	UAE	16 437
Cuba	103 704	Colombia	68 935	Italy	104 908	UK	39 017	Spain	12 537
Netherlands	99 670	Algeria	64 349	Brazil	96 269	UAE	34 899	S.Arabia	12 007
Spain	67 434	UAE	61 163	Kenya	93 116	S.Arabia	31 139	Denmark	11 970
UAE	64 359	Spain	48 197	Mexico	91 712	Jordan	25 539	Jordan	8 046
Italy	63 547	Sudan	46 761	Venezuela	91 479	Italy	23 769	Japan	7 282

Source: FAO.

consumption has been in long-term decline in many world markets. India's per capita consumption decreased from 23 kg in 1961 to 13 kg per year in 2007, while China's fell from 10 kg to 1 kg and Spain's from 9 kg to 5 kg in the same period. Per capita consumption in North America is about 3.5 kg, holding fairly steady over the last 25 years. So while total global pulse consumption is still forecast to rise in coming years due to population increases, growth in demand will potentially be mitigated by falling per capita use.

With this in mind, and as we consider the future of the global pulse industry and demand for pulse foods, the following demographic and market trend factors are relevant considerations to understanding the key challenges and opportunities:

- Global population is projected to increase by 35% over the next 50 years. Over the same period, global demand for food is projected to double as urbanization proceeds and incomes rise (http://www.unfpa.org/public/).
- Processed food represents three-quarters of total world food sales by value. High-income countries account for a majority of processed food sales, with the USA, the EU and Japan alone accounting for more than 60% by value (http://www.who.int/en/)
- Europe released more than 36% of the world's new processed food products in 2007, more than any other region, making it the most innovative and leading edge food industry in the world, followed closely by the USA (http://www.who.int/en/)
- Processed food products using pulses as ingredients represent only about 1% of new launches. The majority of pulses are sold in whole form to low-income countries (http://www.mintel.com/).
- The United Nations estimated that, in 2007, for the first time, the majority of the global population was urban. Urban population is expected to grow 1.8% a year through 2030, almost twice as fast as the global population. In 2005, urbanization in low-income countries was 28%, 44% in middle-income countries and 75% in high-income countries (http://www.unfpa.org/swp/2007/english/notes/indicators/e_indicator2.pdf).
- Urbanization and rising incomes change diets and historically have contributed to reductions in per capita pulse consumption. In contrast, increased meat consumption is directly correlated to rising GDP. Processed food consumption also grows with urbanization and increased GDP. Urbanization is also increasing the share of supermarket retail food sales at the expense of

local or village markets as is seen in Asia and other developing economies.

- In India, the world's largest pulse market, local markets continue to dominate food sales. Yet massive retail change is occurring in India with an estimated 2000 supermarkets added in 2007–2009. While the share of supermarket retail food sales is 12%, annual growth is estimated at 42% with 3200 additional supermarkets expected to be added by 2011 (http://www.who.int/en/).

In summary, the evolution of the food sector in emerging economies toward more "Western style" production, distribution and consumption patterns implies new opportunities, and different challenges for the pulse industry. Moreover, value-added opportunities in developed markets are enormous with pulses significantly under represented compared with other foods. Market development efforts in the future, therefore, must take these realities and the changing environment into consideration as traditional consumers of pulses will be influenced by these trends, leading to the potential for lost market share in some instances, while new consumers will be created in other areas with the potential to generate new and possibly higher value demand.

16.3.2 Health and nutrition

Health and nutrition present one of the most significant market opportunities for the pulse sector. The burden of chronic disease is rapidly increasing worldwide and the role of diet and nutrition in the development of these conditions is well established.

The prevalence of cardiovascular disease (CVD), obesity and diabetes in developed countries has been well documented and reported. About two-thirds of adults in the USA are overweight and almost one-third are obese (http://www.who.int/en/).

CVD is implicated in 42% of US deaths and roughly one-third of Canadian deaths. Approximately 7% of Americans and 6% of Canadians have diabetes. Many European countries are facing similarly dismal health statistics.

But these are not the only countries facing an epidemic of dietary-related illnesses. Increasingly, developing countries are exchanging the diseases of poverty for the diseases of affluence. Diabetes is Mexico's number one cause of death. The cost of providing care for those with diabetes in Mexico is equivalent to 34% of the country's budget for social services and doubles every five years (http://www.who.int/en).

By 2020, developing countries will see 71% of the heart-disease-related deaths in the world, 70% of deaths due to stroke and 70% of deaths due to diabetes. Cardiovascular diseases are now more numerous in India and China than in all the economically developed countries in the world combined. Obesity is becoming a major problem in Asia, Latin America and some parts of Africa, despite the widespread problem of malnutrition (http://www.who.int/en/).

With excellent nutritional benefits, including high levels of dietary fiber and complex carbohydrates, pulses are ideally positioned to provide dietary solutions to the epidemic of chronic disease around the globe. Research on the health benefits of pulses is increasing interest in their potential as a healthy and nutritional food ingredient. Building on the body of evidence that confirms these findings and communicating those messages to a wide audience will be critical to capturing new consumers in markets considered non-traditional pulse consumers, such as North America, and to stemming the tide away from pulse consumption in traditional consuming markets. This can be accomplished by partnering with health professionals and other consumer influencers as well as the food processing and retailing industries to communicate messages on the health benefits of pulses to consumers.

Globally, through further research and innovation to optimize the nutritional value of pulses and integration with other ingredients in food formulations, pulses can be positioned as a prescription for good health. This could be especially valuable in developing nations where nutrition and nourishment and avoidance of the diseases associated with a transition to affluence are key issues that are only going to become even more critical as world population in underdeveloped nations continues to grow and health care costs continue to rise.

The food processing industry is increasingly interested in the potential to incorporate novel ingredients, such as pulses, into food products for nutritional purposes, including their high protein and fiber content, gluten-free status (identified as one of the fastest growing new product trends), low glycemic index, antioxidant levels, as well as functional properties like water binding and fat absorption. Health and nutrition present an enormous opportunity for the pulse sector in coming years, one that will require heavy investment in research and communications activities to capitalize on this demand driver.

16.3.3 Pulse ingredients: processing and functionality

The processing of pulses into ingredients, such as flours and fractions (e.g. protein, starch and fiber), and utilizing them in food products is virtually non-existent in Western style food products, apart from a few specialty or niche markets, and only exists in a limited way in a few other countries. India and some of the surrounding countries, such as Pakistan, are exceptions where chickpea flour is used in a variety of traditional foods. In China, mung bean starch (and now pea starch) is used to make vermicelli noodles. But, generally speaking, pulses have not been widely developed as ingredients in the way that other crops, such as soy, that have invested heavily in research and marketing initiatives to diversify the market base and develop value-added markets.

The North American and EU-25 food industry uses 65 Mt of different flours each year for food products like breads and other baked goods, tortillas, snack foods, cereals, pasta and noodles. Flours used in these products are typically made from wheat, rye, oats, barley, corn and rice as well as from other crops such as soybean and potato. Although bean, pea, lentil and chickpea flours are making small inroads in some food applications, their production and utilization by the food-processing sector remain low due to a limited understanding of their properties and applications in processed foods. As developing countries, such as India, become more Westernized, there is great potential to include pulse ingredients, which are already familiar to consumers in these markets, in Western-style processed food products (e.g. pizza dough) that are increasingly in demand. There are also opportunities to increase the use of pulses in international markets such as China by exploring the use of pulse flours in traditional food products such as steamed bread, noodles and dumplings.

Although food companies are interested in exploring the use of pulse flours and fractions, food manufacturers look to ingredient providers to invest the time and money in R&D and bring "solutions" to the challenges they face. To introduce pulse flours and pulse fractions successfully into the food-processing sector, the pulse industry must invest in pre-commercialization pilot scale utilization and development. This work will provide ingredient manufacturers and food companies with the technology needed to produce and incorporate pulse flours into products that will deliver improved nutrition and health properties, without adversely affecting taste or texture.

The pulse industry must advance its knowledge of the processing of pulses into ingredients and the impact of that processing on the functionality of the ingredients in food product formulations. The optimization of processing in terms of quality and functionality, in addition to other factors, such as yield and energy use, will be needed to introduce successfully more value-added pulse processing and the incorporation of these ingredients into foods. Ultimately, this will open the door to creating new "ingredient" markets for pulses, leading to new food products and reformulated food products that address consumer needs.

16.3.4 Sustainability

Most global food companies now have programs underway to improve sustainability. While many initial programs have focused on reduced/recyclable packaging and reduced energy use in processing, these are gradually giving way to efforts aimed at more substantive change. Leading organizations have used life cycle analysis to examine their supply chains and identify opportunities for reducing impacts. Among the interesting conclusions reached are that roughly 90% of their products' "footprints" come from the agricultural component of their supply chains, and that transportation energy use and greenhouse gas emissions tend to be dwarfed by field emissions, even when products are crossing oceans.

As legumes, pulses have a symbiotic relationship with soil microorganisms which enables them to make their own fertilizer from atmospheric nitrogen, in place of energy-intensive synthetic nitrogen. Preliminary research has revealed that including pulses in a four-year crop rotation with wheat and canola can decrease the rotation's non-renewable energy use by over 20%. Pulses have also been shown to emit less of the powerful greenhouse gas nitrous oxide, relative to non-legume crops, and some studies have indicated that they may increase levels of soil organic matter. Pulses add much needed diversity to crop rotations, and by breaking disease and pest cycles also play an important part of an integrated pest management system.

Food companies are interested in the contribution to environmental sustainability unique to nitrogen-fixing legumes, and the fact that pulses are not genetically modified and are not considered as an ingredient with concerns over food allergies. The positive environmental attributes of pulses can be enhanced through innovation, and will create partnerships between agriculture, non-governmental organizations and food companies focusing on corporate social responsibility.

16.3.5 Feed market opportunities: aquaculture and pet food

Pulses have long been used in the livestock feed sector, particularly in hog rations, although only to a limited extent compared with more traditional feed stocks such as corn, soybean meal, wheat, etc. However, increased demand for pulses in the food sector led to price rises in 2008 and made their use as feed less economic. Since then the use of pulses in the feed sector has been limited, apart from those pulses grown specifically for feed, such as lupins and vetches. However, some higher value market opportunities in the feed sector include aquaculture and pet food. These two sectors are examined below.

16.3.5.1 Aquaculture

Aquaculture is the world's fastest growing source of food with growth rates of 8.0% annually. In the past 20 years, production has quadrupled to an annual production of 48 Mt. Fishmeal and fish oil are the two primary components of aquafeed. However, the production of fishmeal has been capped by diminishing ocean stocks and will not be able to keep pace with the expansion of the "farmed fish" industry. In order for the industry to continue its rapid growth, new protein sources are required. The search for alternative highly digestible protein sources is being carried out in many countries. Certain pulses, such as peas, have proven very digestible for aquaculture species. Market opportunities include the use of peas as a replacement for soybean meal in shrimp diets as well as alternative protein sources for a number of other fish species including trout and tilapia. Other pulse crops, such as faba bean, that have as high or higher protein content may offer another ingredient option into the market.

16.3.5.2 Pet food

The current estimated global size of the pet food industry is estimated to be 14 Mt of dog food and 8.5 Mt of cat food worth an estimated $46 billion per year. Product innovation is based almost entirely on following trends in the human counterparts, e.g. health and nutrition, and environmental sustainability.

Today, most new pet food products released are those with functional health attributes. Diets that are hypoallergenic, contain some mechanism that is beneficial to the gastrointestinal tract of the animals or are "natural" are some of the more popular releases. The gluten-free attributes of pulses allow them to fit well into diets for reduced allergens as many pets are sensitive to the diets that they consume. Pulses have been found to have prebiotic benefits in people and the same can extend into pet food diets. The "natural" products that are released have ingredients that are whole rather than processed. Whole pulse flours will fit well into this category.

16.4 Conclusions

The above review section has identified a few key opportunity areas for the pulse sector in relation to food and feed, but there are other important markets to consider which were not covered. Food service is a huge and growing market where pulses could find new demand, and bio-industrial applications may also prove to be an area of interest in the future. This chapter did not discuss issues around food safety, genetically modified organisms or organics, all of which may present opportunities to add value for industry. And the issues around regulatory environments in relation to labeling, health claims and novel product approval also have a potentially important bearing on the industry. In conclusion, in order to remain viable and profitable in the future, the pulse industry must diversify its market base for pulses and generate value by focusing on consumer drivers such as health, nutrition and environment, and by addressing the food science and food engineering challenges associated with the development of pulse ingredients and their incorporation into new food products. Focused investments in research and communications will be required to educate consumers and the food industry on the value of pulses in terms of nutrition, health, environmental benefits and functionality, and to provide the knowhow to effectively develop new food products with taste and texture that appeal to consumers. The industry will also benefit from support to navigate the regulatory frameworks governing novel food ingredients and health claims.

Index

AACC, *see* American Association of Cereal Chemists

ACE, *see* Acid extraction

Acid extraction (ACE), bioactive compound extraction from by-products, 341–342

AEE, *see* Aqueous alkaline extraction

Air classification, bioactive compound extraction from by-products, 342–345

Alkaloids, pulse content, 40, 369, 390

Allergenicity, pulse proteins, 401–402

American Association of Cereal Chemists (AACC), pulse quality standard parameters, 428–429

Amino acids, *see* Protein, pulse

α-Amylase inhibitor, kidney bean, 396

Amylose, *see* Starch, pulse

ANF, *see* Antinutritional factors

Antinutritional factors (ANF)
elimination, 333–334, 386–387, 391
pulse composition, 369–370, 388
types, 385–387

Antioxidants, *see* Phenolics

Aquaculture, pulse in feeds, 360–461

Aqueous alkaline extraction (AEE), bioactive compound extraction from by-products, 339–341

Broad bean, production, 447

By-products, pulse
extraction of bioactive compounds
acid extraction, 341–342
air classification, 342–345
aqueous alkaline extraction, 339–341
microwave-assisted extraction, 350–351, 353
overview, 338–339
prospects, 353–354
ultrasound extraction, 345–349
water extraction, 351–353
food applications, 329–333
generation, 325–326
milling by-products, 327–329
non-food applications, 333
nutritional value, 333–338

CAC, *see* Codex Alimentarius Commission

Calcium, pulse content, 30

Canada, pulse market, 453–461

Cancer, pulse consumption benefits, 374, 395, 397–398

Carbohydrate, pulse
Fiber, *see* Dietary fiber
glycemic index, 109
non-starch polysaccharides
cellulose, 159–160
content by pulse type, 158–159
gum, 160–161
hemicellulose, 159–160
mucilage, 160–161

Carbohydrate, pulse (*continue*)
  overview, 22–24, 157–158
  pectin, 160–161
  physiological effects, 161–162
  processing effects, 162–165
  starch, *see* Starch, pulse
Carborundum roller mill, 197–199
Cardiovascular disease (CVD),
    legume consumption benefits,
    10, 372–373, 457
Cellulose, pulse, 159–160
CHD, *see* Coronary heart disease
Chickpea
  nutritional composition, 364
  overview of food use, 252
  production, 444–445
  protein composition, 15–16
  roasting, 270–272
Cholesterol, pulse consumption
    benefits in lowering,
    373–374, 397
Codex Alimentarius Commission
    (CAC), pulse quality standard
    parameters
  contaminants, 426–427
  extraneous matter, 425
  hygiene, 426
  labeling, 428
  moisture content, 425
  overview, 422–423
  packaging, 427–428
  pulses standard, 422, 425
  toxic or noxious seeds, 426
Color, quality evaluation, 430–431
Consumption, pulse
  intake and trends, 362–363,
    437–438, 450, 455–456
  statistics, 448–449
Convicine, faba bean, 389
Copper, pulse content, 28
Coronary heart disease (CHD),
    legume consumption benefits,
    10, 372–373, 457
Cowpea
  overview of food use, 249–250
  quick-cook dehydrated pulses,
    257–261
CPI, *see* Cysteine protease inhibitor

CVD, *see* Cardiovascular disease
Cyanogenic glucosides, pulse
    content, 390
Cysteine protease inhibitor (CPI),
    pulse applications, 314

DE, *see* Dehulling efficiency
Dehulling, *see* Milling
Dehulling efficiency (DE),
    calculation, 199–200
Dehulling index, calculation, 200
Dehydration, *see* Drying
DF, *see* Dietary fiber
Diabetes, pulse consumption
    benefits, 370, 372, 396, 457
Diadzein, pulse composition, 368
Dielectric heating
  applications
    microwave heating, 233–235
    radiofrequency heating, 231–232
  principles, 225–226
  pulse properties, 232–233
Dietary fiber (DF)
  composition by pulse type, 20,
    123–125
  definition, 122, 366–367
  factors affecting levels in pulses
    boiling and roasting, 131
    canning, 133
    cooking, 131–133
    drying and dehydration, 133–134
    extrusion cooking, 134–135
    milling and grinding, 129–130
    overview, 127–128
    soaking and fermentation, 130
    thermal processing, 128–129
  health benefits, 20, 122, 144–146,
    367, 384
  insoluble dietary fiber, 125–126
  physicochemical properties
    bulk density, 137–138
    cation exchange capacity, 142–143
    oil-binding capacity, 141–142
    overview, 135–137
    swelling capacity, 138–139
    water-binding capacity, 139–141,
      144
    water-holding capacity, 139–141

proximate composition by legume
    type, 331
soluble fiber, 126–127
Differential scanning calorimetry
    (DSC), pulse starch, 101–103
Dough, *see* Flour
Drying
    advances, 207
    fiber effects, 133–134
    post-harvest system, 174–176
Dry milling, 203–205
DSC, *see* Differential scanning
    calorimetry

EAI, *see* Emulsifying activity index
Emulsifying activity index (EAI),
    pulse proteins, 65–66
Emulsifying stability index (ESI),
    pulse proteins, 65–66
ESI, *see* Emulsifying stability
    index
Extrusion
    dietary fiber effects, 134–135
    food products, 261–263
    snack-based pulse products,
        263–266, 311–312

Faba bean, antinutritional factors,
    389
FAO, *see* Food and Agriculture
    Organization
Fermentation, pulse
    developments, 273–275
    dietary fiber effects, 130
    traditional foods , 253–254
Fiber, *see* Dietary fiber
Flour, pulse
    bakery product applications,
        304–305
    composition, 267–268
    dough system effects of flour
        physical properties
    proximate composition, 296–297
    starch parameters
        amylopectin activity, 299–301
        amylose activity, 299–301
        birefringence, 294
        gelatinization, 297

granule particle size, 294–296
pasting, 298–299
retrodegradation, 298
tempering of flour, 296
extruded snacks, 311–312
fortification with pulse flours,
    212–213, 301–303
fractionation, 215–216
functional properties
    emulsifying activity, 291–292
    foaming properties, 292–293
    least gelling concentration,
        290–291
    oil-absorption capacity, 288–289
    overview, 285, 287–288
    protein solubility index, 290
    water-holding capacity, 289–290
germinated pulse powder
    production, 215
industrial applications, 313–314
meat product applications,
    305–308
milling methods, 208–211
natraceutical ingredients, 403–405
particle size, 211–212
pasta and noodle product
    applications, 308–311
paste product applications,
    312–313
precooked flour and powder
    production, 214
proximate composition, 296
roasting, 213–214
soup applications, 312–313
tortilla acceptance, 267
types
    fiber, 283
    protein flours, concentrates, and
        isolates, 284–285
    starch, 283–284
    whole pulse flour, 282–283
Foaming properties
    pulse flour, 292–293
    pulse proteins, 68–69
Food and Agriculture Organization
    (FAO)
    pulse requirements, 420–421
    statistical data for pulses, 436

Food products, pulse
  extruded products, 261–266
  fermentation developments,
    273–275
  forms
    canned pulses, 251
    dry pulses, 250–251
    fermented products, 253–254
    green beans, 248–249
    sprouted pulses, 252–253
  overview, 247–248
    quick-cook dehydrated pulses,
      256–261
    snack-based products, 263–266
    value added products
    gluten-free products, 270, 272
    noodles, 272–273
    overview, 266–269
    roasting, 269–270

Gelling, *see* Least gelling
  concentration
Genistein, pulse composition, 368
Germinated pulse powder,
  production, 215
GI, *see* Glycemic index
Gluten-free products, pulse foods,
  270, 272
Glycemic index (GI), pulse, 109
Green gram, protein composition,
  16
Grinding, *see* Milling
Gum, pulse, 160–161

Hard-to-cook (HTC), post-harvest
  effects, 182
Hemicellulose, pulse, 159–160
High-pressure (HP) processing
  applications, 237–242
  principles, 229–231
HP processing, *see* High-pressure
  (HP) processing
HTC, *see* Hard-to-cook

Impact milling, 209–211
Intake, *see* Consumption, pulse
International Pulse Quality
  Committee (IPQC), pulse

quality standard parameters,
  421–422, 429
Ionizing radiation (IR), pulse
  treatment, 228, 235–237
IPQC, *see* International Pulse
  Quality Committee
IR, *see* Ionizing radiation
Iron, pulse content, 28, 31, 365
Irradiation, *see* Ionizing radiation
Isoflavones, *see* Phenolics, pulse

Kibble percentage, calculation, 201

Lathyrogens, pulse content, 390
Least gelling concentration (LGC)
  pulse flour, 290–291
  pulse proteins, 71–72
Lectins, cancer inhibition, 397
Legumes, *see* Pulses
Lentil
  nutritional composition, 364
  production, 445–446
LGC, *see* Least gelling
  concentration
Lipid, pulse
  composition, 25–26
  starch content, 18
LPI, *see* Lupin protein isolate
Lunasin, bioactivity, 398
Lupin protein isolate (LPI)
  baked products, 77–78
  meat products, 82–84
Lupin, production, 447–448

MAE, *see* Microwave-assisted
  extraction
Magnesium, pulse content, 30
Meat products, pulse uses, 82–84,
  305–308
Microbes, limits for pulses, 427
Microwave-assisted extraction
  (MAE), bioactive compound
  extraction from by-products,
  350–351, 353
Microwave heating
  applications, 233–235
  dielectric heating principles,
  225–226

Milling
  by-products, 327–329
  dry milling, 203–205
  drying advances, 207
  fiber effects, 129–130
  flour milling
    fractionation, 215–216
    germinated pulse powder
      production, 215
    modern methods, 208–211
    particle size, 211–212
    precooked flour and powder
      production, 214
    roasting, 213–214
    substitution with pulse flours,
      212–213
  geographic distribution of
    dehulling and splitting, 208
  modern dehulling and splitting
    methods
    calculations, 199–201
    carborundum roller mill,
      197–199
    tangential abrasive dehulling
      device, 199–200
    under runner disc sheller,
      196–197, 199
  overview, 193–194
  pre-treatments, 206–207
  seed attributes influencing yield,
    205–206
  traditional methods
    Asia, 194–195
    dehulling and splitting, 196
    dehulling without splitting,
      195–196
  wet milling, 201–203
Minerals, pulse content, 27–31
Mucilage, pulse, 160–161

Niacin, pulse content, 27
Non-starch polysaccharides, *see*
  Carbohydrate, pulse
Noodles, pulse foods, 272–273,
  308–311
Nutraceutical
  antioxidants, *see* Phenolics,
    pulse

bioactive peptides and proteins in
  pulses
    allergenicity, 401–402
    soybean, 396–398
    structure and bioactivity,
      399–401
  definition, 383
  prebiotic properties of pulses,
    391–392
  pulse components as natraceutical
    ingredients, 402–406

OAC, *see* Oil-absorption capacity
Obesity, pulse consumption benefits,
  374–375, 457
Ohmic heating, pulse treatment,
  227–228
Oil-absorption capacity (OAC)
  pulse flour, 288–289
  pulse proteins, 63–64
Oil-binding capacity, fiber, 141–142
Osteoporosis, pulse consumption
  benefits, 375
Oxalate, pulse content, 38–40
Ozone, pulse treatment, 228–229

Pasta, pulse foods, 272–273, 308–311
Pea, nutritional composition, 364
Pectin, pulse, 160–161
PEF, *see* Pulsed electric field
Pesticide, limits for pulses, 427
Pests
  control, 179–180
  detection, 180–181
Pet food, pulse utilization, 461
Phenolics, pulse
  bioavailability, 394–395
  by-products, 336–337
  chemistry, 392–394
  composition by pulse type, 32,
    35–37
  health benefits, 366
  isoflavones, 37–38
  processing effects, 338
  structures, 33–35
Phytic acid
  health benefits, 395–396
  pulse content, 29, 31, 32

Phytosterols, pulse content, 28,
    39–40
Pigeon pea, protein composition,
    14–15
Polysaccharides, *see* Carbohydrate,
    pulse
Post-harvest system, *see also*
    specific technologies
    by-products, *see* By-products,
        pulse
    drying, 174–176
    emerging technologies
        classification, 224
        types, 224–243
    losses, 173–174
    overview of stages, 171–173
    quality impact, 181–186
    storage
        pests
            control, 179–180
            detection, 180–181
            systems, 177–178
Prebiotics, pulses, 391–392
Processing, *see* Post-harvest system
Production, pulses
    area and yield, 440–441
    broad bean, 447
    Canada, 453–461
    chickpea, 444–445
    dry beans, 442–443
    dry peas, 443–444
    global production, 436–437
    lentil, 445–446
    lupin, 447–448
    statistics, 438–440
    sustainability, 459–460
Protein, pulse
    air classification, 342–345
    amino acids
        essential, 13
        composition by pulse type, 11,
            368
    bioactive peptides and proteins in
        pulses
        allergenicity, 401–402
        soybean, 396–398
        structure and bioactivity,
            399–401

composition by pulse type
    chickpea, 15–16
    green gram, 16
    pigeon pea, 14–15
flours, concentrates, and isolates,
    284–285
food applications
    baked products, 75–79
    meat products, 82–84
    pasta products, 79–82
functional properties
    emulsifying properties, 65–67,
        70
    foaming properties, 68–69
    gelation characteristics, 71–74
    oil-absorption capacity, 63–64
    solubility, 62–63
    water-absorption capacity, 63–64
preparation of concentrates and
    isolates, 59–62
structure and classification, 13–14,
    58
Protein–energy malnutrition,
    developing countries, 12
Protein solubility index, pulse flour,
    290
Pulsed electric field (PEF), pulse
    treatment, 226–227
Pulses
    definition, 1
    processing, *see also* Post-harvest
        system
        challenges, 5–6
        overview, 3–4
    types, 1–3

Quality, pulses
    evaluation techniques, 430–431
    extrinsic parameters, 419
    Food and Agriculture Organization
        requirements, 420–421
    improvement, 417–419
    standard parameters
        American Association of Cereal
            Chemists, 428–429
        Codex Alimentarius
            Commission
            contaminants, 426–427

Quality, pulses (*continue*)
    extraneous matter, 425
    hygiene, 426
    labeling, 428
    moisture content, 425
    overview, 422–423
    packaging, 427–428
    pulses standard, 422, 425
    toxic or noxious seeds, 426
    International Pulse Quality
        Committee, 421–422, 429
    list, 431
Quick-cook dehydrated pulses,
    256–261

Radiofrequency heating
    applications, 231–232
    dielectric heating principles,
        225–226
Resistant starch, *see* Carbohydrate,
    pulse
Riboflavin, pulse content, 27
Roasting
    dietary fiber effects, 131
    pulse foods, 269–270

Saponins, pulse content, 38–39,
    369, 389
SCFX, *see* Supercritical fluid
    extrusion
Seed testing, standards, 429
Snacks, extruded pulse products,
    263–266, 311–312
Soybean
    bioactive peptides and proteins,
        396–398
    nutritional composition, 364
    overview of food use, 255
    quick-cook dehydrated pulses,
        256–257
Splits yield (SY), calculation,
    200–201
Squalene, pulse content, 28
Starch, pulse
    air classification, 342–345
    amylose content, 17, 95
    chemical composition, 16–17
    composition by pulse type, 19

digestibility, 108–111
dough system effects of flour
    physical properties
    amylopectin activity, 299–301
    amylose activity, 299–301
    birefringence, 294
    composition, 283–284
    gelatinization, 297
    granule particle size, 294–296
    pasting, 298–299
    retrodegradation, 298
granule structure, 18, 294–296
isolation, 92–94
lipid content, 18
physicochemical properties
    composition, 95–96
    morphology, 94–95
    pasting properties, 105–106
    retrogradation properties,
        106–108
    rheological properties, 103–105
    solubility, 99–101
    structure, 97–99
    swelling power, 99–101
    thermal properties, 101–103
resistant starch, 19–20, 108,
    110–111
yield, 19
Storage
    pests
        control, 179–180
        detection, 180–181
    systems, 177–178
Supercritical fluid extrusion
    (SCFX), pulse treatment, 229
Swelling capacity
    fiber, 138–139
    starch, 99–101
SY, *see* Splits yield

TADD, *see* Tangential abrasive
    dehulling device
Tangential abrasive dehulling device
    (TADD), 199–200
Tannins, pulse by-products,
    335–336, 387
Tempering, flour effects in dough
    systems, 296

Thiamine, pulse content, 27
Tocopherol, pulse content, 28
Trade
    Canada, 453–461
    pulse statistics, 449–454
Trypsin inhibitors, pulse by-products,
        335–336, 387–388, 397

Ultrasound
    bioactive compound extraction
        from pulse by-products,
        345–349
    pulse treatment, 228
Under runner disc (URD) sheller,
        196–197, 199
URD sheller, *see* Under runner disc
        sheller

Vicine, faba bean, 389

Vitamin B_{12}, pulse content, 25–26
Vitamin B_6, pulse content, 27

WAC, *see* Water-absorption capacity
Water extraction, bioactive
        compound extraction from
        by-products, 351–353
Water-absorption capacity (WAC),
        pulse proteins, 63–64
Water-binding capacity (WBC),
        fiber, 139–141, 144
Water-holding capacity (WHC)
    fiber, 139–141
    pulse flour, 289–290
WBC, *see* Water-binding capacity
Wet milling, 201–203
WHC, *see* Water-holding capacity

Zinc, pulse content, 28, 30–31

Food Science and Technology International Series

Maynard A. Amerine, Rose Marie Pangborn, and Edward B. Roessler, *Principles of Sensory Evaluation of Food.* 1965.

Martin Glicksman, *Gum Technology in the Food Industry.*1970.

Maynard A. Joslyn, *Methods in Food Analysis,* second edition.1970.

C. R. Stumbo, *Thermobacteriology in Food Processing*, second edition. 1973.

Aaron M. Altschul (ed.), *New Protein Foods:* Volume 1, *Technology, Part A*—1974. Volume 2, *Technology, Part B*—1976. Volume 3, *Animal Protein Supplies, Part A*—1978. Volume 4, *Animal Protein Supplies, Part B*—1981. Volume 5, *Seed Storage Proteins*—1985.

S. A. Goldblith, L. Rey, and W. W. Rothmayr, *Freeze Drying and Advanced Food Technology.* 1975.

R.B. Duckworth (ed.), *Water Relations of Food.* 1975.

John A. Troller and J. H. B. Christian, *Water Activity and Food.* 1978.

A. E. Bender, *Food Processing and Nutrition.* 1978.

D. R. Osborne and P. Voogt, *The Analysis of Nutrients in Foods.* 1978.

Marcel Loncin and R. L. Merson, *Food Engineering: Principles and Selected Applications.* 1979.

J. G. Vaughan (ed.), *Food Microscopy.* 1979.

J. R. A. Pollock (ed.), *Brewing Science,* Volume 1—1979. Volume 2—1980. Volume 3—1987.

J. Christopher Bauernfeind (ed.), *Carotenoids as Colorants and Vitamin A Precursors: Technological and Nutritional Applications.* 1981.

Pericles Markakis (ed.), *Anthocyanins as Food Colors.* 1982.

George F. Stewart and Maynard A. Amerine (eds), *Introduction to Food Science and Technology,* second edition. 1982.

Hector A. Iglesias and Jorge Chirife, *Handbook of Food Isotherms: Water Sorption Parameters for Food and Food Components.* 1982.

Colin Dennis (ed.), *Post-Harvest Pathology of Fruits and Vegetables.* 1983.

P. J. Barnes (ed.), *Lipids in Cereal Technology.* 1983.

David Pimentel and Carl W. Hall (eds), Food *and Energy Resources.* 1984.

Joe M. Regenstein and Carrie E. Regenstein, *Food Protein Chemistry: An Introduction for Food Scientists.* 1984.

Maximo C. Gacula, Jr. and Jagbir Singh, *Statistical Methods in Food and Consumer Research.* 1984.

Fergus M. Clydesdale and Kathryn L. Wiemer (eds), *Iron Fortification of Foods.* 1985.

Pulse Foods: Processing, Quality and Nutraceutical Applications. DOI: 10.1016/B978-0-1238-2018-1.00007-0

Robert V. Decareau, *Microwaves in the Food Processing Industry*. 1985.

S. M. Herschdoerfer (ed.), *Quality Control in the Food Industry*, second edition. Volume 1—1985. Volume 2—1985. Volume 3—1986. Volume 4—1987.

F. E. Cunningham and N. A. Cox (eds), *Microbiology of Poultry Meat Products*. 1987.

Walter M. Urbain, *Food Irradiation*. 1986.

Peter J. Bechtel, *Muscle as Food*. 1986.

H. W.-S. Chan, *Autoxidation of Unsaturated Lipids*. 1986.

Chester O. McCorkle, Jr., *Economics of Food Processing in the United States*. 1987.

Jethro Japtiani, Harvey T. Chan, Jr., and William S. Sakai, *Tropical Fruit Processing*. 1987.

J. Solms, D. A. Booth, R. M. Dangborn, and O. Raunhardt, *Food Acceptance and Nutrition*. 1987.

R. Macrae, *HPLC in Food Analysis*, second edition. 1988.

A. M. Pearson and R. B. Young, *Muscle and Meat Biochemistry*. 1989.

Marjorie P. Penfield and Ada Marie Campbell, *Experimental Food Science*, third edition. 1990.

Leroy C. Blankenship, *Colonization Control of Human Bacterial Enteropathogens in Poultry. 1991.*

Yeshajahu Pomeranz, *Functional Properties of Food Components*, second edition.1991.

Reginald H. Walter, *The Chemistry and Technology of Pectin*. 1991.

Herbert Stone and Joel L. Sidel, *Sensory Evaluation Practices*, second edition. 1993.

Robert L. Shewfelt and Stanley E. Prussia, *Postharvest Handling: A Systems Approach*. 1993.

Tilak Nagodawithana and Gerald Reed, *Enzymes in Food Processing*, third edition. 1993.

Dallas G. Hoover and Larry R. Steenson, *Bacteriocins*. 1993.

Takayaki Shibamoto and Leonard Bjeldanes, *Introduction to Food Toxicology*.1993.

John A. Troller, *Sanitation in Food Processing*, second edition.1993.

Harold D. Hafs and Robert G. Zimbelman, *Low-fat Meats*. 1994.

Lance G. Phillips, Dana M. Whitehead, and John Kinsella, *Structure-Function Properties of Food Proteins*. 1994.

Robert G. Jensen, *Handbook of Milk Composition*. 1995.

Yrjö H. Roos, *Phase Transitions in Foods*. 1995.

Reginald H. Walter, *Polysaccharide Dispersions*.1997.

Gustavo V. Barbosa-Cánovas, M. Marcela Góngora-Nieto, Usha R. Pothakamury, and Barry G. Swanson, *Preservation of Foods with Pulsed Electric Fields*. 1999.

Ronald S. Jackson, *Wine Tasting: A Professional Handbook*. 2002.

Malcolm C. Bourne, *Food Texture and Viscosity: Concept and Measurement*, second edition. 2002.

Benjamin Caballero and Barry M. Popkin (eds), *The Nutrition Transition: Diet and Disease in the Developing World*. 2002.

Dean O. Cliver and Hans P. Riemann (eds), *Foodborne Diseases,* second edition. 2002. Martin

Martin Kohlmeier, *Nutrient Metabolism*, 2003.

Herbert Stone and Joel L. Sidel, *Sensory Evaluation Practices,* third edition. 2004.

Jung H. Han, *Innovations in Food Packaging*. 2005.

Da-Wen Sun (ed.), *Emerging Technologies for Food Processing*. 2005.

Hans Riemann and Dean Cliver (eds), *Foodborne Infections and Intoxications,* third edition. 2006.

Ioannis S. Arvanitoyannis, *Waste Management for the Food Industries*. 2008.

Ronald S. Jackson*, Wine Science: Principles and Applications*, third edition. 2008.

Da-Wen Sun (ed.), *Computer Vision Technology for Food Quality Evaluation*. 2008.

Kenneth David and Paul Thompson (eds), *What Can Nanotechnology Learn From Biotechnology?* 2008.

Elke K. Arendt and Fabio Dal Bello (eds), *Gluten-Free Cereal Products and Beverages*. 2008.

Debasis Bagchi (ed.), *Nutraceutical and Functional Food Regulations in the United States and Around the World*, 2008.

R. Paul Singh and Dennis R. Heldman, *Introduction to Food Engineering*, fourth edition. 2008.

Zeki Berk, *Food Process Engineering and Technology*. 2009.

Abby Thompson, Mike Boland and Harjinder Singh (eds), *Milk Proteins: From Expression to Food*. 2009.

Wojciech J. Florkowski, Stanley E. Prussia, Robert L. Shewfelt and Bernhard Brueckner (eds), *Postharvest Handling*, second edition. 2009.

Maximo Gacula, Jr., Jagbir Singh, Jian Bi and Stan Altan, *Statistical Methods in Food and Consumer Research*, second edition. 2009.

Takayuki Shibamoto and Leonard Bjeldanes, *Introduction to Food Toxicology*, second edition. 2009.

James BeMiller and Roy Whistler (eds), *Starch: Chemistry and Technology*, third edition. 2009.

Ronald S. Jackson, *Wine Tasting: A Professional Handbook*, second edition. 2009.

Gerald M. Sapers, Ethan B. Solomon and Karl R. Matthews, *The Produce Contamination Problem: Causes and Solutions*. 2009.

Printed and bound by CPI Group (UK) Ltd, Croydon, CR0 4YY

13/05/2025

01869571-0002